AF884917

SOTHIS PRESS

Sothis Press aims to completely reimagine the public domain book with an emphasis on universal access to captivatingly designed print editions. We believe that technology will enable every single book ever published to be perpetually accessible in beautiful print form. Sothis is beauty in ubiquity.

www.sothis.press
1 Waterhouse Square, London, ECIN 2ST

Statistical Mechanics With Applications to Physics and Chemistry

Richard Chace Tolman

SOTHIS PRESS

ISBN 978-1-0365-1268-2

Statistical Mechanics With Applications to Physics and Chemistry is in the public domain.

This edition is published by Sothis Press © 2023. The company is registered in England and Wales. The Company Number is 14900389.

STATISTICAL MECHANICS WITH APPLICATIONS TO PHYSICS AND CHEMISTRY

BY

RICHARD C. TOLMAN, PH.D.

PROFESSOR OF PHYSICAL CHEMISTRY AND MATHEMATICAL PHYSICS
CALIFORNIA INSTITUTE OF TECHNOLOGY

American Chemical Society
Monograph Series

BOOK DEPARTMENT
The CHEMICAL CATALOG COMPANY, *Inc.*
419 FOURTH AVENUE, AT 29TH STREET, NEW YORK, U. S. A.
1927

GENERAL INTRODUCTION

American Chemical Society Series of Scientific and Technologic Monographs

By arrangement with the Interallied Conference of Pure and Applied Chemistry, which met in London and Brussels in July, 1919, the American Chemical Society was to undertake the production and publication of Scientific and Technologic Monographs on chemical subjects. At the same time it was agreed that the National Research Council, in coöperation with the American Chemical Society and the American Physical Society, should undertake the production and publication of Critical Tables of Chemical and Physical Constants. The American Chemical Society and the National Research Council mutually agreed to care for these two fields of chemical development. The American Chemical Society named as Trustees, to make the necessary arrangements for the publication of the monographs, Charles L. Parsons, Secretary of the American Chemical Society, Washington, D. C.; John E. Teeple, Treasurer of the American Chemical Society, New York City; and Professor Gellert Alleman of Swarthmore College. The Trustees have arranged for the publication of the American Chemical Society series of (a) Scientific and (b) Technologic Monographs by the Chemical Catalog Company of New York City.

The Council, acting through the Committee on National Policy of the American Chemical Society, appointed the editors, named at the close of this introduction, to have charge of securing authors, and of considering critically the manuscripts prepared. The editors of each series will endeavor to select topics which are of current interest and authors who are recognized as authorities in their respective fields. The list of monographs thus far secured appears in the publisher's own announcement elsewhere in this volume.

The development of knowledge in all branches of science, and especially in chemistry, has been so rapid during the last fifty years and the fields covered by this development have been so varied that it is difficult for any individual to keep in touch with the progress in branches of science outside his own specialty. In spite of the facilities for the examination of the literature given by Chemical Abstracts and such compendia as Beilstein's Handbuch der Organischen Chemie, Richter's Lexikon, Ostwald's Lehrbuch der Allgemeinen Chemie, Abegg's and Gmelin-Kraut's Handbuch der Anorganischen Chemie and the English and French Dictionaries of Chemistry, it often takes a great deal of time to coördinate the knowledge available upon a single topic. Consequently when men who have spent years in the study of important subjects are willing to coördinate their knowledge and present it in concise, readable form, they perform a service of the highest value to their fellow chemists.

It was with a clear recognition of the usefulness of reviews of this character that a Committee of the American Chemical Society recommended the publication of the two series of monographs under the auspices of the Society.

Two rather distinct purposes are to be served by these monographs. The first purpose, whose fulfilment will probably render to chemists in general the most important service, is to present the knowledge available upon the chosen topic in a readable form, intelligible to those whose activities may be along a wholly different line. Many chemists fail to realize how closely their investigations may be connected with other work which on the surface appears far afield from their own. These monographs will enable such men to form closer contact with the work of chemists in other lines of research. The second purpose is to promote research in the branch of science covered by the monograph, by furnishing a well digested survey of the progress already made in that field and by pointing out directions in which investigation needs to be extended. To facilitate the attainment of this purpose, it is intended to include extended references to the literature, which will enable anyone interested to follow up the subject in more detail. If the literature is so voluminous that a complete bibliography is impracticable, a critical selection will be made of those papers which are most important.

The publication of these books marks a distinct departure in the policy of the American Chemical Society inasmuch as it is a serious attempt to found an American chemical literature without primary regard to commercial considerations. The success of the venture will depend in large part upon the measure of coöperation which can be secured in the preparation of books dealing adequately with topics of general interest; it is earnestly hoped, therefore, that every member of the various organizations in the chemical and allied industries will recognize the importance of the enterprise and take sufficient interest to justify it.

American Chemical Society

MONOGRAPH SERIES

PUBLISHED

The Chemistry of Enzyme Actions (Revised Edition).
By K. George Falk. Price $5.00.

The Chemical Effects of Alpha Particles and Electrons.
By Samuel C. Lind. Price $3.75.

Organic Compounds of Mercury.
By Frank C. Whitmore. Price $7.50.

Industrial Hydrogen.
By Hugh S. Taylor. Price $4.50.

Zirconium and Its Compounds.
By F. P. Venable. Price $4.00.

The Vitamins.
By H. C. Sherman and S. L. Smith. Price $5.50.

The Properties of Electrically Conducting Systems.
By Charles A. Kraus. Price $6.50.

The Origin of Spectra.
By Paul D. Foote and F. L. Mohler. Price $5.50. (Out of print)

Carotinoids and Related Pigments.
By Leroy S. Palmer. Price $6.00.

The Analysis of Rubber.
By John B. Tuttle. Price $3.50.

Glue and Gelatin.
By Jerome Alexander. Price $4.50.

The Chemistry of Leather Manufacture.
By John Arthur Wilson. Price $7.00.

Wood Distillation.
By L. F. Hawley. Price $4.00.

Valence, and the Structure of Atoms and Molecules.
By Gilbert N. Lewis. Price $3.75.

Organic Arsenical Compounds.
By George W. Raiziss and Jos. L. Gavron. Price $9.00.

Colloid Chemistry.
By The Svedberg. Price $4.50.

Solubility.
By Joel H. Hildebrand. Price $4.00.

Coal Carbonization.
By Horace C. Porter. Price $8.00.

The Structure of Crystals.
By Ralph W. G. Wyckoff. Price $7.50.

The Recovery of Gasoline from Natural Gas.
By George A. Burrell. Price $10.00.

The Chemical Aspects of Immunity.
By H. Gideon Wells. Price $5.50.

Molybdenum, Cerium and Related Alloy Steels.
By H. W. Gillett and E. L. Mack. Price $5.50.

[*Continued*]

American Chemical Society

MONOGRAPH SERIES

PUBLISHED

The Animal as a Converter.
By H. P. Armsby and C. Robert Moulton. Price $4.50.

Organic Derivatives of Antimony.
By Walter G. Christiansen. Price $4.50.

Shale Oil.
By Ralph H. McKee. Price $6.00.

The Chemistry of Wheat Flour.
By C. H. Bailey. Price $6.00.

Surface Equilibria of Biological and Organic Colloids.
By P. Lecomte du Noüy. Price $4.50.

The Chemistry of Wood.
By L. F. Hawley and Louis E. Wise. Price $6.00.

Photosynthesis.
By H. A. Spoehr. Price $6.50.

Casein and Its Industrial Applications.
By Edwin Sutermeister. Price $5.00.

Equilibria in Saturated Salt Solutions.
By W. C. Blasdale. Price $4.50.

IN PREPARATION

Thyroxin.
By E. C. Kendall.

The Properties of Silica
By Robert B. Sosman.

The Corrosion of Alloys.
By C. G. Fink and Robert J. McKay.

Piezo-Chemistry.
By L. H. Adams.

Cyanamide.
By Joseph M. Braham.

Liquid Ammonia as a Solvent.
By E. C. Franklin.

Absorptive Carbon.
By N. K. Chaney.

Refining of Petroleum.
By George A. Burrell, *et al.*

The Properties of Metallic Substances.
By Charles A. Kraus.

Phosphoric Acid, Phosphates, and Phosphatic Fertilizers.
By W. H. Waggaman.

Physical and Chemical Properties of Glass.
By Geo. W. Morey.

The Chemistry of the Treatment of Water and Sewage.
By A. M. Buswell.

[*Continued*]

American Chemical Society

MONOGRAPH SERIES

IN PREPARATION

The Rare Gases of the Atmosphere.
By Richard B. Moore.

The Manufacture of Sulfuric Acid.
By Andrew M. Fairlie.

The Biochemistry and the Biological Rôle of the Amino Acids.
By H. H. Mitchell and T. S. Hamilton.

Protective Metallic Coatings.
By Henry S. Rawdon.

Soluble Silicates in Industry.
By James G. Vail.

The Industrial Development of Searles Lake Brines with Equilibrium Data.
By John E. Teeple, *et al.*

Diatomaceous Earth.
By Robert Calvert.

Aromatic Coal Products.
By Alexander Lowy.

Fixed Nitrogen.
By Harry A. Curtis.

Surface Energy and Colloidal Systems.
By W. D. Harkins and T. F. Young.

Nucleic Acids.
By P. A. Levene.

Molecular Rearrangements.
By C. W. Porter.

Catalysis in Homogeneous Systems.
By F. O. Rice.

Nitrocellulose Lacquers.
By Bruce K. Brown and F. M. Crawford.

Hydrochloric Acid and Sodium Sulfate.
By N. A. Laury.

Vapor Phase Catalytic Oxidation of Organic Compounds and Ammonia.
By J. M. Weiss, C. R. Downs and Dorothy A. Hahn.

Absorption Spectra.
By Victor Henri and Emma P. Carr.

The Thermal Decomposition of Organic Compounds.
By Charles D. Hurd.

Carbon Dioxide.
By Elton L. Quinn and Charles L. Jones.

Titanium.
By William M. Thornton, Jr.

Glycerol Production, Properties and Analyses.
By James W. Lawrie.

Noxious Gases.
By Yandell Henderson and Howard W. Haggard.

Preface

It is the purpose of the following book to give in a single place a systematic presentation of the theory of statistical mechanics, together with a survey of most of its fields of application in physics and chemistry. It is hoped that the result will be of assistance to physicists and chemists, especially to the latter since they have to deal for the most part with systems so complicated that statistical methods are specially advantageous.

Certain features of the book may be mentioned in advance.

The attempt has been made to develop the theory of statistical mechanics in a logical fashion, giving special attention to the physical implication of the concepts employed rather than to mathematic subtilities. It is hoped that the frankness with which the ergodic hypothesis is definitely introduced as a postulate will meet with approval, and that the particular form of this postulate, which has appealed most to the writer, will also seem plausible to others.

Chapters have been included which present the elements of classical and quantum mechanics in sufficient measure to serve as a foundation for statistical mechanics. The treatment of quantum mechanics is of course only partially satisfactory owing to the rapidity of development in this field of investigation.

Chapter 15, on general principles regarding molecular states and molecular processes, contains a new treatment of the principle of microscopic reversibility which may appear complicated but will perhaps assist in removing earlier illogicalities in the treatment of this principle.

Other original contributions in the book include the deduction of a general equipartition law which holds in cases where the principle of equipartition of energy breaks down, an extension of Boltzmann's H-theorem to include radiation, and a discussion of the possibility of introducing the quantrum theory by changing the dependence of energy on the coördinates and momenta instead of by concentrating μ-weight on the allowed quantrum states. It is hoped that these tentative additions to the theory may prove useful.

Special attention has been given to the application of statistical mechanics to rates of physical-chemical change, since this is one of the fields where statistical mechanics most clearly demonstrates its superiority to thermodynamics. It is thought that the treatment in a single place of transport problems both for gases at moderate concentrations and at high dilutions will prove useful. Apology should possibly be offered for presenting a treatment of chemical reaction rate which gives perhaps too much attention to the author's own interests and point of view in that field. The fault is, however, a natural one.

In the preparation of the book, the author has often had to make use of Boltzmann's "Gas Theorie", Gibbs' "Statistical Mechanics", Jeans' "Dynamical Theory of Gases" and the treatment of P. and T. Eherenfest "Mechanik der aus sehr zahlreichen diskreten Teilen bestehende Systeme" and specific reference to these works has for the most part been omitted. References to newer literature have, however, often been inserted with the hope that they will prove useful.

Professor Roscoe G. Dickinson has been good enough to read all the proof and to make many valuable suggestions for which the writer is very grateful.

The writer would be glad to have his attention called to errors in the text.

RICHARD C. TOLMAN.

California Institute of Technology,
August 30, 1926.

Table of Contents

The Classical Statistical Mechanics.

Introduction of the Quantum Theory.

Conclusion.

STATISTICAL MECHANICS WITH APPLICATIONS TO PHYSICS AND CHEMISTRY

Chapter I

The Methods of Statistical Mechanics

1. Introduction.—In the case of very complicated systems such as those studied by the chemist and the biologist, it is very difficult if not impossible either to observe experimentally or predict theoretically the *precise* behavior of all the elements of any given system. Even if the chemist were told at a certain instant the exact locations, orientations, and velocities of all the atoms and molecules of which a reacting mixture is composed, and given the precise laws governing their individual behavior, the complexities of calculation would prevent him from predicting, even a fraction of a second later, the positions and chemical combinations of these elements. And for the biologist with his even more complicated systems, which consist not of a single reacting mixture but of an integrated combination of reaction zones, with an interplay between them of the physical-chemical processes taking place therein, the problem would be even more impossible of solution.

Nevertheless, in spite of the impossibility of precise observation and prediction, the chemist and the biologist have not been deterred from their labors. Even if they can not follow nor predict the *exact* behavior of individual elements, they can observe and predict with remarkable surety the *gross* behavior of their systems as a whole. Each time that the chemist touches a match to a soap bubble full of hydrogen and oxygen, he obtains an explosion; each time the bacteriologist autoclaves his glassware, he kills the bacteria therein; and each time there is a war, the sociologist can predict that rich men will profiteer and poor men die. These complicated phenomena do not lie beyond the realm of law and predictability.

To investigate the laws that do describe the gross behavior of systems containing many molecules, the science of *statistical mechanics* has been specially devised. The essence of its method consists in abandoning the attempt to predict the exact behavior of each molecule of a given system, and determining by statistical methods what on the average will be the behavior of a great number of systems of similar constitution to the one of interest. The appeal thus made to the methods of statistics in the realm of physical-chemistry is hence quite similar to the use of statistics in the biological sciences.

It is the purpose of the following book to develop in a logical manner the principles of statistical mechanics and apply them to the elucidation of a number of physical and chemical phenomena. Special stress will be laid on applications in the field of chemistry, since the chemist always deals with such a complicated collection of atoms that other methods are hard to apply. More and more the chemist must turn to the methods of statistical mechanics in order to continue the theoretical development of his science.

2. Comparison of the Methods of Ordinary and Statistical Mechanics.—A better idea of the methods of statistical mechanics will be obtained if we compare them first with those of ordinary mechanics and then with those of thermodynamics.

To take a concrete illustration, let us consider a gas composed of a large number of similar molecules moving about in all directions inside the walls of a container. And for simplicity let us take the molecules as small rigid elastic spheres.

To apply the methods of *ordinary mechanics* to such a gas we should have to know at some given instant the positions and velocities of all the molecules. We could then foresee the collisions that were about to take place among the molecules themselves and between the molecules and walls of the container, and by applying the principles of the conservation of energy and momentum to these collisions could predict their result. We should thus theoretically be able to follow the further behavior of the gas and predict its condition at any future time. Practically, however, we should be lost in our calculations before the lapse of a thousandth of a second.

The difficulty of thus using the methods indicated by ordinary mechanics led in the older kinetic theory of gases to the introduction of various artificial simplifications, which are not now necessary, such as the assumption that all the molecules had the same velocity distributed in random directions, thus neglecting the effect of collisions between the molecules themselves in changing the velocity. With the help of such simplifications the methods of ordinary mechanics can lead to a number of valuable approximate results.

To apply the methods of *statistical mechanics* to determine the behavior of a gas, such as that postulated above, we do not try to investigate the behavior of a single sample of the gas. Instead we consider an infinite number or *ensemble* of samples of the gas, started off with their molecules having all possible combinations of positions and velocities. Even though we can not predict the precise behavior of any single sample of the gas, nevertheless, surprising as it may seem, we are able by the application of the principles of mechanics to obtain information as to the *statistical behavior* of the whole ensemble, and thus to draw conclusions as to the *probable* behavior of a single sample. Thus, for example, we find that of all the samples in the ensemble only an infinitesimal fraction of them will have the velocities of their molecules distributed in a manner appreciably different from that given by the Maxwell distribution law, and hence conclude that any given sample of gas will probably have the Maxwell distribution of velocities and if displaced therefrom will tend to assume that distribution of velocities when left to itself.

At first sight it seems surprising that the indicated method should be so powerful. A single sample of the gas with its many molecules was so complicated a system that we could not directly predict its future behavior, and yet by going to the much more complicated system of an infinite ensemble of similar samples we appear to obtain such predictions as to behavior. The apparent paradox, however, is dependent on the circumstance that in the first case with the single sample we attempted to predict the *actual* behavior of this single sample; with our new method on the other hand we content ourselves with a determination of the *statistical* behavior of the ensemble and hence a prediction as to the probable behavior of a single sample.

A similar situation arises in the biological sciences. Thus a given man between the ages of thirty-four and thirty-five is such a complicated system of atoms and molecules that it is impossible to predict whether or not he will die within the coming year. Nevertheless, in a given country we have very reliable statistics as to the number of men per thousand between the ages of thirty-four and thirty-five that die per annum, and hence we can make a very satisfactory prediction as to the probability or not that the man in question will live throughout the year. In this biological case the prediction is based on statistical laws obtained as a result of observation, while in the physical-chemical cases which will be of interest to us the statistical laws will be obtained as the result of deduction from the laws of mechanics. Nevertheless the reason for the appeal to statistical methods is the same in both cases, the complexity of the systems making the use of more direct methods difficult.

The impression of course must not be gained that statistical mechanics can be profitably applied only to systems which are too complicated to permit the application of ordinary mechanical methods. Also in the case of a system of only a few degrees of freedom governed by simple equations of motion, we can consider an ensemble of such systems and discuss the statistical behavior of this ensemble.[1] An application of this kind can be of importance in case the initial values of the coordinates and velocities of the system at some given instant are unknown or not of interest. Nevertheless in the present book we shall develop the application of statistical mechanics to physical-chemical systems with many degrees of freedom corresponding to their many molecules, and our chief interest will be the practical one of elucidating their behavior by a method that has elements of superiority over any other.

3. Comparison of the Methods of Statistical Mechanics and Thermodynamics.—For the theoretical treatment of the behavior of complicated physical-chemical systems the method of thermodynamics has been used in the past even more than that of statistical mechanics. A comparison of the nature of the two methods and some estimate of their relative advantages will be of interest.

If we first compare the two sciences as to the certainty of their foundation and the rigor of their development, the advantage most certainly lies with thermodynamics.

The whole of classical thermodynamics is based entirely upon two postulates, the first and second laws of energy, and must be regarded merely as the logical system of theorems which can be drawn from these postulates by applying them to different types of physical-chemical systems. Moreover, the truth of the two fundamental laws, at least as far as the macroscopic behavior of physical-chemical systems is concerned, can hardly be questioned; deviations from the laws have never been found, and the whole past history of successful application furnishes a vindication of the starting point. We must regard the theorems of thermodynamics as among the most certain possessions of science.

On the other hand, when we turn to statistical mechanics we find a much less certain foundation, and for parts of the science a less rigorous method of development.

The first underlying assumption, in the application of the science, is the hypothesis of the atomic and molecular structure of matter. This general hypothesis no longer has to withstand the attacks of the older school of physical chemists, influenced by Ostwald's point of view that the scientist should give attention only to observable phenomena. Nevertheless, just at the present moment, the physicists themselves, greatly influenced by the development of the general theory of relativity, which concerns itself only with observable space-time coincidences, are tempted to follow Heisenberg's lead in hoping that the difficult problems of the quantum theory can be solved by neglecting the atomic constitution of matter and concentrating attention solely on the observed relations as to

[1] See E. B. Wilson, *Annals of Math.*, **10**, 129, 149 (1909).

the frequencies and intensities of spectral lines. In any case, even though we feel justified in accepting as one of our starting points the general theory of atomic structure, we shall find in the course of our development of statistical mechanics many times a necessity for specific hypotheses as to the precise nature of atoms and molecules, which at the best can be regarded only as necessary simplifying assumptions.

Having started with the hypothesis of the atomic structure of matter, the method of statistical mechanics then consists in applying the laws of dynamics to determine the behavior of an ensemble of systems, each system itself composed of atoms and molecules. In the time of Gibbs and Boltzmann there would have been few to question this procedure. The laws of dynamics, as codified for example in the form of Hamilton's principle, are certainly a valid abstraction drawn from actual experiments on the behavior of moving bodies of macroscopic size, and the danger of applying these same laws to the motions of molecules was then but dimly appreciated. Today, however, the development of quantum theory has indicated that the most satisfactory treatment of atomic and molecular behavior may involve a complete revision of the laws of dynamics and perhaps even of our notions as to the continuity of space and time.

In the present book we shall try to solve this difficulty by first developing what may be called the classical statistical mechanics based on the older laws of dynamics. We shall then regard this as a limiting case, valid when high frequencies of periodic motion are not involved, and shall proceed by introducing somewhat arbitrarily such modifications as seem most suitable for the treatment of the great group of quantum phenomena.

One further lack of rigor in the method of statistical mechanics must be mentioned. Very early in the development, even of its classical applications to physical-chemical systems, we are forced to introduce the famous ergodic hypothesis or principle of continuity of path. The introduction of this hypothesis is certainly justified by its results, and the reasons for its introduction in a suitably modified form can, as we shall see, be made plausible. Nevertheless, this hypothesis still presents a baffling problem for further study.

This concludes our statement of the sources of lack of rigor in the method of statistical mechanics. They are:—first, the assumption of the atomic hypothesis itself, not so much the general assumption of the atomic structure of matter as specific assumptions as to atomic properties which must be made from time to time in the course of the development; second, the application of classical dynamics to describe the behavior of atomic systems, arbitrarily modified to take account of quantum phenomena rather than revivified by the newer dynamics which is yet to come, even though already heralded by birth pangs; and third, the introduction of the ergodic hypothesis which has to be made in order to obtain the important applications to physical-chemical systems that we have in mind.

It should be remarked that this lack of rigor is not resident in the science of statistical mechanics *per se,* but only arises when we come to actual applications to the problems of physics and chemistry. Nevertheless, we are interested in comparing the methods of thermodynamics and statistical mechanics as applied to actual physical-chemical systems, and for this purpose there is no question that thermodynamics at the present time is superior as to rigor.

4. If we turn our attention now, however, to the results that can be achieved by the two methods, the balance is all in the favor of the newer science.

In the first place, as will be shown in a later chapter of the present book, the two laws of thermodynamics may themselves be regarded as derivable from the principles of statistical mechanics and interpretable as the necessary result of atomistics. Hence all the possible conclusions of thermodynamics can in this way at least be indirectly obtained from statistical mechanics, although there are often advantages in their direct derivation since the theorems then obtain a more intimate atomistic interpretation.

In the second place, the properties of matter enter the equations of thermodynamics merely as empirical parameters whose values have to be determined by actual measurements applied to macroscopic systems. In the treatments of statistical mechanics, however, the actual values of these quantities are often deducible from their atomistic interpretation. Thus, for example, in thermodynamic deductions specific heats of gases and solids, heats of reaction, and equilibrium constants enter the equations and are related one with another. Their absolute values remain, however, unknown. In statistical mechanical treatments, however, values of specific heats can be obtained from a consideration of the number of degrees of freedom of the system in question and the extent of their excitation; values of heats of reaction are gradually being related to the energy levels of atoms and molecules as determined from spectroscopy; and absolute values of equilibrium constants are obtained by considering the relative probability of different arrangements of the atoms to form molecules.[8]

In the third place, even in the field of equilibrium relations, for the treatment of which thermodynamics is specially fitted, laws arise which thermodynamics must use but can not derive. Thus in particular the gas laws and the equations of state for matter in any form can for thermo-

[8] The *third* law of thermodynamics essays this same task of determining absolute values of equilibrium constants. The correct formulation of the third law, however, involves statistical mechanical interpretation.

dynamics be only empirical facts, but for statistical mechanics they are already either deduced or in any case deducible by methods that we can now foresee.

In the fourth place, for a system which is not in equilibrium, thermodynamics can only tell us that those changes are permitted which involve an increase in entropy. It can give us no information as to the velocities with which the permitted changes will take place and hence in actuality usually gives us no prediction as to what will really happen in a system which is not in equilibrium. Thus such questions as rates of diffusion and evaporation, rates of thermal and electrical conduction, rate of momentum transfer, and rates of thermal and photochemical reactions all lie beyond the power of thermodynamic reasoning. They are, however, as we shall see, in the present book, amenable to statistical mechanical treatment.

Finally, when phenomena arise which are essentially atomistic in character, thermodynamics can of course have absolutely no help to give. The Brownian motion of particles, the fluctuations in the density of a fluid, the differences in the velocities of electrons from a hot filament, and the distribution of thermal radiation among its different frequencies all furnish problems which statistical mechanics may treat, but which a non-atomistic thermodynamics must actually fail to recognize.

This long recital does not exhaust the achievements of statistical mechanics. It will give some idea, however, of the power of the method and the scope of its attack. The future of theoretical chemistry is dependent on its applications and there will be a mutual and advantageous interplay in the development of these two sciences.

Chapter 2

The Elements of Classical Mechanics

5. Introduction.—In the following brief chapter we shall present some of the principles of classical dynamics which will be necessary for an understanding of the foundations of statistical mechanics.

Any general treatment of dynamics is of course beyond the scope of a single short chapter. We shall try, however, in the first place to present the major part of what is necessary for future reference, and in the second place to give a correct feeling as to the fundamental nature of the methods employed in dynamical investigations.

6. Generalized Coordinates and Velocities.—The essence of any dynamical problem always consists in the calculation of the future condition of some system of interest from a knowledge of its initial condition at some chosen time. The condition of the system at any instant can always be given by a statement of the coordinates which determine the positions of its parts and the velocities which determine the rate of change of these coordinates with the time.

In the case of a system of point particles, or bodies which can be treated as such, it will often be natural to specify the positions of the parts of the system by giving the values of the Cartesian coordinates x, y and z for each particle, and their velocities by giving the values of the component velocities $\dot{x}$, $\dot{y}$ and $\dot{z}$ for each particle.

In many cases, however, some choice of other than Cartesian coordinates for the specification of positions may be more convenient or natural. Thus for the position of a swinging pendulum bob, it is more satisfactory to give the single angle ϑ which the pendulum makes with the vertical, than to give the values of the two Cartesian coordinates x and y which could also be used to determine its position. For an electron rotating in a circular hydrogen orbit, the angle ϕ made by the radius vector from nucleus to electron is the natural coordinate. And for an electron, moving inside an atom subjected to an external electric field as in the Stark effect, the parabolic coordinates furnished by a system of parabolæ with the nucleus at their focus and their axes parallel to the field are found to be convenient.

Such variables are called ***generalized coordinates*** and may be chosen in a great variety of ways depending on their convenience for the problem at hand. Corresponding to each of these coordinates will be a *generalized velocity*, giving the rate of change with the time of the particular coordinate in question. For the generalized coordinates of a system we shall use the letters $q_1\ q_2\ \ldots\ q_n$, and for the corresponding generalized velocities the symbols $\dot{q}_1\ \dot{q}_2\ \ldots\ \dot{q}_n$. It has been one of the most useful achievements of dynamics to obtain its laws in forms which are independent of the particular nature of the coordinates that are employed.

7. The number of coordinates chosen must of course be at least sufficient for the complete specification of the configuration of the system, and it is most convenient when it does not exceed that number. If the coordinates so chosen are all of them capable of independent displacement without violating the constraints of the structure the system is said to be *holonomic* and the number of coordinates is called the ***number of degrees of freedom of the system.*** If the minimum number has been chosen and they are not all capable of independent variation, but their displacement connected by non-integrable relations,[1] the system is said to be *non-holonomic*. The number of degrees of freedom is then equal to the number of coordinates capable of independent displacement.

8. Hamilton's Principle.—The search for a single general dynamical principle which would apply to the motions of any system has had a considerable history, which starts with the theological arguments of Maupertuis in favor of the *principle of least action* in 1744, includes the general formulation of this principle by Lagrange in 1760-61, and the introduction of the *principle of least curvature* by Gauss in 1829,[2] and having found a certain temporary culmination in the discovery of *Hamilton's principle* in 1834, is perhaps still continuing at the present time in the investigation of different variation principles suitable for use in the general theory of relativity.

For the needs of the present book we shall take Hamilton's principle as the starting point for our discussions. For ordinary mechanical systems all the different principles lead to the same conclusions, but Hamilton's principle seems to be the most satisfactory formulation for our purposes. The principle is to be regarded as a postulate, justified by the correspondence between experimental facts and the conclusions that can be deduced from it.

[1] If the relations were integrable a reduction in the number of coordinates chosen would be possible.

[2] Later developed by Hertz.

9. Hamilton's principle can be stated in the form

$$\delta \int_{t_1}^{t_2} L dt = 0 \tag{1}$$

The equation applies to the motion of any conservative system [3] from the configuration it has at time t_1 to the configuration it assumes at time t_2. The quantity L is the so-called Lagrangian function of the generalized coordinates and velocities

$$L = L\,(q_1\, q_2 \ldots q_n\, \dot{q}_1\, \dot{q}_2 \ldots \dot{q}_n) \tag{2}$$

and the sign of variation indicates that the integral $\int L dt$ has a minimum [4] value for the actual path by which the system travels from its initial to its final configuration, as compared with any neighboring slightly displaced path by which it could be made to go in the *same time interval* from the initial to the final point. The requirement that the time interval should not be changed means that the limits t_1 and t_2 are to be kept constant in performing the indicated variation.

The principle is a wonderfully compact formulation to serve as the single axiom for an immense body of science. It has, however, the disadvantage that it is difficult to get an intuitional grasp of its significance and inevitability as seems to be possible, for example, in the case of the laws of the conservation of energy and momentum, and the law of the increase of entropy.

10. Applications of the principle can of course not be made until we have information as to the way the Lagrangian function L depends on the generalized coordinates and velocities $(q_1 \ldots . \dot{q}_n)$ as indicated by equation (2). For ordinary mechanical systems involving low velocities, the Lagrangian function is the difference between the kinetic and potential energies of the system

$$L = T - V \tag{3}$$

For a system of particles involving high velocities of the order of the speed of light, such as are considered in the special theory of relativity, the Lagrangian function has the value

$$L = \Sigma m_0 c^2 (1 - \sqrt{1 - u^2/c^2}) - V \tag{4}$$

where m_0 is the mass of a particle at rest, c is the velocity of light, u the

[3] Systems which are not conservative owing to the action of external forces can always be made so by including in the system "reservoirs" of potential energy which will exert these forces.

[4] For extended motions $\int L dt$ is not necessarily a minimum.

velocity of the particle and the summation is to be taken over all the particles of the system.

For other systems including non-mechanical ones, the Lagrangian function may assume a variety of forms. It should be regarded as a matter of experimental science to determine the actual form of the function L for any particular application. This can indeed be considered one of the major problems of experimental physics which has, however, been successfully solved for a wide variety of phenomena.

11. It may seem somewhat surprising that the Lagrangian function should contain only the generalized coordinates and their first derivatives with respect to the time, since this would indicate that the future motion of a system is never affected by the initial values of the accelerations or higher derivatives. In the case of a system of moving electrons, for example, this would seem to lead to difficulties, since we know that the classical theory makes the emission of energy dependent on the accelerations of the electrons and hence that we should not be able to predict their further motion solely from a knowledge of their initial positions and velocities. The problem of a system of moving electrons is, nevertheless, made amenable to the principles of dynamics by the device of including in the system to be considered the energy of the electromagnetic fields surrounding the electrons, and hence including in the Lagrangian function L what amounts to a coordinate and a velocity for all three directions at every point in space. Since this makes the Lagrangian function contain, as well as the coordinates and velocities for the electric charges, an infinite number of further coordinates and velocities, it is not surprising that the treatment then becomes possible.

The possibility of throwing an exceedingly wide variety of problems into the form demanded by dynamics may in part be due to the great flexibility that is open to us in the choice of the form of the Lagrangian function L, as illustrated by the example just cited. The advantages, however, in treating as many phenomena as possible by equations having the same mathematical form will of course be apparent.

12. **Derivation of the Equations of Motion in the Lagrangian Form.**—From Hamilton's principle we may now proceed to obtain the equations for the motion of a system, first in the Lagrangian form and then in the canonical form due to Hamilton himself. These forms of the equations can be made the starting point for all of dynamics that we shall need to use.

To obtain the Lagrangian equations from Hamilton's principle, we have merely to perform the indicated variation given in equation (1). We obtain

$$\delta\int_{t_1}^{t_2} L dt = \int_{t_1}^{t_2}\sum_{i=1}^{i=n}\left(\frac{\partial L}{\partial q_i}\delta q_i + \frac{\partial L}{\partial \dot{q}_i}\delta \dot{q}_i\right)dt = 0 \tag{5}$$

where δq_i and $\delta \dot{q}_i$ are the variations in the values of the i'th coordinate and velocity in changing from the actual to a neighboring displaced path. The variation is carried out solely behind the integral sign since Hamilton's principle requires that the time of passage shall not be changed.

The second term behind the integration sign can easily be transformed by substituting

$$\delta\dot{q}_i = \dot{q}_i + \delta\dot{q}_i - \dot{q}_i = \frac{d}{dt}(q_i + \delta q_i) - \frac{dq_i}{dt} = \frac{d}{dt}(\delta q_i) \tag{6}$$

giving us

$$\int_{t_1}^{t_2}\frac{\partial L}{\partial \dot{q}_i}\delta\dot{q}_i dt = \int_{t_1}^{t_2}\frac{\partial L}{\partial \dot{q}_i}\frac{d}{dt}(\delta q_i)dt = \int_{t_1}^{t_2}\frac{\partial L}{\partial \dot{q}_i}d(\delta q_i)$$

which on integration by parts gives

$$\left|\frac{\partial L}{\partial \dot{q}_i}\delta q_i\right|_{t_1}^{t_2} - \int_{t_1}^{t_2}\frac{d}{dt}\left(\frac{\partial L}{\partial \dot{q}_i}\right)\delta q_i dt$$

The first of the two terms, however, is zero, since by hypothesis the configurations are the same in the actual and varied motions at the initial and final times t_1 and t_2, so that we finally obtain

$$\int_{t_1}^{t_2}\frac{\partial L}{\partial \dot{q}_i}\delta\dot{q}_i dt = -\int_{t_1}^{t_2}\frac{d}{dt}\left(\frac{\partial L}{\partial \dot{q}_i}\right)\delta q_i dt \tag{7}$$

This expression can now be profitably substituted in equation (5) to give an expression which no longer contains the variations of the velocities; we obtain

$$\int_{t_1}^{t_2}\sum_{i=1}^{i=n}\left(\frac{d}{dt}\frac{\partial L}{\partial \dot{q}_i} - \frac{\partial L}{\partial q_i}\right)\delta q_i dt = 0 \tag{8}$$

For holonomic systems, however, the different variations δq_i are entirely arbitrary, and furthermore the limits of integration t_1 and t_2 are also arbitrary, hence the truth of equation (8) can be preserved in general only if we have for each of the n expressions

$$\frac{d}{dt}\frac{\partial L}{\partial \dot{q}_i} - \frac{\partial L}{\partial q_i} = 0 \tag{9}$$

These are the equations of motion in the Lagrangian form for a holonomic, conservative system.[5]

13. In applying these equations to a system in which L is the difference between the kinetic energy T or some other function solely of the velocities, and the potential energy V given solely as a function of the coordinates, we can write

$$\frac{\partial L}{\partial q_i} = -\frac{\partial V}{\partial q_i} \tag{10}$$

and regard this change of potential energy with the coordinate as a *generalized force*. We are then at liberty if we desire to take any part of this generalized force Q_i, which it may be convenient to think of as due to external parts of the system, and transpose to the other side of the equation, giving us

$$\frac{d}{dt}\frac{\partial L}{\partial \dot{q}_i} - \frac{\partial L}{\partial q_i} = Q_i \tag{11}$$

where the quantities Q_i are the *external* generalized forces.

14. Definition of the Generalized Momenta and the Hamiltonian Function.—We may next define the generalized momenta $p_1 \ldots . p_n$ corresponding to the n generalized coordinates and velocities by the equations

$$p_i = \frac{\partial L}{\partial \dot{q}_i} \qquad (i = 1, 2, 3, \ldots . n) \tag{12}$$

For the case of a particle whose position is given by Cartesian coordinates, these generalized momenta are readily seen to become identical with the ordinary components of momentum $m\dot{x}$, $m\dot{y}$, $m\dot{z}$ for slow velocities (see equation 3), and equal to the relativity expressions for momentum $\frac{m_0\dot{x}}{\sqrt{1-u^2/c^2}}$ etc., for high velocities (see equation 4).

We may further define the Hamiltonian function H by the equation

$$H = p_1\dot{q}_1 + \ldots . + p_n\dot{q}_n - L \tag{13}$$

[5] For a non-holonomic system the variations δq_i would not all be arbitrarily independent, and the treatment would have to be changed. It is possible, however, to reduce non-holonomic systems to holonomic ones by introducing forces which do no work, but preserve the postulated lack of freedom in the arbitrary displacement of the coordinates. As remarked above, systems which are not conservative because of the action of external forces can be made conservative by including in the system "reservoirs" of potential energy which will exert these forces. Hence Lagrange's equation can be regarded as having a very wide application.

15. Derivation of the Equations of Motion in the Canonical Form.—Making use of these definitions, we are now ready to derive the equations of motion in the Hamiltonian or so-called canonical form.

Differentiating equation (13) we obtain

$$\begin{aligned} dH = & \quad p_1 d\dot{q}_1 + \ldots + p_n d\dot{q}_n \\ & + \dot{q}_1 dp_1 + \ldots + \dot{q}_n dp_n \\ & - \frac{\partial L}{\partial q_1} dq_1 - \ldots - \frac{\partial L}{\partial q_n} dq_n \\ & - \frac{\partial L}{\partial \dot{q}_1} d\dot{q}_1 - \ldots - \frac{\partial L}{\partial \dot{q}_n} d\dot{q}_n \end{aligned}$$

and referring to equations (12) we see that the first and last lines cancel giving us

$$\begin{aligned} dH = & \quad \dot{q}_1 dp_1 + \ldots + \dot{q}_n dp_n \\ & - \frac{\partial L}{\partial q_1} dq_1 - \ldots - \frac{\partial L}{\partial q_n} dq_n \end{aligned} \qquad (14)$$

This differential form shows that H can be obtained solely as a function of the q's and p's, the coefficients of the differentials in (14) being the partial derivatives of H with respect to these quantities, in accordance with the expressions

$$\frac{\partial H}{\partial p_i} = \dot{q}_i \quad \text{and} \quad \frac{\partial H}{\partial q_i} = -\frac{\partial L}{\partial q_i} \qquad (15)$$

The quantity $\frac{\partial L}{\partial q_i}$, however, in accordance with Lagrange's equation (9) and the definitions (12) is given by

$$\frac{\partial L}{\partial q_i} = \frac{d}{dt}\frac{\partial L}{\partial \dot{q}_i} = \frac{d}{dt} p_i = \dot{p}_i$$

So that equations (15) can now be written in the symmetrical form

$$\frac{\partial H}{\partial q_i} = -\dot{p}_i \; ; \quad \frac{\partial H}{\partial p_i} = \dot{q}_i \qquad (i = 1, 2, 3 \ldots n) \qquad (16)$$

where H is to be taken as a function of the q's and p's.

These are the equations of motion in the canonical form. A part of their importance for dynamical investigations arises from the circumstances that we have in this manner replaced the n second-order differential equations of Lagrange by $2n$ first-order equations, which may facilitate the integration. These equations will form the starting point for our investigation of the properties of statistical ensembles.

16. Interpretation of the Hamiltonian Function as the Energy of the System.—The quantity H which was originally defined by equation (13) can now be shown to have the important property of being a constant independent of time. To demonstrate this we may evidently write

$$\frac{dH}{dt}=\frac{\partial H}{\partial q_1}\dot{q}_1+\ldots.+\frac{\partial H}{\partial q_n}\dot{q}_n+\frac{\partial H}{\partial p_1}\dot{p}_1+\ldots.+\frac{\partial H}{\partial p_n}\dot{p}_n$$

and by substitution of the canonical equations (16) obtain

$$\begin{aligned}\frac{dH}{dt}&=-\dot{p}_1\dot{q}_1-\ldots-\dot{p}_n\dot{q}_n+\dot{q}_1\dot{p}_1+\ldots.+\dot{q}_n\dot{p}_n\\&=0\end{aligned}\tag{17}$$

We have thus obtained one integral of the equations of motion, namely

$$H=\text{const.}\tag{18}$$

This quantity whose value is conserved throughout the motion of the system may now be defined as the total *energy* of the system (expressed of course as a function of the q's and p's), and from our present point of view this is the most natural manner for the concept of energy first to enter into our considerations.

For a system of particles moving with low velocities where L in accordance with equation (3) is the difference between the kinetic and potential energies it is easy to show that H is equal to the ordinary expression for energy, since we can write

$$L=T-V=\sum_i \tfrac{1}{2}\,m_i\dot{q}_i^2-V$$

$$p_i=\frac{\partial L}{\partial \dot{q}_i}=m_i\dot{q}_i$$

and substituting in equation (13) obtain

$$H=\sum_i m_i\dot{q}_i^2-\sum_i \tfrac{1}{2}\,m_i\dot{q}_i^2+V=T+V=E\tag{19}$$

It is to be noted, however, that the Hamiltonian H is for *all* systems the energy, expressed as a function of the coordinates and momenta, even when the Lagrangian function does not assume the form $T-V$ of ordinary slow velocity mechanics.

This brief treatment of dynamics is all that will be necessary for our further development of statistical mechanics.

Chapter 3

The Properties of Statistical Ensembles

17. Introduction. Ensemble and Phase.—We are now ready to proceed with the development of statistical mechanics. As already stated, we shall not discuss the behavior of a single system, but rather the behavior of a collection or *ensemble* of systems, containing an enormous number of sample systems identical in nature with the one of interest. From the behavior of this collection of samples we shall then be able to draw conclusions as to the probable behavior of a single system.

The different systems in the ensemble will be of the same structure, that is, composed of the same numbers of atoms of each of the different kinds enclosed in similar containers, but will differ in the positions and velocities of their component elements. In the language introduced by Gibbs they will be said to differ in *phase.*

If each system in the ensemble has m generalized coordinates $Q_1 \ldots . Q_m$ and the corresponding m momenta $P_1 \ldots P_m$, then the instantaneous phase of a system could be specified by giving the momentary values of these m coordinates and momenta. The reason for choosing the generalized coordinates and momenta rather than coordinates and velocities will later become apparent.

18. Phase Spaces.—To follow the behavior of such a system, that is, its change in phase with time, it is very convenient to think of its phase as given by the position of a representative point (phase point) in a $2m$-dimensional space (phase space) corresponding to the $2m$ coordinates and momenta whose values are to be specified. As time proceeds, the phase point for a given system will describe a trajectory in this $2m$-dimensional phase space, and the phase points for all the systems of an ensemble can be regarded as describing stream lines in the space. This device will have the advantage of providing a ready made geometrical language for discussing the behavior of the ensemble.

If our system is composed of molecules, it will also often be convenient to represent the values of the coordinates and momenta for a single molecule, in a phase space of a smaller number of dimensions than that necessary for the whole system. Thus the phase of a single molecule having the $2n$ coordinates and momenta $q_1 \ldots\ldots q_n p_1 \ldots\ldots p_n$ could be represented by the position of a point in a space of $2n$ dimensions.

In order to avoid the possibility of confusing these two different phase spaces, it will sometimes be convenient to employ the nomenclature of Ehrenfest and call the phase-space used for representing the condition of the whole system or "gas" a γ-space, and the phase space for a single molecule a μ-space.

If our system can be regarded as built up from N *monatomic* atoms, the γ-space would have $2m = 6N$ dimensions, and the μ-space for one atom would have $2n = 6$ dimensions. Furthermore, the $6N$-dimensional γ-space could evidently be regarded as composed of the N different 6-dimensional spaces belonging to the individual atoms.

19. Distribution of the Ensemble in Phase.—Each system in the ensemble will give us a point in the γ-space. We may initially distribute these points in any way we desire, and then follow their motion as they describe stream lines in this space. We shall define the density of distribution ϱ at any position in the phase space as the number of points per unit $2m$-dimensional volume at the position in question.

Different methods of initially distributing the points in the phase space can be used for different purposes. The three distributions that have been most employed are the *canonical ensemble* introduced by Gibbs and used by him and others

$$\varrho = N e^{\frac{\Psi - E}{\Theta}} \tag{20}$$

where N is the total number of systems in the ensemble, E is the energy of a single system, and Ψ and Θ are constants having the dimensions of energy; the *microcanonical ensemble* used by Gibbs, Boltzmann, Jeans and others

$$\begin{array}{ll} \varrho = \text{const.} & \text{(When the energy lies between } E \text{ and } E + dE\text{)} \\ \varrho = 0 & \text{(For other values of the energy)} \end{array} \tag{21}$$

and the *surface ensemble* used by Ehrenfest, in which the phase points are distributed on a surface in the γ-space of constant energy, with the surface density

$$\sigma = \frac{\text{const.}}{\sqrt{\left(\frac{\partial E}{\partial Q_1}\right)^2 + \ldots\ldots + \left(\frac{\partial E}{\partial P_m}\right)^2}} \tag{22}$$

All three of these distributions can be used for drawing conclusions as to the probable behavior of a single system.

The surface ensemble would at first sight seem to be the most suitable for this purpose, since all the systems of the ensemble could be given the same energy as that of the system of interest. From the point of view of mathematical simplicity, however, the surface distribution is not as easy to handle as the others, and we shall not employ it in the present book.

From many points of view, the canonical ensemble of Gibbs is mathematically the simplest and theoretically the most illuminating. Its use, for predicting the behavior of a single system, is dependent on the fact that most of the systems in the ensemble will be found to have an energy sensibly equal to the average energy for all the systems in the ensemble, the fluctuations in energy being of the order that might be expected for a system placed in a thermostat. Hence a system picked at random from the ensemble may be regarded as a fair sample of a single system left to itself. In Chapter 23 we shall consider some of the properties of this distribution.

The microcanonical ensemble is the one which has been most employed, and we shall find it the most satisfactory for our purpose. By taking the energy range dE, within which the phase-points are distributed in the γ-space, as small as we desire, we can give all the systems sensibly the same energy, and shall find that we can then regard a system, picked at random from this ensemble, as a fair sample of a single system left to itself with an energy content within this small range. Most of our later work will be based on the behavior of microcanonical ensembles.

20. Change in Density of Distribution with the Time. Liouville's Theorem.—We must now investigate for *any arbitrary* distribution of the density ϱ as a function of the Q's and P's, the change of the density ϱ with the time.

Let us consider some particular cubical element $dQ_1 \ldots dQ_m dP_1 \ldots dP_m$ in the 2 m-dimensional phase space and let

$$N = \varrho dQ_1 \ldots\ldots dQ_m dP_1 \ldots\ldots dP_m \tag{23}$$

be the number of phase points within that element of volume. In general the number of points within the element will be changing with the time, since the phase points will be moving in accordance with the laws of motion governing the behavior of the systems, and a different number of points may enter any face of the cube than leave by the opposite face. If we consider the two parallel faces of the cube perpendicular to the Q_1 axis, each having the area $dQ_2 \ldots\ldots dP_m dP_1 \ldots\ldots dP_m$ the number of points entering the cube through the first face per second will evidently be $\varrho \dot{Q}_1 dQ_2 \ldots\ldots dQ_m dP_1 \ldots\ldots dP_m$ and the number leaving through the opposite face

$$\left(\varrho + \frac{\partial \varrho}{\partial Q_1} dQ_1\right)\left(\dot{Q}_1 + \frac{\partial \dot{Q}_1}{\partial Q_1} dQ_1\right) dQ_2 \ldots\ldots dQ_m \, dP_1 \ldots\ldots dP_m \,.$$

Hence considering all the pairs of parallel faces, we can write for the total rate of change of the number of points within the cube

$$\frac{\partial N}{\partial t} = \varrho \dot{Q}_1 dQ_2 \ldots\ldots dP_m - \left(\varrho + \frac{\partial \varrho}{\partial Q_1} dQ_1\right)\left(\dot{Q}_1 + \frac{\partial \dot{Q}_1}{\partial Q_1} dQ_1\right) dQ_2 \ldots\ldots dP_m$$

$$\begin{array}{cccccccccccc} \cdot & \cdot & \cdot & \cdot & \cdot & \cdot & \cdot & \cdot & \cdot & \cdot & \cdot \\ \cdot & \cdot & \cdot & \cdot & \cdot & \cdot & \cdot & \cdot & \cdot & \cdot & \cdot \end{array}$$

$$+ \varrho \dot{P}_m dQ_1 \ldots\ldots dP_{m-1} - \left(\varrho + \frac{\partial \varrho}{\partial P_m} dP_m\right)\left(\dot{P}_m + \frac{\partial \dot{P}_m}{\partial P_m} dP_m\right) dQ_1 \ldots\ldots dP_{m-1}$$

or simplifying and omitting second-order terms

$$\frac{\partial N}{\partial t} = -\varrho\left(\frac{\partial \dot{Q}_1}{\partial Q_1} + \frac{\partial \dot{P}_1}{\partial P_1} + \ldots\ldots + \frac{\partial \dot{Q}_m}{\partial Q_m} + \frac{\partial \dot{P}_m}{\partial P_m}\right) dQ_1 \ldots\ldots dP_m$$
$$- \left(\dot{Q}_1 \frac{\partial \varrho}{\partial Q_1} + \dot{P}_1 \frac{\partial \varrho}{\partial P_1} + \ldots\ldots + \dot{Q}_m \frac{\partial \varrho}{\partial Q_m} + \dot{P}_m \frac{\partial \varrho}{\partial P_m}\right) dQ_1 \ldots\ldots dP_m$$

We have, however, from the canonical equations of motion (16) Chapter 2, the relations

$$\frac{\partial \dot{Q}_1}{\partial Q_1} = \frac{\partial}{\partial Q_1}\frac{\partial H}{\partial P_1} \quad \text{and} \quad \frac{\partial \dot{P}_1}{\partial P_1} = \frac{\partial}{\partial P_1}\left(-\frac{\partial H}{\partial Q_1}\right) \quad \text{etc.} \tag{24}$$

and since the order of differentiation is immaterial can see that pair by pair all the terms in the first parenthesis will cancel, and hence since $\partial N/\partial t$ divided by the element of volume $dQ_1 \ldots\ldots dP_m$ is the rate of change of the density ϱ, we can write

$$\frac{\partial \varrho}{\partial t} = -\left(\dot{Q}_1 \frac{\partial \varrho}{\partial Q_1} + \dot{P}_1 \frac{\partial \varrho}{\partial P_1} + \ldots\ldots + \dot{Q}_m \frac{\partial \varrho}{\partial Q_m} + \dot{P}_m \frac{\partial \rho}{\partial P_m}\right) \tag{25}$$

for the rate at which the density of distribution is changing at any point in our phase space.

21. It is immediately evident that this rate will be zero if the points themselves are uniformly distributed, since the quantities $\partial \rho/\partial Q_1$, etc., will then all individually be zero. Furthermore, since a given system cannot change its energy in the course of its motion, it is evident that the density of distribution will remain unchanged with time if the phase points are distributed as any function of the energy, but uniformly in each region between the limits E and $E + dE$. So that we may write

$$\frac{\partial \varrho}{\partial t} = 0 \tag{26}$$

whenever the distribution is merely a function of the energy of the systems.

Such ensembles, in which the density of distribution does not change with the time, are said to be in *statistical equilibrium*. It will be noted that the canonical, and micro-canonical distributions, mentioned in Section **19**, would give ensembles which are in statistical equilibrium. This can also be shown to be true for the surface distribution given in that section.[1]

22. Equation (25) gives us a general expression for the rate at which the density of distribution of an ensemble is changing at *a particular point* in the phase space, and to indicate the fact that we were interested in the change of density at some such chosen stationary point we have used the symbols of partial differentiation $\partial \varrho / \partial t$. We can also, however, follow the motion of some selected phase point and consider the change of density in the neighborhood of that moving point. To do this we merely have to transpose the expression in brackets to the other side of the equation and write

$$\left(\frac{\partial \varrho}{\partial t}\right)_{Q,P} + \left(\dot{Q}_1 \frac{\partial \varrho}{\partial Q_1} + \dot{P}_1 \frac{\partial \varrho}{\partial P_1} + \dots\dots + \dot{Q}_m \frac{\partial \varrho}{\partial Q_m} + \dot{P}_m \frac{\partial \varrho}{\partial P_m}\right) = 0 \quad (27)$$

The result gives us a theorem of wonderful simplicity. Since the density is a function of the time t and the coordinates and momenta $Q_1 \dots . P_m$, its total rate of change with the time will evidently be obtained by adding to the partial rate of change with the time, partials of the density with respect to the other quantities multiplied by their rate of change with the time. Hence the full expression gives us the total rate of change of the density for any distribution of the ensemble provided we follow a moving point in the phase space. So that using total differentials we can write in general

$$\frac{d\varrho}{dt} = 0 \quad (28)$$

This result is sometimes called Liouville's theorem, or the ***principle of the conservation of density in phase.***

The possibility of obtaining this significant result is dependent on our choice of generalized coordinates and momenta as the $2m$ axes for our phase space. If we had plotted our representative points in a phase space in which the axes were the coordinates and velocities, we should not in general have been able to achieve so simple a result, since we could not then have obtained the simplification, arising from the introduction of equations (24), which are dependent on the use of the equa-

[1] See P. and T. Ehrenfest, *Encykl. der Math. Wiss.*, IV 2, II Heft 6, page 29.

tions of motion in the Hamiltonian form involving the generalized momenta. This is the reason why the equations of motion in the canonical Hamiltonian form are so important for statistical mechanics and why in our future work momenta rather than velocities will practically always appear in the expressions we obtain. It is to Boltzmann that we owe the clear recognition of the importance of using momenta in statistical mechanical considerations.

23. With the help of equation (28) we can obtain a further conclusion as to the motion of phase points of great importance. Suppose we consider a small but finite element of volume $\delta Q_1 \ldots \delta P_m$ in the phase space. It will evidently contain the number of phase points

$$N = \varrho \delta Q_1 \ldots\ldots \delta P_m \tag{29}$$

Let us now follow the motion of this element containing these phase points, through the phase space, as the points themselves move. Since no systems are created or destroyed we shall evidently have

$$\frac{dN}{dt} = \frac{d\varrho}{dt}\delta Q_1 \ldots \delta P_m + \varrho \frac{d}{dt}\,(\delta Q_1 \ldots \delta P_m) = 0 \tag{30}$$

The first of the two terms, however, is equal to zero in accordance with equation (28), and hence we obtain the important conclusion

$$\frac{d}{dt}(\delta Q_1 \ldots \delta P_m) = 0 \tag{31}$$

This is called the *principle of the conservation of extension in phase.* In accordance with this principle, if we draw a boundary surface around the points in a given region of the phase space, and follow the motion of these points as they describe stream lines in the space, the shape of the bounding surface may change, but its volume will not alter.

24. On the Properties of a Microcanonical Ensemble.—Most of our applications of statistical mechanics to molecular systems will be based, as we shall later show, on the device of using a microcanonical ensemble of systems with energies in the range E to $E + dE$ to give an idea of the probable behavior of a single system having an energy in that same range. We shall first need to consider some of the simpler properties of such an ensemble.

First of all it should be noted that the considerations of Section **21**, see equation (26), have shown us that the uniform density of phase points in a microcanonical ensemble, lying in the shell between E and $E + dE$, will not alter with the time. Hence, if at any time we pick at random some system from the ensemble, it will have the same chance of coming from any one of the equal small regions $dQ_1 \ldots dP_m$ into which we could divide the region between E and $E + dE$.

To put this result in somewhat more convenient language, we shall define the *microscopic state* of a system as being determined by the particular small region $dQ_1 \ldots . dP_m$ in which the representative point for the system happens to lie.[2] We may then state that in a *microcanonical ensemble all the different microscopic states have the same probabilities.*

Two further conclusions can immediately be drawn from this principle.

In the first place, in most of our statistical considerations we shall not be primarily interested in the probability of only a *single* microscopic state, but rather in the combined probability of any one of a collection of microscopic states all having some feature of interest. Thus to give an example, when we are considering the vapor pressure of a crystal we shall be interested in the total probability of the collection of all possible states in which a given number of the molecules are in the crystal, the rest being in the form of vapor. It is immediately evident from the principle stated above that in a microcanonical ensemble the combined probability of such a collection of microscopic states is proportional to the total volume in the phase space corresponding to the specified properties.

In the second place, we shall often be interested in the average of some property either for all the systems in the ensemble or for those in a group having some feature in common. It is immediately evident, however, since all the microscopic states have the same probability that in a microcanonical ensemble an average can be taken by summing the values of the property in question for all the microscopic states and dividing by the number of such states in the group.

25. Application to Molecular Systems.—The foregoing simple results provide practically all that is necessary as a fundamental apparatus for the applications of statistical mechanics, and we are now ready to discuss the behavior of actual systems of molecules.

Let us consider a system of molecules enclosed in a suitable container and having the energy content E. In general this energy content, and the components of linear and angular momenta of the whole system, which we shall take as zero, will be the only dynamical quantities of which we have any knowledge, since the motions of the individual particles are too complex for us to follow.[3] Hence, since we do not know the precise molecular configurations and velocities of our system, we must consider all possible phases that such a system with the known

[2] We shall usually find it convenient to regard the regions $dQ_1 \ldots dP_m$ as small but finite, so that there will be only a finite number of microscopic states.

[3] The fact that they are beyond the powers of our observational methods is a secondary matter. The methods would be just as necessary in the case of a system of equal complexity even if composed of particles of size large enough to observe easily.

energy content E might assume, and to do this we can use a microcanonical ensemble of systems similar to the one of interest with their representative points distributed uniformly throughout a shell which lies between the surfaces determined in the $2m$-dimensional space by E and $E + dE$.

26. Introduction of the Ergodic Hypothesis.—So far we have proceeded without the introduction of any hypotheses, except that the system obeys the laws of dynamics in the canonical form. To go further, however, we must now introduce the assumption that such a microcanonical ensemble of systems having energies between E and $E + dE$ gives a fair representation of the different phases which will be assumed in the course of time by a single system having an energy content between E and $E + dE$. To state this more specifically we shall have to assume that the results obtained by selecting a collection of systems at random from such an ensemble are practically the same as would be obtained by considering a given system at random times.

A certain empirical justification for this assumption is given by the fact that our later developments based on its use are found to meet the test of experiment.

The theoretical justification for the assumption is more difficult and less certain, though nevertheless plausible.

27. In the first place it should be noted that the maintenance of uniform density in a microcanonical ensemble, guaranteed by equation (26), means that the representative points for the systems have no tendency to crowd into any particular region of the phase space. If such a crowding did take place, the representative point for any given system would at any arbitrary time be more likely to lie in such a region of compression than elsewhere in the phase space. In the absence of such crowding, however, we can definitely state that a *given system taken at any random time has an equal chance of being in any one of the different microscopic states through which it actually passes in the course of its history.*

This definitely valid conclusion is of course, however, not identical with the statement that the system has an equal chance of being in any one of all the microscopic states which lie in the phase space between the surfaces determined by E and $E + dE$, as is true for a microcanonical ensemble, since a given system may not in the course of its complete history be able to reach all the regions between E and $E + dE$.

28. To meet this the *ergodic hypothesis,* as it was called by Boltzmann, or the *principle of continuity of path,* as it was called by Maxwell, makes the definite assumption that a system of molecules actually does pass through all possible microscopic states consonant with its energy content before completing its cycle of motions. This hypothesis, if true, would of course make the selection of random samples from a microcanonical ensemble absolutely identical with the consideration of a given system at random times and hence would justify the further procedure we wish to undertake.

The ergodic hypothesis in its simple form, however, is not even plausible. The reentrant orbits of astronomy have made us quite familiar with the possibility of an infinite number of closed paths for a given dynamical system all of them having the same energy content. Indeed no single example of a dynamical system of more than one degree of freedom has ever been discovered in which the motion carries the system through all possible phases consonant with its energy content.

29. It seems evident that we must reject the ergodic hypothesis in its simple form as not having sufficient plausibility and investigate other similar assumptions which would lead to equivalent results. We may consider three further simple possibilities.

a. It might be assumed that for a system of molecules with a given energy content there are a number of closed paths possible, but that one of them is infinitely longer than all the others put together, so that we could neglect the possibility of the system being on any path but the longest, and the microscopic states lying along this path would include practically all the microscopic states in the region between E and $E + dE$. Such a possibility was apparently first suggested by Jeans, but we must agree with him in rejecting it, as being no more plausible than the assumption of a single closed path itself.

b. A second possibility would be to assume that for a given energy content a system of molecules can move on any one of a number of closed paths, but cannot of itself get from one of these closed paths to another. It might, however, be fortuitously removed from one path to another by external agencies in such a way as to have an equal probability of being in any microscopic state belonging to the corresponding microcanonical ensemble. This point of view has also been suggested by Jeans, who has further indicated the plausibility of regarding one of the molecules themselves as a fortuitous external disturbing agency which acts on the system which is now regarded as composed of all the other molecules present. This assumption would of course lead to the desired results.

c. A third possibility, which appears to the present writer most satisfactory for the purposes at hand, is to grant that there may indeed be more than one closed path for the motion of a system of molecules, with a given energy content, perhaps indeed a great many such paths.

Lying along each closed path will be a succession of microscopic states, through which the system must pass before it rearrives at the starting point; this, however, it will inevitably do, since we shall take the regions $dQ_1 \ldots . dP_m$ which specify the different microscopic states as small but nevertheless finite, so that there will not be an infinite number of possible states.

Let us now consider the motion of a given system when left to itself. Its representative point will have to lie on one of the possible paths which correspond to its energy content. We cannot say on which path, but, as we saw above, we can say that at any random time it has an equal chance of being in any one of the microscopic states that lie along this particular path.

So far we have introduced no assumption. Let us now consider, however, the probable nature of the different closed paths. It is evident that for an actual system containing many molecules most of these closed paths would have to be very long. Consider for example a gas composed of many molecules moving back and forth between the walls of its container. Even if the molecules collided only with these walls and not among themselves, it is evident that we should have to wait an exceedingly long time before we should again find every individual molecule back at its starting point, and when we allow for collisions among the molecules, the time is enormously longer yet. Hence at least the vast majority of the paths will contain an enormous number of microscopic states taken out from those that lie in the total shell between E and $E + dE$.

It hence seems reasonable, since so many states have to be taken, to assume, by and large, that the microscopic states lying along the particular closed path on which the system of interest happens to lie are themselves a fair sample of the different kinds of states present in the whole region of phase space between the limits E and $E + dE$. If this be admitted, any large collection of samples chosen at random from the whole microcanonical ensemble between E and $E + dE$ will have practically the same properties as the collection obtained by observing the given system at random times, and the use of the microcanonical ensemble as giving a fair representation of the succession of states arrived at in time by a single system is justified.

To illustrate this conclusion by a specific example, we shall presently find that the vast majority of all the states in a microcanonical ensemble have distributions of molecular velocities that do not differ appreciably from Maxwell's law. We shall then feel justified in believing that this is also probably true for the vast majority of the states lying along the particular path on which a given system is to be found.

We shall henceforth proceed on the assumption that a microcanonical ensemble may be taken as giving a fair representation of the successive states of an actual molecular system.

30. Nature of the Systems to be Considered.—Our further considerations will usually deal especially with systems containing a large number of identical elements of one or more kinds, for example molecules, atoms, electrons, modes of vibration or what-not, and for short we shall use for the present the general name molecules as applying to such elements.

Evidently the total number of degrees of freedom of such a system will be equal to the sum of the degrees of freedom belonging to the individual molecules of which the system is composed. Furthermore, we can consider the coordinates and momenta used for describing the system as a whole as being the total collection of coordinates and momenta which belong to the individual molecules. Thus, for example, if we have a system composed of N molecules each having n degrees of freedom, we can identify the coordinates and momenta $Q_1 \ldots . P_m$ belonging to the system as a whole with the N sets of variables $q_1 \ldots . q_n p_1 \ldots . p_n$ belonging to the N molecules.

In the case of systems containing many identical molecules, it should be noted that we must think of these molecules as being separately identifiable so that a specification of the microscopic state of the system gives us the location and motion of each individual molecule, and a new microscopic state would result even if all we did was merely to swap some pair of molecules with regard to their coordinates and momenta. It is evident, however, that for the purposes of predicting the further gross behavior of the system, such a swap would be of no importance. This is one of the reasons why we shall usually be interested in a collection of microscopic states rather than in a single microscopic state.

31. γ-Weight and μ-Weight.—In concluding this chapter we may mention a nomenclature that we shall often find very convenient.

As already stated, in most of our statistical considerations we shall not be primarily interested in the probability of only a *single* microscopic state, but rather in the total probability of a collection of microscopic states, all having some common feature of interest. It is evident that the total probability of such a collection of microscopic states will be proportional to the total number of microscopic states involved, since they are all of equal probability. Each microscopic state, however, corresponds to a particular one of the equal regions $dQ_1 \ldots . dP_m$ into which we have divided our γ-space. Hence the total probability of our collection of microscopic states will be proportional to the total volume in the γ-space which corresponds to the collection.

Adopting the convenient language of Ehrenfest and Trkal, we shall call the collection of microscopic states of the system in which we are interested *the state* of the system, and the corresponding volume in the phase space the γ-weight of the state. We shall represent the γ-weight of a state by the symbol γ and the probability of the state will be proportional to this quantity.

Similarly we shall define the μ-weight μ of a given state of a particular molecule as the volume of the μ-space in which the representative point for the molecule falls. In the case of a dilute system, we shall find it possible to calculate the γ-weight of a given state by considering the μ-weights of the molecules of which the system is composed.

Chapter 4

Derivation of the Maxwell-Boltzmann Distribution Law

32. Specification of a Given State of the System.—Our first application of the principles discussed in the preceding chapter will be to determine, for a physical-chemical system which has come to equilibrium, how many molecules of each particular kind have values for their coordinates and momenta which fall in any particular range. The result which we shall obtain is called the Maxwell-Boltzmann distribution law for the coordinates and momenta of a system of molecules.

Let A, B, C, etc., be the numbers of molecules of each of the various kinds which go to make up the complete system, a, b, c, etc., being the number of degrees of freedom respectively for each kind of molecule.

In order to specify the condition of any one of these molecules, for example of the kind A, we shall make use of a μ-space of $2a$-dimensions corresponding to the $2a$ coordinates and momenta belonging to each one of these molecules. The position of a representative point in this phase space will evidently give the condition of the molecule.

We shall divide this μ-space into equal infinitesimal regions having the volume

$$d\sigma_A = dq_1 \ldots . dq_a dp_1 \ldots . dp_a \tag{32}$$

and number them off by the designations $1A$, $2A$, $3A$, etc., in order to distinguish them one from another. In general the first three coordinates q_1, q_2, q_3 will be taken as the Cartesian coordinates x, y, z which locate the center of gravity of the molecule, and the first three momenta p_1, p_2 p_3 will then be the components of linear momentum $m\dot{x}$, $m\dot{y}$ and $m\dot{z}$ for the motion of the molecule as a whole. The remaining coordinates and momenta will refer to the internal configuration and motions of the molecule, for example its rotation. Hence the statement that the phase-point for a given molecule falls in a particular region $d\sigma_A$ will give us the location and motion of the molecule in space, as well as its internal configurations and motions.

We can now specify any given state of the system by giving the numbers of molecules A_1, A_2, A_3, ... B_1, B_2, B_3, ... C_1, C_2, C_3, etc., of the different kinds which have coordinates and momenta lying in the different infinitesimal regions Nos. $1A$, $2A$, $3A$, ... $1B$, $2B$, $3B$, ... $1C$, $2C$, $3C$, etc., into which we have divided the μ-spaces for the different kinds of molecules.

33. Probability of a Given State of the System.—We must now consider the probability of any given state of the system, since we shall presently inquire as to what state has the greatest probability.

To do this we must first note that *a state* of the system, specified in the manner just described, will evidently correspond to many different microscopic states, since if we had any exemplar of the system in a microscopic state with the above distribution of molecules, we could evidently obtain a new microscopic state with the same distribution of molecules by swapping any pair of molecules of the same kind that had their coordinates and momenta in different regions of the μ-space. Thus if we had 2 molecules of the kind A in each of the regions Nos. $1A$ and $2A$, we could rearrange these 4 molecules in 6 different ways and each time have two molecules in each of the regions. Hence when we consider the great number of molecules of each kind available for such swaps, it is evident that there will ordinarily be an enormous number of individual microscopic states corresponding to *the* state of the system.

34. We are now ready to investigate the probability of such a state of the system.

In the last chapter we saw that the probability of any state of the system could be taken proportional to the volume of the γ-space corresponding to that state, or as we shall say to the γ-weight of the state.

If now we consider any single microscopic state that belongs to the total collection of microscopic states that go to make up *the* state in which we are interested, it is evident that each molecule will have to fall in a specified region in its own μ-space, so that the representative point for the whole system will have to fall in a perfectly definite region in the γ-space, having a hyper-volume

$$d\sigma = (d\sigma_A)^A\ (d\sigma_B)^B\ (d\sigma_C)^C \ldots . \tag{33}$$

which is obtained by multiplying together the volumes of μ-space assigned to each of the $A + B + C +$, etc., molecules in the system. The result will perhaps be more easily understood, if it is appreciated that each molecule contributes its own axes in the μ-space to the total number of axes necessary for the γ-space, and thus $d\sigma$ as given by equation (33), is the product of infinitesimal distances along each of these $2aA + 2bB + 2cC +$, etc., axes. Specifying a given region $d\sigma$ in the γ-space thus tells the values of the coordinates and momenta for each separate molecule.

Since the γ-weight of any state of the system is the volume of γ-space corresponding to it, the γ-weight of the above particular microscopic state of the system will be

$$\gamma = do \tag{34}$$

35. We are, however, interested in the probability or γ-weight of the total collection of microscopic states which correspond to *the* state of the system in which A_1, A_2, A_3, etc., B_1, B_2, B_3, etc., C_1, C_2, C_3, etc., etc., molecules have coordinates and momenta lying in the regions Nos. $1A$, $2A$, $3A$, etc., $1B$, $2B$, $3B$, etc., $1C$, $2C$, $3C$, etc., etc. Evidently the total γ-weight of this collection can be found by multiplying the γ-weight of one microscopic state, do, by the total number of microscopic states in the collection. We obtain for the γ-weight of *the* state

$$\gamma = \frac{\underline{|A}\ \underline{|B}\ \underline{|C} \ldots\ldots}{\underline{|A_1}\ \underline{|A_2}\ \underline{|A_3} \ldots\ \underline{|B_1}\ \underline{|B_2}\ \underline{|B_3} \ldots\ \underline{|C_1}\ \underline{|C_2}\ \underline{|C_3} \ldots} do \tag{35}$$

where the factor which multiplies do is the number of ways in which we can rearrange all the molecules of the system so as to get microscopic states belonging to the collection. To show this latter point, we note that the total number of ways in which we could arrange A molecules in a linear series would evidently be $A(A-1)(A-2) \ldots (2)(1) = \underline{|A}$. This total number of permutations, however, is greater than the number in which we are interested, since some of the different permutations would result merely from changes in the order of the molecules in some particular region do_A and this would not give us a new microscopic state. Hence the total number of permutations $\underline{|A}$ is equal to the number of permutations in which we are interested multiplied by $\underline{|A_1}\ \underline{|A_2}\ \underline{|A_3} \ldots$ which is the number of permutations that do not involve the transfer of a molecule from one region to another. Hence dividing $\underline{|A}$ by $\underline{|A_1}\ \underline{|A_2}\ \underline{|A_3} \ldots$ we obtain the number of ways of rearranging the molecules of kind A to obtain different microscopic states, and by extension to include all the different kinds of molecules obtain the expression given.

36. We shall almost always find it easiest to work with the logarithm of the γ-weight, and for purposes of simplification shall introduce for each of the factorial numbers occurring in equation (35) the Sterling approximation

$$\underline{|N} = \sqrt{2\pi N}\left(\frac{N}{e}\right)^N \tag{36}$$

This formula is valid only for large numbers, so that the possibility of using it in equation (35) is dependent on the assumption that the number of molecules $A_1\ A_2\ A_3 \ldots B_1\ B_2\ B_3 \ldots$ etc., in each of the regions into which we have divided the μ-spaces for the different kinds of molecules are themselves large. Since we have, however, an enormous number of molecules of each kind, this can be made true for the regions of interest in the μ-space by not taking the infinitesimals $d\sigma_A$, $d\sigma_B$, etc., too small. This point will be more fully discussed in a later section.

Taking the logarithm of (36) we obtain

$$\log \underline{|N} = N \log N - N + \log \sqrt{2\pi N} \tag{37}$$

where on the basis of our assumptions the last term can be neglected in comparison with the other two.

37. Taking the logarithm of the expression given by equation (35) and substituting (37) we obtain

$$\begin{aligned}\log \gamma = \; & A \log A \quad + B \log B \quad + C \log C \quad + \dots \\ & -A \qquad\quad -B \qquad\quad -C \qquad\quad - \dots \\ & -A_1 \log A_1 - A_2 \log A_2 - A_3 \log A_3 - \dots \\ & -B_1 \log B_1 - B_2 \log B_2 - B_3 \log B_3 - \dots \\ & -C_1 \log C_1 - C_2 \log C_2 - C_3 \log C_3 - \dots \\ & \cdot \quad \cdot \quad \cdot \quad \cdot \quad \cdot \quad \cdot \quad \cdot \\ & +A_1 \qquad +A_2 \qquad +A_3 \qquad + \dots \\ & +B_1 \qquad +B_2 \qquad +B_3 \qquad + \dots \\ & +C_1 \qquad +C_2 \qquad +C_3 \qquad + \dots \\ & \cdot \quad \cdot \quad \cdot \quad \cdot \quad \cdot \quad \cdot \quad \cdot \\ & + \log d\sigma\end{aligned}$$

It should be noted, however, that the total number of molecules of any kind must be equal to the sum of the numbers in the individual regions, so that we have for example

$$A = A_1 + A_2 + A_3 \dots\dots$$

Hence we may simplify the above to obtain as our final expression for the logarithm of the γ-weight of the specified state

$$\begin{aligned}\log \gamma = \; & A \log A + B \log B + C \log C + \dots \\ & - \sum_i A_i \log A_i - \sum_j B_j \log B_j - \sum_k C_k \log C_k - \dots \\ & + \log d\sigma\end{aligned} \tag{38}$$

where the summations Σ are to be taken over all the infinitesimal regions i, j, k, etc., for the μ-spaces corresponding to each kind of molecule.

38. State of Maximum Probability and Derivation of the Maxwell-Boltzmann Distribution Law.—We shall now be specially interested in the state of maximum probability when the system has a given volume, energy content, and composition. The condition for maximum probability will evidently be

$$-\delta \log \gamma = \sum_i (\log A_i + 1)\delta A_i + \sum_j (\log B_j + 1)\delta B_j + \ldots . = 0 \quad (39)$$

where the variation δ is brought about by a change in the numbers of molecules having coordinates and momenta in the different infinitesimal regions. This variation, however, cannot be carried out entirely arbitrarily, since we desire to know the most probable condition for a system having a given volume, energy and composition, and hence must maintain the volume, energy and number of molecules of each kind in the system constant, when the variation is performed.

39. The maintenance of constant volume is satisfied if our variation does not involve the introduction of molecules into regions corresponding to values of the coordinates which lie outside of the container, and this will obviously be taken care of since we shall consider only the same infinitessimal regions Nos. $1A$, $2A$ .... $1B$, $2B$ .... etc., in our original and varied states of the system.

The maintenance of constant energy can be achieved if we introduce the additional equation

$$\delta E = \sum_i \varepsilon_i \delta A_i + \sum_j \varepsilon_j \delta B_j + \sum_k \varepsilon_k \delta C_k + \ldots = 0 \quad (40)$$

where ε_i for example is the rate of change in the energy of the system per molecule of the kind A introduced into the i'th infinitesimal region, as given by the equation[1]

$$\varepsilon_i = \frac{\partial E}{\partial A_i} \quad (41)$$

Finally the maintenance of constant composition will be taken care of if we introduce the further equations

$$\begin{aligned} \delta A &= \sum_i \delta A_i = 0 \\ \delta B &= \sum_j \delta B_j = 0 \\ \delta C &= \sum_k \delta C_k = 0 \\ &\text{etc.} \end{aligned} \quad (42)$$

40. We now have the set of simultaneous equations (39), (40), and (42), all of which must be satisfied. The equations can be most easily treated by the well-known method of undetermined multipliers.

[1] For dilute systems ε_i will be the actual energy of a molecule in region i. For concentrated systems ε_i can depend on the mutual energy of the molecules.

Multiplying equation (40) by an undetermined quantity μ and equations (42), by undetermined quantities λ_A, λ_B, λ_C, etc., and adding to (39) we obtain,

$$\sum_i(\log A_i + 1 + \mu\varepsilon_i + \lambda_A)\delta A_i + \sum_j(\log B_j + 1 + \mu\varepsilon_j + \lambda_B)\delta B_j + \ldots. = 0 \qquad (43)$$

and this equation must be true for arbitrary variations δA_i, δB_j, etc., so that we obtain as the solution

$$\begin{aligned} &\log A_i + 1 + \mu\varepsilon_i + \lambda_A = 0 \\ &\log B_j + 1 + \mu\varepsilon_j + \lambda_B = 0 \\ &\text{etc.} \end{aligned} \qquad (44)$$

It will be noted that μ is the same quantity for all the different kinds of molecules in the system, but that there is no relation between the quantities λ_A, λ_B, λ_C, etc., for the different kinds of molecules.

Equations (44) may now be solved for the numbers of molecules A_i, B_j, etc., and written in the more convenient form

$$\begin{aligned} A_i &= a_A\, e^{-\mu\varepsilon_i} \\ B_j &= a_B\, e^{-\mu\varepsilon_j} \\ &\text{etc.} \end{aligned} \qquad (45)$$

where a_A, a_B, etc., are themselves constants for each kind of molecule, corresponding to the constants λ_A, λ_B, etc., in equations (44).

41. These equations are as a matter of fact an expression of the *Maxwell-Boltzmann distribution law* for the molecules of a system when it is in its state of *maximum probability*. They show that the number of molecules of any kind, for example A, having coordinates and momenta falling in i'th infinitesimal region of the phase space appropriate for plotting the condition of molecules of that kind, is determined by the energy increase of the system ε_i, which is brought about by the introduction of a molecule into that region. Since the total energy of the system is limited, it is evident that μ is a positive quantity, and that the number of molecules per infinitesimal region in the phase space decreases as ε increases.

42. Critique of the Method of Derivation.—Before proceeding to the expression of these equations in the more familiar forms ordinarily used in making applications of the Maxwell-Boltzmann law, we must consider the effect resulting from the introduction of the Sterling approximation in the course of the derivation.

Since the Sterling formula for factorial N becomes rapidly more nearly true as N is increased, we might expect equations (45) to be substantially true for those regions in the phase space where the number of molecules is large, that is, where ε is small. On the other hand, for large values of ε we might expect the equations to fail. This is, moreover, indicated by the form of equations (45) themselves, since for large enough values of ε we should evidently arrive at fractional numbers of molecules.

To show that equations (45) can really be taken as correct for those regions in the μ-space where the numbers of molecules are large, it will be sufficient to consider a simple system containing N molecules all of the same kind.

In analogy with equation (35) in Section **35** we can write for the γ-weight of a particular distribution of the N molecules of the system

$$\gamma = \frac{\underline{|N}}{\underline{|N_1}\ \underline{|N_2}\ \underline{|N_3}\ldots\ldots} do \tag{46}$$

or

$$\log \gamma = \log \underline{|N} - \sum_i \log \underline{|N_i} + \log do \tag{47}$$

where N_i is the number of molecules in the i'th region of the μ-space. For the variation in γ-weight with variations in the arrangement we shall then obtain, considering only regions which do contain large numbers of molecules,

$$\delta \log \gamma = -\sum_i \frac{\delta \log \underline{|N_i}}{\delta N_i} \delta N_i = 0 \tag{48}$$

and this is set equal to zero as the condition for a maximum. In carrying out variations of the arrangement, however, we shall wish to keep the total energy E constant and the number of molecules N in the system constant, in accordance with the equations

$$\sum_i \varepsilon_i \delta N_i = 0 \tag{49}$$

and

$$\sum_i \delta N_i = 0 \tag{50}$$

Treating equations (48), (49) and (50) as before by the method of undetermined multipliers, we easily obtain the result

$$\frac{\delta \log \underline{|N_i}}{\delta N_i} + \mu \varepsilon_i + \lambda = 0 \tag{51}$$

which will be true even for regions containing a small number of molecules N_i.

43. For regions which *do* contain large numbers of molecules, however, we can use the *Sterling approximation* for factorial N, giving us

$$\frac{\delta}{\delta N_i} \log \sqrt{2\pi N_i} \left(\frac{N_i}{e}\right)^{N_i} + \mu\varepsilon_i + \lambda = 0$$

or

$$\frac{1}{2N_i} + \log N_i + \mu\varepsilon_i + \lambda = 0$$

and since N_i is large the first term can be neglected, and the equation solved in the form

$$N_i = ae^{-\mu\epsilon_i} \tag{52}$$

which agrees with equations (45).

We thus see that the Maxwell-Boltzmann distribution law does hold for all the regions which contain large numbers of molecules, and this is all that is necessary for our further purposes, since we can neglect the effect of molecules in the remaining regions.

44. Probability of States Which Deviate from the Maxwell-Boltzmann Distribution.—The Maxwell-Boltzmann law gives us the distribution of the coordinates and momenta of the molecules when the system is in its state of *maximum* probability. We must now investigate the probability of states which deviate from the Maxwell-Boltzmann distribution. We shall find that the Maxwell-Boltzmann distribution is not only the most probable but very much more probable than distributions which differ even moderately from it.

For simplicity let us take a system containing N molecules of only a single kind. Restricting ourselves to the consideration of regions containing large numbers of molecules, we can write in accordance with equation (38) of Section **37** for the γ-weight of any distribution

$$\log \gamma = N \log N - \sum_i N_i \log N_i + \log d\sigma \tag{53}$$

and for the γ-weight of a neighboring different distribution

$$\log (\gamma + \Delta\gamma) = N \log N - \sum_i (N_i + \Delta N_i) \log (N_i + \Delta N_i) + \log d\sigma \tag{54}$$

where $\Delta\gamma$ is the change in γ-weight and ΔN_i is the change in the number of molecules in i'th region when we pass from the original to the neighboring state. Subtracting (53) from (54), we obtain

$$\log \frac{\gamma + \Delta\gamma}{\gamma} = -\sum_i (N_i + \Delta N_i) \log (N_i + \Delta N_i) + \sum_i N_i \log N_i$$

$$= -\sum_i \left\{ N_i \log\left(1 + \frac{\Delta N_i}{N_i}\right) + \Delta N_i \log\left(1 + \frac{\Delta N}{N_i}\right) + \Delta N_i \log N_i \right\}.$$

Taking now the changes ΔN_i small enough so that we can expand the logarithm of $(1 + \Delta N_i/N_i)$ into a series and neglect higher terms, we obtain

$$\log \frac{\gamma + \Delta\gamma}{\gamma} = -\sum_i \left[N_i \left\{ \frac{\Delta N_i}{N_i} - \frac{1}{2}\left(\frac{\Delta N_i}{N_i}\right)^2 + \ldots \right\} \right.$$

$$\left. + \Delta N_i \left\{ \frac{\Delta N_i}{N_i} - \ldots \right\} + \Delta N_i \log N_i \right]$$

$$= -\sum_i \frac{1}{2} N_i \left(\frac{\Delta N_i}{N_i}\right)^2 - \sum_i \Delta N_i - \sum_i \Delta N_i \log N_i \; .$$

Let us now take the original distribution corresponding to equation (53) as the state of maximum probability agreeing with the Maxwell-Boltzmann law. We can then substitute for N_i in the last term the expression given by equation (52), and write

$$\log \frac{\gamma}{\gamma_{\text{max.}}} = -\frac{1}{2} \sum_i N_i \left(\frac{\Delta N_i}{N_i}\right)^2 - \sum_i \Delta N_i - \log a \sum_i \Delta N_i + \mu \sum_i \varepsilon_i \Delta N_i$$

where $\gamma_{\text{max.}}$ is the probability of the Maxwell-Boltzmann state, and γ that of the state which deviates therefrom. It is immediately evident, however, that the last three terms in this equation must be zero, since we must preserve the total number of molecules and total energy of the system constant. We obtain as our final expression

$$\log \frac{\gamma}{\gamma_{\text{max.}}} = -\frac{1}{2} \sum_i N_i \left(\frac{\Delta N_i}{N_i}\right)^2 \qquad (55)$$

and note as should evidently be true that any change from the Maxwell-Boltzmann distribution will lead to a decrease in probability dependent on the second power of the deviations.

45. It is at once evident from equation (55) that the γ-weight is very sensitive to deviations from the Maxwell-Boltzmann distribution law, whenever large numbers of molecules are involved. Consider for example that we had two similar regions in the μ-space each containing in the most probable state as small a number as 1000 molecules, then the transfer of *one* per cent of these molecules from one of these regions to the other would alone reduce the probability of the configuration by *ten* per cent. Deviations involving larger numbers of molecules would have an even greater effect. Thus if we had a cubic centimeter of gas containing 2×10^{19} molecules, and considered a state with an average deviation

$$\frac{\Delta N_i}{N_i} = 0.01$$

for all regions, we should obtain for the ratio of probabilities

$$\frac{\gamma}{\gamma_{\text{max.}}} = e^{-10^{15}}$$

so that a state with an average deviation of one per cent from the normal number of molecules in all regions of the μ-space would have almost no chance of occurring.

We may conclude that in a system containing a large number of molecules appreciable deviations from the Maxwell-Boltzmann distribution will have a very small probability of occurrence. Hence we may take the Maxwell-Boltzmann distribution law as substantially representing the state of the molecules in a system which has come to equilibrium, and may proceed to deduce from it the properties of such systems.

46. Value of the Constant a Occurring in the Maxwell-Boltzmann Distribution Law.—We shall now return to the equations which we obtained in Section **40** for the Maxwell-Boltzmann distribution law

$$\begin{aligned} A_i &= a_A e^{-\mu\epsilon_i} \\ B_j &= a_B e^{-\mu\epsilon_j} \\ &\text{etc.} \end{aligned} \tag{56}$$

or in general

$$N_i = a e^{-\mu\epsilon_i} \tag{57}$$

and show the methods of transformation into the more familiar expressions which are convenient for actual application to physical-chemical problems. To do this we shall first need to obtain more rational expressions for the constants a and μ which have so far been undetermined.

47. The quantity N_i occurring in equation (57) is the number of molecules at equilibrium whose coordinates and momenta have values falling in the i'th one of the equal infinitesimal regions

$$do = dq_1 \ldots\ldots dp_n \tag{58}$$

into which we have divided the $2n$-dimensional μ-space used for plotting the values of these coordinates and momenta. It will obviously be natural to take this number proportional to the total number of molecules N of the kind in question and to the volume $d\sigma$ of the infinitesimal region. This we can do by writing

$$a = NCdq_1 \ldots . dp_n \tag{59}$$

where C is a new constant. Substituting into (57) and changing the notation to that corresponding to the differential form, we can then write the Maxwell-Boltzmann distribution law in the more familiar way

$$dN = NCe^{-\mu\epsilon}\, dq_1 \ldots . dp_n \tag{60}$$

where dN is the number of molecules whose coordinates and momenta have values falling in an infinitesimal range $dq_1 \ldots . dp_n$, the quantity ϵ being the energy increase in the system per molecule introduced into this region.

48. This quantity ϵ is evidently a function of the coordinates and momenta $q_1 \ldots p_n$, and when the form of this function is known it then becomes possible to determine the value of the constant C by integration. To do this we may write

$$\int dN = \int \ldots . \int NCe^{-\mu\epsilon}\, dq_1 \ldots . dp_n$$

and if this integration is carried out over all the regions of the μ-space, we evidently obtain, since N and C are constants,

$$N = NC \int \ldots . \int e^{-\mu\epsilon}\, dq_1 \ldots . dp_n$$

or

$$C = \frac{1}{\int \ldots . \int e^{-\mu\epsilon}\, dq_1 \ldots . dp_n} \tag{61}$$

so that C can be evaluated whenever the indicated integration can be performed. Since the dependence of ϵ on the q's and p's will be different for different kinds of molecules, it is evident that the constant C will also depend on the kind of molecule, corresponding to the lack of relation between the different constants α_A, α_B, etc., already noted in Section **40.**

49. Value of the Constant μ Occurring in the Maxwell-Boltzmann Distribution Law.—To obtain the value of the constant μ occurring in the Maxwell-Boltzmann distribution law, it is necessary to apply the law to the calculation of some actual macroscopic property of the system. To do this we shall consider that a part of the system consists of a container of volume V, enclosing N molecules of a monatomic gas so dilute that the forces between the molecules of the gas can be neglected; in other words N molecules of a perfect monatomic gas. We shall then calculate the pressure of this gas and find it related to the quantity μ. On the other hand the pressure of such a gas is related by the gas laws to the temperature T of the whole system, so that we shall finally obtain what proves to be a very simple relation between μ and T.

Consider a gas consisting of N non-attracting point particles of mass m. For such a system the generalized coordinates may be taken as the Cartesian coordinates x, y and z locating the particles, and the corresponding momenta as the components of linear momentum $m\dot{x}$, $m\dot{y}$ and $m\dot{z}$, so that the Maxwell-Boltzmann distribution law as given by equation (60) can be written in the form

$$dN = NCe^{-\mu\epsilon}\, dxdydzmd\dot{x}md\dot{y}md\dot{z} \tag{62}$$

where dN is the number of particles having coordinates falling in the region $dx\, dy\, dz$ and velocities in the range $d\dot{x}\, d\dot{y}\, d\dot{z}$. Or integrating over the whole volume of the container V, we obtain for the total number of molecules which have velocities in the range in question

$$dN = NCVm^3e^{-\mu\epsilon}\, d\dot{x}d\dot{y}d\dot{z} \tag{63}$$

and denoting constant terms by the single symbol A may write

$$dN = NAe^{-\mu\epsilon}\, d\dot{x}d\dot{y}d\dot{z} \tag{64}$$

50. This result can immediately be used to obtain an expression for the pressure of the gas. Consider an element of surface dS, for convenience perpendicular to the X-direction, and let the pressure of the gas be P so that the force exerted on the surface dS by particles which are coming up to it along the X-direction and then rebounding is PdS.

Now by equation (64) the total number of particles in the container having a component of velocity in the X-direction between $\dot{x}$ and $\dot{x} + d\dot{x}$ is evidently

$$\int_{-\infty}^{+\infty}\int_{-\infty}^{+\infty}\int_{\dot{x}}^{\dot{x}+d\dot{x}} NAe^{-\mu\epsilon}\, d\dot{x}d\dot{y}d\dot{z}$$

But $\dot{x}dS$ gives us the volume which contains the number of particles having the velocity $\dot{x}$ which will reach the surface dS per second. Hence the fraction of the above molecules which impinge on dS per second will be $\dot{x}dS/V$ and the actual number will be

$$\frac{\dot{x}dS}{V}\int_{-\infty}^{+\infty}\int_{-\infty}^{+\infty}\int_{\dot{x}}^{\dot{x}+d\dot{x}} NAe^{-\mu\epsilon}\, d\dot{x}d\dot{y}d\dot{z}$$

Finally, the momentum which these particles give to the surface, allowing for the effect of the rebound, will be

$$2m\dot{x}\frac{\dot{x}dS}{V}\int_{-\infty}^{+\infty}\int_{-\infty}^{+\infty}\int_{\dot{x}}^{\dot{x}+d\dot{x}} NAe^{-\mu\epsilon}\,d\dot{x}d\dot{y}d\dot{z}$$

The force PdS, however, which the gas exerts on the element of surface will be the total momentum given to the element per second by the rebounding molecules, and this can be calculated by integrating the above expression for all possible values of $\dot{x}$ between zero and infinity. We obtain

$$PdS = \frac{2dS}{V}\int_{-\infty}^{+\infty}\int_{-\infty}^{+\infty}\int_{0}^{+\infty} NAe^{-\mu\epsilon}\,m\dot{x}^2d\dot{x}d\dot{y}d\dot{z}$$

Cancelling dS, transposing V, and noting that symmetry considerations will permit us to change the limits of integration, we obtain

$$PV = \int_{-\infty}^{+\infty}\int_{-\infty}^{+\infty}\int_{-\infty}^{+\infty} NAe^{-\mu\epsilon}\,m\dot{x}^2d\dot{x}d\dot{y}d\dot{z} \tag{65}$$

51. The value of this integral can, however, be obtained by a simple artifice. Returning to equation (64) and substituting for the energy ϵ its evident value in terms of the velocities $m/2(\dot{x}^2+\dot{y}^2+\dot{z}^2)$, we can write

$$N = \int_{-\infty}^{+\infty}\int_{-\infty}^{+\infty}\int_{-\infty}^{+\infty} NAe^{-\mu\frac{m}{2}(\dot{x}^2+\dot{y}^2+\dot{z}^2)}\,d\dot{x}d\dot{y}d\dot{z}$$

or performing the integration with respect to $\dot{x}$ by parts, we obtain

$$N = \left|\int_{-\infty}^{+\infty}\int_{-\infty}^{+\infty} NAe^{-\mu\frac{m}{2}(\dot{x}^2+\dot{y}^2+\dot{z}^2)}\,\dot{x}d\dot{y}d\dot{z}\right|_{\dot{x}=-\infty}^{\dot{x}=+\infty}$$
$$+\int_{-\infty}^{+\infty}\int_{-\infty}^{+\infty}\int_{-\infty}^{+\infty} NAe^{-\mu\frac{m}{2}(\dot{x}^2+\dot{y}^2+\dot{z}^2)}\,\mu m\dot{x}^2d\dot{x}d\dot{y}d\dot{z}$$

Owing to the negative sign of the exponent, it is evident, however, that on the substitution of the limits the first term on the right-hand side of this equation becomes zero, and we obtain

$$N = \int_{-\infty}^{+\infty}\int_{-\infty}^{+\infty}\int_{-\infty}^{+\infty} NAe^{-\mu\epsilon}\,\mu m\dot{x}^2 d\dot{x}d\dot{y}d\dot{z} \tag{66}$$

Comparing now with equation (65), we see that we have finally obtained the simple result

$$PV = \frac{N}{\mu} \tag{67}$$

In accordance with the gas laws, however, we know experimentally that the pressure volume product for such a gas will be given by the equation

$$PV = NkT \tag{68}$$

where the Boltzmann constant k is the ordinary gas constant R divided by the number of molecules in one mol. of gas. We thus obtain from (67) and (68) the desired relation

$$\mu = \frac{1}{kT} \tag{69}$$

52. This equation relates the quantity μ to the temperature T of the perfect monatomic gas which we assumed as part of our equilibrium system. The quantity has as we saw, however, in Section **40** the same value for all the kinds of molecules in the system, and furthermore, the macroscopic quantity T is obviously the same throughout such an equilibrium system, so that the relation is a general one. It should also be noted that the inclusion of a container holding a perfect gas as part of the system is analogous to the experimental use of a gas thermometer for measuring the temperature of the system, and will not in any harmful way affect the behavior of the system.

53. Expression of the Maxwell-Boltzmann Distribution Law in Different Forms.—Having thus obtained the familiar expressions for the values of the constants C and μ, we can now write the Maxwell-Boltzmann distribution law in the usual forms which prove convenient for applications. Substituting the new value for μ into equation (60), we may write as our first fundamental form of the law

$$dN = NCe^{-\frac{\epsilon}{kT}}\,dq_1 \ldots dp_n \tag{70}$$

where in accordance with equation (61), the constant C has the value

$$C = \frac{1}{\int \ldots \int e^{-\frac{\epsilon}{kT}}\,dq_1 \ldots dp_n} \tag{71}$$

the integration to be taken over all regions of the μ-space. Combining (70) and (71) we may eliminate C and obtain the distribution law in the form

$$dN = \frac{Ne^{-\frac{\epsilon}{kT}} dq_1 \dots . dp_n}{\int \dots . \int e^{-\frac{\epsilon}{kT}} dq_1 \dots . dp_n} \qquad (72)$$

Still another form of the distribution law which is sometimes useful is obtained by the introduction of a constant ψ defined by the equation

$$NC = e^{\frac{\psi}{kT}} \qquad (73)$$

which leads to the expression

$$dN = e^{\frac{\psi - \epsilon}{kT}} dq_1 \dots . dp_n \qquad (74)$$

54. Method of Obtaining Averages from the Maxwell-Boltzmann Distribution Law.—Since the Maxwell-Boltzmann expression for dN gives the number of molecules having coordinates and momenta in the range $dq_1 \dots . dp_n$, it is evident that the average value of any property P of the molecules which is dependent on the coordinates and momenta can be calculated from the equation

$$P_{av.} = \frac{1}{N} \int PdN = \int \dots . \int Ce^{-\frac{\epsilon}{kT}} Pdq_1 \dots dp_n \qquad (75)$$

where in performing the integration the quantity P is to be regarded as a function of the coordinates and momenta $q_1 \dots . p_n$, and the integration is to be taken over the whole or a part of the μ-space, according as the average is desired for the whole or a part of the molecules.

55. As an illustration of the usefulness of this method of averaging, we can use it to show the possibility of obtaining a simple expression for the temperature coefficient of the quantity C, occurring in the Maxwell-Boltzmann expression, even in the absence of a knowledge as to its absolute value. Returning to the expression for C given by equation (71) and differentiating with respect to the temperature we obtain

$$\frac{dC}{dT} = -\frac{1}{\left(\int \dots \int e^{-\frac{\epsilon}{kT}} dq_1 \dots dp_n\right)^2} \int \dots \int e^{-\frac{\epsilon}{kT}} \frac{\varepsilon}{kT^2} dq_1 \dots dp_n.$$

Substituting equation (71) itself, however, this may be rewritten in the form

$$\frac{1}{C}\frac{dC}{dT} = -\int \dots \int Ce^{-\frac{\epsilon}{kT}} \frac{\varepsilon}{kT^2} dq_1 \dots dp_n$$

and in accordance with equation (75). this gives us

$$\frac{d \log C}{dT} = -\frac{\bar{\varepsilon}}{kT^2} \tag{76}$$

where $\bar{\varepsilon}$ is the average value of the energy ε.

56. Change of Variables.—In the application of the Maxwell-Boltzmann law to further problems, it is sometimes necessary to make a change in the variables $q_1 \ldots . p_n$. To do this we can make use of a theorem due to Jacobi which connects the expressions for an infinitesimal area in different sets of coordinates. Consider a change from the variables x and y to u an v, where

$$u = u(x, y) \quad \text{and} \quad v = v(x, y)$$

We can then substitute

$$J\left|\frac{u, v}{x, y}\right| dxdy = dudv \tag{77}$$

or

$$J\left|\frac{x, y}{u, v}\right| dudv = dxdy \tag{78}$$

where the Jacobian J is defined as the functional determinant

$$J\left|\frac{u, v}{x, y}\right| = \begin{vmatrix} \frac{\partial u}{\partial x} & \frac{\partial u}{\partial y} \\ \frac{\partial v}{\partial x} & \frac{\partial v}{\partial y} \end{vmatrix} \tag{79}$$

The method can be extended by substituting in this manner for co-ordinates a pair at a time.

We are now ready to undertake the application of the Maxwell-Boltzmann distribution law to the study of actual physical-chemical problems.

Chapter 5

Application of the Maxwell-Boltzmann Distribution Law to Problems Involving Molecular Velocities

57. Introduction.—It often happens that we are not primarily interested in the values of all the coordinates and momenta that determine the condition of a molecule, but only in the three coordinates which give its location in space and the three components of momentum which are related to its motion as a whole. If we concentrate our attention on these three degrees of freedom, we can obtain Maxwell's law for the distribution of velocities among the molecules of a system, which may be regarded as a special case of the more general distribution law, and is indeed the one which was first deduced.

In the present chapter, we shall first obtain Maxwell's law for the distribution of velocities in a variety of convenient forms, and shall then use it to investigate the pressure volume relations of gases, and to determine the number of collisions which occur between the molecules of a gas.

58. Maxwell's Law as a Special Case of the General Distribution Law.—Consider a system containing N similar molecules each having n degrees of freedom. Then the Maxwell-Boltzmann distribution law for the molecules of the system can be written in the general form

$$dN = NCe^{-\frac{\varepsilon}{kT}}\, dq_1 \ldots . dp_n \tag{80}$$

Let us now assume, however, that the energy increase of the system ε, per molecule introduced into the various regions of the μ-space, can be taken as the sum of three terms

$$\varepsilon = \varepsilon\,(x, y, z) + \frac{m}{2}\,(\dot{x}^2 + \dot{y}^2 + \dot{z}^2) + \varepsilon_i \tag{81}$$

where the first term gives the potential energy of the whole molecule as a function of its coordinates of position x, y and z, the second term its kinetic energy as a whole as determined by its mass m and components of velocity $\dot{x}$, $\dot{y}$ and $\dot{z}$, and the last term gives the remaining energy which it contributes to the system as a function solely of the remaining coordinates and momenta $q_4 \ldots . q_n p_4 \ldots . p_n$. Whenever the system is of such a nature that this separation can be made the considerations which are to follow will be justified. It is evident that this will be true for dilute gases, and perhaps closely true for a wider variety of other systems than has sometimes been thought the case, provided always of course that quantum restrictions do not enter the problem.

Since the momenta corresponding to the three coordinates x, y and z will now evidently have the values $m\dot{x}$, $m\dot{y}$ and $m\dot{z}$, we can now rewrite the Maxwell-Boltzmann distribution law in the form

$$dN = N(Ce^{-\frac{\epsilon_i}{kT}} dq_4 \ldots dq_n\, dp_4 \ldots dp_n m^3)e^{-\frac{\epsilon(x,y,z) + m/2(\dot{x}^2 + \dot{y}^2 + \dot{z}^2)}{kT}} dxdydzd\dot{x}d\dot{y}d\dot{z} \quad (82)$$

where for convenience we have collected inside the brackets quantities which are not dependent on the positional coordinates or velocities of the molecules.

As we are now interested for our present purposes in all the molecules which have positions and velocities falling in the range $dx\ dy\ dz\ d\dot{x}\ d\dot{y}\ d\dot{z}$, without reference to the value of other variables determining the condition of the molecule, we may integrate over all possible values of the coordinates and momenta which occur inside the brackets, and thus obtain the simple expression

$$dN = NAe^{-\frac{\epsilon(x,y,z) + m/2(\dot{x}^2 + \dot{y}^2 + \dot{z}^2)}{kT}} dxdydzd\dot{x}d\dot{y}d\dot{z} \quad (83)$$

where dN is the number of molecules falling in the range $dx\ dy\ dz\ d\dot{x}\ d\dot{y}\ d\dot{z}$ and A represents the quantity resulting from the integration. The equation has, we see, exactly the same form for all different kinds of molecules, monatomic or polyatomic, and indeed applies quite rigorously to any kind of molecules, whenever the quantity ϵ can be separated into three terms as discussed above.

59. From the form of equation (83) it is immediately evident that the number of molecules in any region of the container $dx\ dy\ dz$ is determined by the potential energy of a molecule in that region, and this can be used, for example, to determine the variation in the density of a gas with height in a gravitational field. Furthermore, from the quadratic and symmetrical form of the kinetic energy we see that the velocities of the molecules will be distributed uniformly in all directions.

60. Let us now turn our attention primarily to this distribution of velocities. To do this we may take the potential energy of the molecules as zero independent of their location and integrate over all the regions in the total volume of the container V and obtain

$$dN = NVAe^{-\frac{m}{2}\frac{(\dot{x}^2+\dot{y}^2+\dot{z}^2)}{kT}}d\dot{x}d\dot{y}d\dot{z} \tag{84}$$

for the number of molecules having velocities in the range $d\dot{x}\ d\dot{y}\ d\dot{z}$ and this is Maxwell's distribution law. It can, however, be expressed in rather more convenient forms.

Let us first introduce polar coordinates with the help of the evident expressions (see Section **56** in the last chapter)

$$c^2 = \dot{x}^2 + \dot{y}^2 + \dot{z}^2 \tag{85}$$

and

$$d\dot{x}d\dot{y}d\dot{z} = c^2 \sin\vartheta d\vartheta d\phi dc \tag{86}$$

where c is the total velocity and ϑ and ϕ the angles determining the direction of motion. Substituting in (84) we obtain

$$dN = NVAe^{-\frac{mc^2}{2kT}}c^2 \sin\vartheta d\vartheta d\phi dc \tag{87}$$

and integrating over all values of the longitudinal angle ϕ from 0 to 2π, and over all values of the latitudinal angle ϕ from 0 to π we obtain

$$dN = 4\pi NVAe^{-\frac{mc^2}{2kT}}c^2dc \tag{88}$$

The value of the constant A occurring in these equations can now, however, easily be determined by integration over all possible values of c from zero to infinity. We evidently obtain

$$N = 4\pi NVA\int_0^\infty e^{-\frac{mc^2}{2kT}}c^2dc = 4\pi NVA\left\{\frac{1}{4}\sqrt{\frac{8\pi k^3T^3}{m^3}}\right\}$$

or

$$VA = \left(\frac{m}{2\pi kT}\right)^{3/2} \tag{89}$$

61. By substitution of equation (89) in (84), (87) and (88), we can now obtain Maxwell's law in the very useful forms

$$dN = N\left(\frac{m}{2\pi kT}\right)^{3/2} e^{-\frac{m}{2}\frac{(\dot{x}^2+\dot{y}^2+\dot{z}^2)}{kT}}d\dot{x}d\dot{y}d\dot{z} \tag{90}$$

$$dN = N\left(\frac{m}{2\pi kT}\right)^{3/2} e^{-\frac{mc^2}{2kT}}c^2 \sin\vartheta d\vartheta d\phi dc \tag{91}$$

$$dN = 4\pi N\left(\frac{m}{2\pi kT}\right)^{3/2} e^{-\frac{mc^2}{2kT}}c^2dc \tag{92}$$

Finally one further form of Maxwell's law will be found useful where the range is expressed in terms of the kinetic energy of the molecules

$$\varepsilon = \frac{mc^2}{2} \tag{93}$$

Introducing in (92) we obtain

$$dN = N\frac{2}{\sqrt{\pi}(kT)^{3/2}}\, e^{-\frac{\varepsilon}{kT}} \sqrt{\varepsilon}\, d\varepsilon \tag{94}$$

62. Average Values of the Velocity.—In a number of applications it is convenient to have expressions for different average velocities which can be obtained from equation (92).

We easily obtain the *mean velocity* by multiplying (92) by c, integrating over all values and dividing by N, in accordance with the principles of averaging discussed in Section **54** of the last chapter. We obtain the result

$$\Omega = \sqrt{\frac{8kT}{\pi m}} = 14{,}500 \sqrt{\frac{T}{M}}\ \frac{\text{cm.}}{\text{sec.}} \tag{95}$$

For the *root-mean-square-velocity,* which is the square root of the mean value of c^2 we obtain

$$G = \sqrt{\frac{3kT}{m}} = 15{,}800 \sqrt{\frac{T}{M}}\ \frac{\text{cm.}}{\text{sec.}} \tag{96}$$

and for the *most probable velocity* we easily obtain by the conditions for a maximum

$$W = \sqrt{\frac{2kT}{m}} = 12{,}900 \sqrt{\frac{T}{M}}\ \frac{\text{cm.}}{\text{sec.}} \tag{97}$$

It will be noticed that these different average velocities do not differ greatly from one another owing to the sharp convergence of the velocities around the most probable value. In the numerical expressions for the averages, the temperature is in degrees centigrade absolute, and the mass is the molecular weight of the substance.[1]

63. The Virial.—Our first application of this knowledge as to the distribution of molecular velocities will be to obtain a general expression for the pressure of a gas. As a preliminary to this we shall first find it necessary to discuss an expression called the virial which was originally introduced by Clausius.

Let x, y and z be the coordinates of the center of gravity of a molecule of mass m and X, Y, Z the components of force acting on it. Then the equations of motion will be of the form

[1] Numerical values taken from Dushman, "High Vacuum," Schenectady, 1922.

$$m \frac{d^2x}{dt^2} = X \text{ etc.} \tag{98}$$

We may evidently write, however,

$$\frac{d}{dt}\left(mx\frac{dx}{dt}\right) = m\left(\frac{dx}{dt}\right)^2 + mx\frac{d^2x}{dt^2} \tag{99}$$

and substituting (98) and rearranging obtain

$$\left\{m\left(\frac{dx}{dt}\right)^2 + xX\right\}dt = d\left(mx\frac{dx}{dt}\right) \tag{100}$$

Let us now integrate this equation over a time interval between t_1 and t_2, and divide by the interval so as to obtain an average value for the expression on the left hand side. We obtain

$$\overline{m\left(\frac{dx}{dt}\right)^2} + \overline{xX} = \frac{1}{t_2 - t_1}\left\{\left(mx\frac{dx}{dt}\right)_{t_2} - \left(mx\frac{dx}{dt}\right)_{t_1}\right\} \tag{101}$$

where averages are indicated by the bars. If we take the time interval long, however, it is evident that the quantities on the left hand side of the equation will approach steady values, while the quantity on the right hand side will approach zero since infinite values of the terms in the brackets do not occur. Hence averaged over a long time interval, or averaged at a given time over all the molecules of the system we can write

$$\overline{m\left(\frac{dx}{dt}\right)^2} + \overline{xX} = 0$$

and adding the similar expressions for the other axes, summing for all the molecules and dividing by 2 we obtain the remarkable theorem

$$\Sigma\, \tfrac{1}{2}\, \overline{mc^2} + \Sigma\, \tfrac{1}{2}\, (\overline{xX} + \overline{yY} + \overline{zZ}) = 0 \tag{102}$$

where the first term is the total kinetic energy of the molecules and the second is the virial of Clausius.

64. Application of the Virial to Calculate the Pressure of a Gas. In order to apply this equation to determine the pressure of a gas we note that the terms in the virial will arise from two causes,—the forces between the molecules and the walls which lead to the pressure, and the forces between the molecules themselves.

To obtain the first or external contribution, let us for simplicity assume a rectangular container having the sides Ox, Oy and Oz. Since the area of the two faces perpendicular to the X-axis is yz and the x coordinates are 0 and x, the contribution of this pair of faces will evidently be

$-\frac{P}{2}xyz$, so that the total contribution of the walls to the virial will be

$$V_e = -\frac{3}{2}PV \tag{103}$$

To obtain the contribution of the internal forces in a convenient form, let us represent by $f(r)$ the force (taken positive for repulsion) acting between two molecules at the distance r from each other, having the respective positions x, y, z and x', y', z'. The force acting on the first molecule evidently has the components

$$X = f(r)\frac{x-x'}{r}, \quad Y = f(r)\frac{y-y'}{r} \quad \text{and} \quad Z = f(r)\frac{z-z'}{r}$$

and hence contributes to the virial the term

$$\frac{f(r)}{2r}\{x(x-x') + y(y-y') + z(z-z')\}$$

while for the second molecule the contribution will be

$$\frac{f(r)}{2r}\{x'(x'-x) + y'(y'-y) + z'(z'-z)\}$$

or adding we obtain

$$\frac{f(r)}{2r}\{(x'-x)^2 + (y'-y)^2 + (z'-z)^2\}$$

so that for the total contribution of the internal forces to the virial, we have

$$V_i = \tfrac{1}{2}\Sigma\Sigma\, r f(r) \tag{104}$$

where the summation is to be taken over all pairs of molecules.

65. Substituting the two contributions given by (103) and (104) into (102) and rearranging we obtain the beautiful relation

$$PV = \tfrac{1}{3}\Sigma\, \overline{mc^2} + \tfrac{1}{3}\Sigma\Sigma\, r f(r) \tag{105}$$

The first term on the right-hand side of this equation has, however, in accordance with equation (96) the value NkT, so that we obtain the gas laws in the form

$$PV = NkT + \tfrac{1}{3}\Sigma\Sigma\, r f(r) \tag{106}$$

and at once see that for a gas dilute enough so that we can neglect the internal forces between the molecules the laws for a perfect gas will hold.

66. Equation (106) can now be made the starting point for a number of important investigations which unfortunately cannot be included in the present work. Thus the calculation of the internal part of the virial due to the Coulomb forces between charged ions in an electrolytic solution has led to the Milner-Debye theory of interionic attraction, and by making suitable assumptions as to the forces between the molecules of a gas it has been possible to arrive at reasonable expressions for the equations of state of compressed gases and liquids.

67. As a single illustration of the latter possibility we may show the method of obtaining Van der Waals' equation, introducing the assumptions that he makes, only as a possible first approximation. To carry out the calculation, let us divide the term $\frac{1}{3}\Sigma\Sigma\, r f(r)$ into two terms $\frac{1}{3}\Sigma\Sigma\, r f_1(r)$ which arises from the repulsion of molecules when they are close together and $\frac{1}{3}\Sigma\Sigma\, r f_2(r)$ which arises from the residual attraction when they are far apart.

With regard to the first term we shall assume that the molecules behave as rigid spheres so that the force of repulsion always acts at the same distance r when the two molecules collide. The number of collisions, however, can evidently be taken proportional to $\sqrt{T}/V$ and the force which is dependent on the relative velocity as also proportional to $\sqrt{T}$ (see Section 70), so that the first term can be written as

$$\frac{1}{3}\Sigma\Sigma\, r f_1(r) = \frac{BT}{V} \tag{107}$$

where B is a proportionality constant.

With regard to the second term, we shall assume with Van der Waals that the forces of attraction can be regarded as balancing out in the interior of the fluid, and need only be considered at the boundary where they will manifest themselves as a cohesion pressure P_i, so that using the same considerations that led to equation (103), we can write

$$\frac{1}{3}\Sigma\Sigma\, r f_2(r) = -P_i V \tag{108}$$

Or if we assume again with Van der Waals that this cohesion pressure is inversely proportional to V^2, since it is dependent both on the number of molecules at the boundary and the number within, which are doing the attracting, we can write

$$\frac{1}{3}\Sigma\Sigma\, r f_2(r) = -\frac{a}{V} \tag{109}$$

Substituting (107) and (109) into (106), we obtain

$$PV = NkT + \frac{BT}{V} - \frac{a}{V}$$

or referred to one mol of gas

$$\left(P+\frac{a}{V^2}\right)V=RT\left(1+\frac{b}{V}\right) \tag{110}$$

where b is a new constant, which can be shown to be related to the volume of the molecules themselves.

For small values of the ratio b/V, this can be written in the Van der Waals' form

$$\left(P+\frac{a}{V^2}\right)(V-b)=RT \tag{111}$$

The result is of course not entirely satisfactory, because of the assumptions.

68. Still making the hypothesis that the repulsive forces may be regarded as arising from the collision of rigid spheres, but making more general assumptions as to the residual attractive forces and not neglecting their action within the interior of the fluid, Reinganum [1] obtains the equation

$$PV=RT\left(1+\frac{be^{\frac{C}{T}}}{V}\right)-\frac{a\phi(T)}{V} \tag{112}$$

where the exponential term arises from the action of the attractive forces within the interior of the fluid and the form of $\phi(T)$ depends on the law of attraction.

69. The Number of Collisions between Molecules.—If we treat the molecules of a gas as rigid elastic spheres, it is possible to calculate the number of collisions between them which occur per second. Although such a treatment must be regarded as a first approximation, it nevertheless seems to be a reasonably good one, and the results will be of considerable later interest, especially in connection with the rate of chemical reactions.

Let us examine the chances of collision between two molecules having the masses m_1 and m_2, and diameters σ_1 and σ_2, enclosed in a container having a volume of one cubic centimeter. At the instant of contact the center of the second molecule will evidently have to lie on a sphere drawn about the center of the first molecule, with a radius σ_{12}, which is the mean

[1] Reinganum, *Ann. der Phys.*, **6**, 533 (1901). For another equation of state, with extensive empirical tests, see the work of Keyes,—Proc. Nat. Acad., **3**, 323 (1917); **11**, 38 (1925); Journ. Amer. Chem. Soc., **41**, 589 (1919); **43**, 1452 (1921); **46**, 1584 (1924); Amer. Soc. Refrigerat. Eng. Journ., **7**, 371 (1921); Journ. Math. Phys., Mass. Inst. Tech., **1**, 211 (1922).

of the two diameters σ_1 and σ_2. Consider especially now collisions in which the angle between the direction of the relative velocity V and the line of centers at the instant of contact lies in the range between ϑ and $\vartheta + d\vartheta$. Under these circumstances the center of the second molecule at the instant of contact will evidently have to lie on a ring of area $2\pi\sigma_{12}^2 \sin\vartheta d\vartheta$ cut out of the sphere of radius σ_{12}. And since the component relative velocity, parallel to the line of centers and hence perpendicular to the surface of this ring is $V\cos\vartheta$, it is evident that the center of the second molecule must lie in a volume $2\pi\sigma_{12}^2 \sin\vartheta d\vartheta V \cos\vartheta dt$, if such a collision is to occur within a short time interval dt. Neglecting the space occupied by the molecules themselves, however, the chance that the center of a molecule will lie in this small volume is equal to the volume itself divided by the total volume unity, so that we may write for the chance of such collisions *per second* $2\pi\sigma_{12}^2 V \sin\vartheta \cos\vartheta d\vartheta$.

Taking now the total number of molecules per cubic centimeter of the two kinds as being N_1 and N_2, we can use the equation (90) to give us the number of each kind having component velocities in the ranges of interest from $\dot{x}_1$ to $\dot{x}_1 + d\dot{x}_1$, $\dot{x}_2$ to $\dot{x}_2 + d\dot{x}_2$, etc., and hence can write for the total number of collisions per second of the kind described

$$dZ = 2\pi N_1 N_2 \sigma_{12}^2 \left(\frac{m_1 m_2}{4\pi^2 k^2 T^2}\right)^{3/2} e^{-\frac{m_1 c_1^2 + m_2 c_2^2}{2kT}} V \sin\vartheta \cos\vartheta d\dot{x}_1 d\dot{y}_1 d\dot{z}_1 d\dot{x}_2 d\dot{y}_2 d\dot{z}_2 d\vartheta \quad (113)$$

where V is determined by the values of $\dot{x}_1$, $\dot{y}_1$, $\dot{z}_1$, $\dot{x}_2$, $\dot{y}_2$ and $\dot{z}_2$.

To treat this expression let us introduce new variables u, v, w and α, β, γ defined by the equations

$$u = \frac{m_1 \dot{x}_1 + m_2 \dot{x}_2}{m_1 + m_2} \text{ etc.} \qquad \alpha = \dot{x}_2 - \dot{x}_1 \text{ etc.} \qquad (114)$$

so that u, v and w are the components of the velocity of the center of gravity of the two molecules and α, β and γ the components of their relative velocity V. Then if we put

$$c^2 = u^2 + v^2 + w^2 \qquad \text{and} \qquad V^2 = \alpha^2 + \beta^2 + \gamma^2 \qquad (115)$$

we can write in accordance with a well-known relation

$$m_1 c_1^2 + m_2 c_2^2 = (m_1 + m_2) c^2 + \frac{m_1 m_2}{m_1 + m_2} V^2 \qquad (116)$$

as can be verified by the substitution of equations (114). Also in accordance with Jacobi's theorem (see Section **56** in the last chapter), we may substitute for the area $d\dot{x}_1\, d\dot{x}_2$ by using the equation

$$J \left| \frac{u, \alpha}{\dot{x}_1, \dot{x}_2} \right| d\dot{x}_1 d\dot{x}_2 = du d\alpha$$

and since the Jacobian is easily found to have the value 1, we see that by considering all three pairs of coordinates we can write

$$d\dot{x}_1 d\dot{y}_1 d\dot{z}_1 d\dot{x}_2 d\dot{y}_2 d\dot{z}_2 = du\ dv\ dw\ d\alpha\ d\beta\ d\gamma \qquad (117)$$

Substituting in equation (113) we obtain

$$dZ = 2\pi N_1 N_2 \sigma_{12}^2 \left(\frac{m_1 m_2}{4\pi^2 k^2 T^2}\right)^{3/2} e^{-\frac{(m_1+m_2)c^2 + \frac{m_1 m_2}{m_1+m_2}V^2}{2kT}} V \sin\vartheta \cos\vartheta\, du\ dv\ dw\ d\alpha\ d\beta\ d\gamma\ d\vartheta \qquad (118)$$

On integrating $du\ dv\ dw$ over all possible directions in space for the velocity of the center of gravity c, we evidently obtain the volume $4\pi c^2 dc$ and similarly integrating $d\alpha\ d\beta\ d\gamma$ over all possible directions for the relative velocity V we obtain $4\pi V^2 dV$ so that expression (118) may now be written in the form

$$dZ = 32\pi^3 N_1 N_2 \sigma_{12}^2 \left(\frac{m_1 m_2}{4\pi^2 k^2 T^2}\right)^{3/2} e^{-\frac{(m_1+m_2)c^2 + \frac{m_1 m_2}{m_1+m_2}V^2}{2kT}} c^2 V^3 \sin\vartheta \cos\vartheta\, dc\ dV\ d\vartheta \qquad (119)$$

And finally, since we shall not usually be interested in the velocity of the center of gravity we may integrate this over all values of c from zero to infinity and obtain

$$dZ = 8 N_1 N_2 \sigma_{12}^2 \left(\frac{\pi\, m_1^3 m_2^3}{8k^3 T^3 (m_1+m_2)^3}\right)^{1/2} e^{-\frac{m_1 m_2}{m_1+m_2}\frac{V^2}{2kT}} V^3 \sin\vartheta \cos\vartheta\, dV d\vartheta \qquad (120)$$

This is the number of collisions per second in which the relative velocity lies in the range dV and the angle between the relative velocity and line of centers at the instant of contact lies in the range $d\vartheta$.

70. *The total number* of *all collisions* between the two kinds of molecules may now be obtained by integrating over all possible values of ϑ from 0 to $\pi/2$, and all possible values of V from zero to infinity. We obtain

$$Z = 2N_1 N_2 \sigma_{12}^2 \left(\frac{2\pi kT(m_1+m_2)}{m_1 m_2}\right)^{1/2} \qquad (121)$$

71. On the other hand, we may be interested in the number of collisions in which the relative velocity V has a certain value. By integrating equation (120) for all values of ϑ we easily obtain for the number of collisions in which the *relative velocity* lies in the *range* dV [2]

[2] See Jeans, "Dynamical Theory of Gases," Section **363**, Second Edition, Cambridge, 1916.

$$dZ = 4N_1N_2\sigma_{12}^2\left(\frac{\pi\, m_1^3m_2^3}{8k^3T^3(m_1+m_2)^3}\right)^{1/2} e^{-\frac{m_1m_2}{m_1+m_2}\frac{V^2}{2kT}}\,V^3dV \quad (122)$$

and by integrating this between all values of V from V_0 to ∞, obtain for the number of collisions in which the *relative velocity* is *greater than* V_0

$$Z(V_0) = Z\left(1+\frac{m_1m_2}{m_1+m_2}\,\frac{V_0^2}{2kT}\right) e^{-\frac{m_1m_2}{m_1+m_2}\frac{V_0^2}{2kT}} \quad (122a)$$

where Z is the total number of collisions.

72. Finally in our application of rates of chemical reaction, we shall be specially interested in the number of collisions in which the component of relative velocity parallel to the line of centers

$$R = V\cos\vartheta \quad (123)$$

has a certain value.

To treat this problem we shall find it simplest to return to equation (120) and first determine the number of collisions in which $R = V\cos\vartheta$ is greater than some particular value R_0. For a given value of $V > R_0$, this will evidently be true when $\cos\vartheta$ lies between unity and R_0/V. Hence we may write for this number of collisions

$$Z(R_0) = \int_{R_0}^{\infty}\int_{\cos\vartheta=1}^{\cos\vartheta=R_0/V} 8N_1N_2\sigma_{12}^2\left(\frac{\pi\, m_1^3m_2^3}{8k^3T^3(m_1+m_2)^3}\right)^{1/2} e^{-\frac{m_1m_2}{m_1+m_2}\frac{V^2}{2kT}}\,V^3\sin\vartheta\cos\vartheta\, d\vartheta\, dV$$

and performing the integrations obtain for the number of collisions in which the *component of relative velocity parallel to the line of centers is greater than* R_0.

$$Z(R_0) = Ze^{-\frac{m_1m_2}{m_1+m_2}\frac{R_0^2}{2kT}} \quad (123a)$$

where Z is again the total number of collisions.[3]

And if we are interested in the number of collisions in which the *component R lies in a particular range dR,* we can obtain by the differentiation of (123a), the equation

$$dZ = Z\frac{m_1m_2}{m_1+m_2}\,\frac{1}{kT}\,e^{-\frac{m_1m_2}{m_1+m_2}\frac{R^2}{2kT}}\,RdR \quad (123b)$$

73. The foregoing results will be found of great value in interpreting the rates of chemical reactions.

In case we desire to use the formulæ to calculate the number of col-

[3] See Langevin and Rey, *Le Radium,* **10**, 142 (1913).

lisions between like molecules, they can evidently be modified by putting $m_1 = m_2 = m$, $\sigma_{12} = \sigma$, and replacing $N_1 N_2$ by $N^2/2$.

For use in numerical calculations, the following formulæ for *total* numbers of collisions will be found convenient[4]

For unlike molecules

$$\frac{Z}{A} = 10^{28.4427} C_1 C_2 \sigma_{12}^2 \sqrt{T \frac{M_1 + M_2}{M_1 M_2}} \tag{124}$$

For like molecules

$$\frac{Z}{A} = 10^{28.2922} C^2 \sigma^2 \sqrt{\frac{T}{M}} \tag{124a}$$

where A is Avogadro's number, so that Z/A is the number of collisions per cubic centimeter expressed in mols colliding, C_1, C_2 and C are concentrations in mols per cubic centimeter, σ_{12} and σ are in centimeters, T is in degrees centigrade absolute, and M_1, M_2 and M are ordinary molecular weights.

[4] See Dushman, *Journ. Amer. Chem. Soc.*, **43**, 407 (1921).

Chapter 6

Application of the Maxwell-Boltzmann Distribution Law to the Theory of Energy Partition and to Specific Heats

74. Introduction.—In the present chapter, we shall use the Maxwell-Boltzmann distribution law to investigate the distribution of energy among the different degrees of freedom of a system. We shall first derive a very general equipartition law, which shows that there is a general function of the coordinates and momenta of a system which has the average value kT for each coordinate or momentum involved, and which reduces to the principle of the equipartition of energy for those coordinates and momenta which contribute quadratic terms to the total energy of the system. We shall then use the principle of the equipartition of energy to obtain values for the specific heat of gases and solids, provided the substances are in temperature ranges such that the results are not affected by quantum restrictions.

We shall specially stress the fact that, even in the absence of quantum restrictions, the principle of the equipartition of energy holds *only* for those coordinates and momenta which contribute quadratic terms to the expression for total energy, and shall investigate the actual partition of energy in certain cases of interest, where equipartition does not obtain. This mode of presentation differs from those usually adopted, which have by implication often given the impression that the principle of the equipartition of energy was always a necessary consequence of the classical dynamics, when in reality it is primarily a consequence of the assumption that the energy of a system is a homogeneous quadratic function of the coordinates and momenta.

75. The General Equipartition Law.—Starting with the Maxwell Boltzmann distribution law, it is possible to derive a general equipartition law by carrying out a partial integration. In accordance with the Maxwell-Boltzmann distribution law, we may evidently write

$$\int \ldots \int C e^{-\frac{\epsilon}{kT}} dq_1 \ldots dp_n = 1 \qquad (125)$$

where the integrations are to be taken over all possible values of the coordinates and momenta. Suppose now we perform a partial integration with respect to some one of the coordinates or momenta $q_1 \ldots . p_n$, say for simplicity with respect to q_1. We shall evidently obtain in this way the following equation

$$\left| \int \ldots \int Ce^{-\frac{\epsilon}{kT}} q_1 dq_2 \ldots dp_n \right|_{q_1 = \text{lower limit}}^{q_1 = \text{upper limit}} + \int \ldots . \int Ce^{-\frac{\epsilon}{kT}} \frac{q_1}{kT} \frac{\partial \varepsilon}{\partial q_1} dq_1 \ldots dp_n = 1 \qquad (126)$$

76. In interpreting the above equation we usually encounter two sorts of cases, either ε is independent of the variable q_1 in question, or else depends upon ε in such a way as to become infinite when q_1 becomes infinite.

An example of the first possibility is given when q_1 is a positional coordinate for a molecule of a dilute gas in the absence of an external field. In such a case the second term in the equation becomes zero since $\partial\varepsilon/\partial q_1$ is zero, and the expression reduces to an equality of the first term with unity, an evident result under the circumstances.

In the cases of interest for our present considerations ε is dependent on q_1 and becomes infinite when q_1 is infinite. Furthermore at the lower and upper limits q_1 itself assumes either the values zero and infinity or the values minus and plus infinity.

This will be true for example when q_1 is a displacement with which potential energy is associated, the amount of potential energy increasing without limit with the displacement. It will also be true in the very usual case when the variable in question is a momentum, an infinite value of kinetic energy being associated with the infinite positive or negative value assumed by the momentum at its upper and lower limits.

77. Under the conditions assumed, the first term of equation (126) becomes zero, at both limits either because q_1 itself is zero at the limit, or because the infinite value of the negative exponent of e outweighs the infinite value of q_1. The equation can then be rewritten in the form

$$\int \ldots \int Ce^{-\frac{\epsilon}{kT}} \left(q_1 \frac{\partial \varepsilon}{\partial q_1} \right) dq_1 \ldots dp_n = kT \qquad (127)$$

In accordance, however, with the principles of averaging discussed in Section **54** of Chapter 4 the left-hand term of this equation gives us the average value of the quantity in brackets, so that we obtain the *general equipartition law*

$$\left[q_1 \frac{\partial \varepsilon}{\partial q_1} \right]_{\text{av.}} = \ldots . . = \left[p_n \frac{\partial \varepsilon}{\partial p_n} \right]_{\text{av.}} = kT \qquad (128)$$

applying in all cases where ε depends in the way stated on the coordinate or momentum involved.

78. The Equipartition of Energy.—There are many cases in which a variable q_i contributes merely a term of the form

$$\varepsilon_i = \tfrac{1}{2} a_i q_i^2 \tag{129}$$

to the energy of the system, where a_i is a constant. This will be true when q_i is a displacement with an elastic force which obeys Hooke's law (*ut tenseo sic vis*), or when the variable is an ordinary linear or angular momentum. In such cases the general law (128) evidently reduces to

$$[\varepsilon_i]_{\text{av.}} = \tfrac{1}{2} kT \tag{130}$$

and we have the law of the equipartition of energy for such coordinates or momenta.

79. The Specific Heat of Gases.—We may immediately use this result to determine the specific heat of gases.

Since monatomic molecules have only the kinetic energy corresponding to the three components of linear momentum, we may evidently write in accordance with equation (130) for the energy and heat capacity of one mol of a *monatomic* gas

$$E = \frac{3}{2} NkT = \frac{3}{2} RT \qquad C_v = \frac{3}{2} R \tag{131}$$

where N is Avogadro's number.

For *diatomic* gases, where the molecules can be regarded as *rigid* dumbbells capable of angular momentum around the two principal axes, we have to add the energy corresponding to these two further momenta, and obtain

$$E = \frac{5}{2} NkT = \frac{5}{2} RT \qquad C_v = \frac{5}{2} R \tag{132}$$

For polyatomic gases, provided the molecules could be regarded as *rigid,* we should have kinetic energy corresponding to still one more mode of rotation and should obtain

$$E = \frac{6}{2} NkT = 3RT \qquad C_v = 3R \tag{133}$$

80. On comparison with experiment, it is found that the above conclusions with regard to the specific heat of *monatomic gases* are in excellent agreement with the facts. In the case of *diatomic* and *polyatomic* gases, however, the simple results expressed in equations (132) and (133) have to be modified in two ways. At low temperatures the actual specific heat can fall below that given by the equations, since as we shall later see, quantum restrictions must be considered, which act in such a way as to make it more difficult for the molecule to acquire rotational energy than we assumed. On the other hand, at high temperatures the specific heat may rise above that given by equations (132) and (133), since the molecules begin to pick up vibrational energy and can no longer be regarded as rigid structures. These matters will be more fully discussed after we have introduced the further restrictions which quantum theory imposes on statistical mechanics.

81. The Specific Heat of Solids.—Let us now consider one mol of a solid, composed of N atoms and hence having $3N$ degrees of freedom, and let the atoms be bound to their equilibrium positions by elastic forces which obey Hooke's law. The total energy of this system will evidently be a homogeneous quadratic function of the $3N$ variables giving the positions of the atoms and the corresponding $3N$ linear momenta. Applying the principle of the equipartition of energy as given by equation (130), we shall then evidently obtain for the energy and specific heat of the solid.

$$E = \frac{6}{2} NkT = 3RT \qquad C_v = 3R \tag{134}$$

This result is in accordance with the well-known law of Dulong and Petit for the specific heat of solid elements.

82. Here too, however, we find deviations from the simple law, due to the interference of quantum restrictions, which will be thoroughly discussed in a later chapter. In general, at low temperatures the specific heat falls below that given by equation (134), owing to the fact that the vibrational energy of the system must be quantized; so that equation (134) becomes a limiting law usually holding fairly well when we get up to room temperatures. On the other hand in the case of strongly metallic solid elements, the specific heat often rises above the value $3R$ owing presumably to the participation of electrons in the thermal motion.

83. The Partition of Energy in General.—From the above discussion in Section **78,** it is evident that the law of the equipartition of energy is a special case holding when the energy depends in a particular way on the coordinates and momenta. In the more general case of any system of such a nature that ε may be regarded as being the energy of an individual molecule, we may evidently obtain in accordance with principles of averaging the average energy of a molecule from the equation

$$\varepsilon_{\text{av.}} = \int \ldots \int C e^{-\frac{\varepsilon}{kT}} \varepsilon \, dq_1 \ldots dp_n \tag{135}$$

which, introducing the value of C, given by equation (174) in Chapter 4, can evidently be rewritten in the form

$$\varepsilon_{\text{av.}} = \frac{\int \dots \int e^{-\frac{\varepsilon}{kT}} \varepsilon \, dq_1 \dots dp_n}{\int \dots \int e^{-\frac{\varepsilon}{kT}} dq_1 \dots dp_n} \tag{136}$$

And, in case it is possible to separate out the portion ε_i of the energy associated with a single variable q_i, we can write

$$[\varepsilon_i]_{\text{av.}} = \frac{\int e^{-\frac{\varepsilon_i}{kT}} \varepsilon_i \, dq_i}{\int e^{-\frac{\varepsilon_i}{kT}} dq_i} \tag{137}$$

These equations will permit a general treatment of the partition of energy, provided the dependence of ε on the coordinates and momenta $q_1 \dots p_n$ is known.

84. In some cases the dependence of ε upon the q's and p's is such as to make the integrations indicated in equation (136) very difficult or possible only by approximation. It is then often convenient to make use of the general equipartition law for the quantity $[q_i \partial\varepsilon/\partial q_i]$ as given by equation (128). By comparing the value of this quantity with the value of the energy, it may then be possible to obtain, in a very easy way, at least an idea of the actual partition.

Using one or the other of these methods of attack, we may now consider the partition of energy in a number of cases where equipartition does not prevail.

85. Energy Proportional to Any Power of the Variable.—Consider for example a case where the energy associated with the ith coordinate is given by the expression

$$\varepsilon_i = a q_i^n \tag{138}$$

which could be the form in a hypothetical *non*-harmonic oscillator. Substituting this expression in equation (137) and performing the integrations it can easily be shown that we obtain in general

$$[\varepsilon_i]_{\text{av.}} = \frac{kT}{n} \tag{139}$$

which reduces to the familiar $\frac{1}{2}kT$ when the energy is proportional to the square of the coordinate.

86. Oscillators not Obeying Hooke's Law.—Actual oscillating systems usually differ from ideal harmonic oscillators in that the elastic restoring force falls below proportionality to the displacement as the latter gets larger. Hence the energy does not continue to build up proportionally to the square of the displacement when the displacement becomes large, and indeed usually falls off increasingly as the displacement continues to increase. This could be allowed for by using, for the potential energy ε_i corresponding to the displacement q_i, the following expression which may be regarded as an empirical formula,

$$\varepsilon_i = aq_i^2 - bq_i^4 - cq_i^6 - \ldots . \qquad (140)$$

a, b, c, etc., being positive quantities. Applying our general equipartition law as given by equation (128), we obtain

$$[q_i(2aq_i - 4bq_i^3 - 6cq_i^5 - \ldots)]_{\text{av.}} = kT \qquad (141)$$

Dividing by two and combining with (140) we obtain

$$[\varepsilon_i]_{\text{av.}} = \tfrac{1}{2}kT + [bq_i^4 + 2cq_i^6 + \ldots]_{\text{av.}} \qquad (142)$$

The thing of interest is to note that the deviation from the usual equipartition value is in the direction of increased energy content and increased heat capacity.

Some care must be used in connection with the above treatment, since our method of deriving the general equipartition law evidently limits the application to cases where ε_i may be regarded as approaching infinity with q_i, and this assumption might cease to be justifiable at temperatures so high that very large displacements became frequent. In addition the form of the empirical equation (140) would not be suitable for large displacements.

If the deviation from Hooke's law is in the opposite direction, we shall evidently obtain lower energies and heat capacities than the usual equipartition values. This possibility may be of importance in the case of large oscillations of the atoms of a crystal.

Too much stress, however, must not be laid on these possibilities, until after we have investigated the very large deviations from equipartition which are due to the quantum theory.

87. Particles Obeying the Relativity Laws of Motion.—Another interesting case of deviation from equipartition arises when we use the exact relativity laws for the motion of the molecules of a gas instead of the approximate Newtonian laws which are true only at low velocities.

In relativity mechanics the total energy of a particle is given by the expression

$$\varepsilon = mc^2 \left(\frac{1}{\sqrt{1 - u^2/c^2}} \right) \tag{143}$$

where m is the mass of the particle at rest, u is its velocity and c the velocity of light; and the three components of momenta are given by the equations

$$p_x = \frac{m\dot{x}}{\sqrt{1 - u^2/c^2}} \; ; \; p_y = \frac{m\dot{y}}{\sqrt{1 - u^2/c^2}} \; ; \; p_z = \frac{m\dot{z}}{\sqrt{1 - u^2/c^2}} \tag{144}$$

Noting that

$$u^2 = \dot{x}^2 + \dot{y}^2 + \dot{z}^2 \tag{145}$$

and defining what may be called the total momentum by the equation

$$p^2 = p_x^2 + p_y^2 + p_z^2 \tag{146}$$

we can obtain by substitution in (143) the following expression for the energy in terms of the momentum

$$\varepsilon = c\sqrt{p^2 + m^2c^2} \tag{147}$$

We may now write the Maxwell-Boltzmann distribution law in the form

$$dN = NCe^{-\frac{c\sqrt{p^2 + m^2c^2}}{kT}} \, dx \, dy \, dz \, dp_x dp_y dp_z \tag{148}$$

or integrating over the total volume of the vessel V and changing for convenience to polar coordinates we obtain

$$dN = NCVe^{-\frac{c\sqrt{p^2 + m^2c^2}}{kT}} \, p^2 \sin\vartheta d\vartheta d\phi dp \tag{149}$$

and integrating for all directions

$$dN = 4\pi NCVe^{-\frac{c\sqrt{p^2 + m^2c^2}}{kT}} \, p^2 dp \tag{150}$$

Referring to equation (136), we can now write for the average energy of a particle

$$\varepsilon_{av.} = \frac{\int_0^\infty e^{-\frac{c\sqrt{p^2 + m^2c^2}}{kT}} c\sqrt{p^2 + m^2c^2} \; p^2 dp}{\int_0^\infty e^{-\frac{c\sqrt{p^2 + m^2c^2}}{kT}} \, p^2 dp} \tag{151}$$

Unfortunately there appear to be no general algebraic expressions for the above integrals and we must adopt some method of approximation.

88. Energy Partition in Accordance with Relativity Theory in the Case of Very High Temperatures or Very Small Masses.—Since the molecules tend to assume larger and larger momenta the higher the temperature, we may neglect the constant quantity m^2c^2 in comparison with p^2 in carrying out our integrations provided the temperature is exceedingly *high*. Under these circumstances, equation (151) reduces to

$$\varepsilon_{av.} = \frac{\int_0^\infty e^{-\frac{cp}{kT}} cp^3 dp}{\int_0^\infty e^{-\frac{cp}{kT}} p^2 dp} \qquad (152)$$

in terms of well-known integrals. Evaluating we obtain

$$\varepsilon_{av.} = 3kT \;\; ; \;\; E = 3RT \;\; ; \;\; C_v = 3R \qquad (153)$$

We thus obtain the interesting result that at *extremely high* temperatures, the heat content and specific heat of a monatomic gas are twice as large as we should expect on the basis of the Newtonian mechanics.

It is also interesting to note that the reduction to the form given by equation (152) would be true at all temperatures, provided the mass of the particle at rest becomes zero. We obtain the surprising result, that a mol of particles having zero mass at the absolute zero, would acquire the energy $3RT$ and the mass $3RT/c^2$ at higher temperatures. Thus a mol of particles which approach zero mass at the absolute zero would have a mass of

$$\frac{3 \times 8.31 \times 10^7 \times 300}{10^{21}} = 7 \times 10^{-11}$$

grams at room temperature (300° absolute). This suggests a fascinating and very likely profitless field of speculation.

89. Energy Partition in Accordance with Relativity Theory at Any Temperature.—In order to obtain an idea of the energy partition at any temperature, it is simplest to make use of our general equipartition law as given by equation (128). Referring to equations (146) and (147) for momentum and energy, we evidently obtain from the general equipartition law

$$\left[p_x \frac{\partial \varepsilon}{\partial p_x}\right]_{av.} = \left[p_x\, c(p^2 + m^2c^2)^{-\frac{1}{2}}\, p_x\right]_{av.} = kT \qquad (154)$$

Substituting (144) and introducing the similar equations for the y and z components we obtain

$$\left[\frac{m\dot{x}^2}{\sqrt{1-u^2/c^2}}\right]_{av.} = \left[\frac{m\dot{y}^2}{\sqrt{1-u^2/c^2}}\right]_{av.} = \left[\frac{m\dot{z}^2}{\sqrt{1-u^2/c^2}}\right]_{av.} = kT \quad (155)$$

or by combination

$$\left[\frac{mu^2}{\sqrt{1-u^2/c^2}}\right]_{\text{av.}} = 3kT \tag{156}$$

We may compare this result, which we have obtained from the general equipartition law, with the expression which we must have on the basis of the relativity theory for the average *kinetic* energy of a particle, namely [1]

$$\varepsilon_{\text{av.}} = \left[\frac{mc^2}{\sqrt{1-u^2/c^2}} - mc^2\right]_{\text{av.}} \tag{157}$$

If we expand both expressions (156) and (157) in the form of series and combine we easily obtain

$$\varepsilon_{\text{av.}} = \frac{3}{2}kT + mc^2\left[\frac{1}{8}\frac{u^4}{c^4} + \frac{1}{8}\frac{u^6}{c^6} + \frac{15}{128}\frac{u^8}{c^8} + \ldots\right]_{\text{av.}}$$

For the case of low enough temperatures, the velocities of the molecules will become so low that the bracketed terms become negligible, and the expression reduces to the familiar $3/2\ kT$ given in the Newtonian mechanics. For the case of somewhat higher temperatures we may introduce a first approximation by substituting

$$[u^4]_{\text{av.}} = 15\left(\frac{kT}{m}\right)^2$$

which is the value it has in the Newtonian mechanics. We obtain

$$E = \frac{3}{2}RT + \frac{15}{8M}\left(\frac{RT}{c}\right)^2 + \ldots \tag{158}$$

where M is the molecular weight.

It will be noted as might be expected that the correction term is larger the smaller the mass of the particles of the gas, and the higher the temperature. If we go to a gas with a molecular weight as low as that of the electron, about 1/1800 that of monatomic hydrogen, we find at room temperature that the correction term would only be 6×10^{-6} per cent of the first term and even in a star having temperature of 30,000° it would still be only 6×10^{-4} per cent. Hence for all ordinary temperatures we may expect the partition of energy to be that given by the Newtonian mechanics for all particles with a mass as great as that of the electron.

90. Gas Laws and Relativity Theory.—These deviations from the Newtonian theory for particles which obey the relativity laws of motion

[1] To obtain the kinetic energy, the "internal" energy of the particle at rest mc^2 must be subtracted from the total energy.

suggest an inquiry as to the validity of the gas laws in relativity mechanics. It can easily be shown, however, that also in the relativity theory the laws of Boyle, Charles and Avogadro will be true.

Since the X-component of momentum for a particle of mass m and component of velocity $\dot{x}$ is $m\dot{x}/\sqrt{1-u^2/c^2}$, it is easy to see in accordance with the methods of Section **50** Chapter **4** that the pressure will be given by the equation

$$P = \frac{N}{V}\left[\frac{m\dot{x}^2}{\sqrt{1-u^2/c^2}}\right]_{\text{av.}}$$

where N is the total number of particles in the volume V. In accordance with the general equipartition law, however, the quantity in the brackets has the average value kT as shown in equation (155), so that we obtain at once the complete expression embodying all three of the gas laws

$$PV = NkT$$

Chapter 7

The Elements of Quantum Theory Mechanics

91. Introduction.—It would be unwise to proceed further with the development of statistical mechanics without first introducing the hypotheses of the quantum theory, since too many of our results would need modification before they could be applied to actual conditions, if we based them solely on our present foundation. Indeed, as indicated in the last chapter, some of the results we have already obtained need to be modified under certain conditions. However, since all the conclusions of the classical mechanics are subject to revision in the light of the quantum theory, we need feel no chagrin because of the necessity of modifying the results of the older statistical mechanics.

Interestingly enough, the first deviations from the predictions of the classical mechanics were actually encountered in the very field of statistical mechanics itself, in connection with predictions as to the distribution of radiant energy in an enclosure which has come to temperature equilibrium with the walls. These deviations led to the first introduction of a quantum hypothesis by Planck in the year 1900. Since that time a large number of further facts have been discovered not only in the field of statistical mechanics but more especially in connection with the mechanics of individual atoms or molecules, all of which indicate even more clearly the necessity of some such modification as proposed by Planck.

These later facts as to the mechanics of single atoms form of course a superior starting point for the foundation of a quantum theory, than could possibly be furnished by facts in such a complicated field as statistical mechanics itself. Hence we shall first consider the modifying postulates which must be introduced into ordinary mechanics in order to account for the behavior of *individual* atoms, and then extend the application of these postulates to the field of statistical mechanics in order to treat the behavior of systems containing many atoms or molecules.

92. Present Status of the Quantum Theory.—At the time the manuscript for this chapter was first prepared, the quantum theory had reached a stage in its development which, although evidently not entirely satisfactory, nevertheless had great definiteness and apparently a certain degree of permanency. The status at the time was that so excellently presented in the fourth edition of Sommerfeld's "Atombau und Spektrallinien" (1924) or in Van Vleck's report on "Quantum Principles and Line Spectra" (1926).

At the present moment, however, the quantum theory is in a very rapid state of new development, owing principally to the work of Heisenberg,[1] Born and Jordan,[2] and Schrödinger[3] in devising new forms of dynamics which are specially adapted for treating motions that are subject to quantum restrictions. This present rapid flux is unfortunate from the point of view of our immediate necessity of giving an outline of quantum theory mechanics, upon which to base the further development of our statistical mechanical considerations. The older definite form of the quantum theory is sure to be greatly modified in the future, and yet the precise nature of the modifications, which will prove most satisfactory, is still uncertain.

Under the circumstances it seems wisest to present our outline of quantum theory mechanics in its older definite form, and we shall offer no further apologies for so doing. It may be remarked, however, that the features of the older quantum theory which are most certain to be modified are the use of the Wilson-Sommerfeld rule of quantization for the theoretical calculation of the energies of the quantized states, and the use of the Bohr correspondence principle for the theoretical calculation of the intensities of spectral lines. Nevertheless, since the energy levels of the quantized states and the intensities of spectral lines are both of them subject to direct experimental determination, mere changes in the methods of their theoretical prediction will not affect the general standpoint that we must assume in adapting statistical mechanics to take account of quantum phenomena. On the other hand, there is no tendency in the newer dynamics which would indicate the necessity of abandoning either the Bohr relation between the energy levels of an atom and the frequency of the radiation which they emit or absorb, or Einstein's general relations connecting the probabilities of absorption and emission, or the procedure of assigning an area in the phase space of magnitude h to each quantized degree of freedom; and these are the features of quantum mechanics which are essential for the extension of statistical mechanics which will be presented.

Let us now turn to the discussion of the new postulates which have

[1] Heisenberg, *Zeitschr. für Phys.*, **33**, 879 (1925).
[2] Born and Jordan, *Zeitschr. für Phys.*, **34**, 858 (1925); Born, Heisenberg and Jordan, *Zeitschr. für Phys.*, **35**, 557 (1926).
[3] Schrödinger, *Ann. der Phys.*, **79**, 361, 489, 734 (1926).

been introduced in the attempt to modify classical dynamics so as to take account of quantum phenomena.[4]

93. The Quantum States.—The first modification of classical views which must be introduced is to recognize the existence of definite internal states of atoms or molecules, in which the atom can persist for an appreciable length of time without loss of energy. These are the so-called quantum states of the atom. Since they correspond to different quantities of internal energy they give the atom a series of energy levels, and it is found that the atom can exist, at least for an appreciable time, only at one or another of these levels, and may then change suddenly to some other level.

We now have very definite experimental evidence for the existence of these energy levels, obtained by the method of electronic impact. In using this method we arrange to bombard the atoms of a dilute gas with electrons coming from a hot filament with a definite velocity and kinetic energy corresponding to a fall through a known potential drop.

If the energy, obtained from this potential drop, is less than that necessary to raise the atoms from their normal energy level to the next higher level, we find that only *elastic* collisions can take place between the atoms and electrons. This can be shown for example by determining the magnitude of an opposing potential drop that the electrons can traverse after they have made many such collisions. If, however, we increase the initial energy of the electrons to a value slightly greater than that between the first two levels of the atom, inelastic impacts immediately take place, in which the kinetic energy of the electron is used in raising the atom to the higher level. These inelastic impacts are immediately made evident by the inability of electrons, which have made such collisions, to traverse electric fields which are easily overcome by electrons which start out initially with less energy. Hence it seems as if the existence of definite energy levels or quantum states could be taken as an undoubted fact.

94. As to the different internal configurations of the atom or molecule which correspond to different quantum states, there can be a wide variety. In the case of monatomic hydrogen, the simplest of all atoms, composed merely of a single electron and a single proton, we believe that the different quantum states correspond to different orbits in which the electron can rotate around the proton. In more complicated systems, such as a molecule of hydrogen chloride, we have different states corre-

[4] In the final theory, the classical dynamics may take its place as a special limiting case of a more general dynamics. In the status of the theory, which we are presenting, the quantum postulates enter as restrictions on classical dynamics when motions involving high frequencies are involved.

sponding to different velocities of rotation of the molecules as a whole, and to different velocities of oscillation of the two nuclei along the line joining their centers. Finally as an extreme we may regard the different isomeric or tautomeric forms of an organic molecule as different quantum states, say for example propyl and isopropyl alcohol, or the enol and keto forms of acetoacetic ester.

95. The development of a theoretical method of calculating the internal configurations and motions of a molecule which correspond to different quantum states has formed one of the main problems of the quantum theory. The method employed in the status of the theory which we are presenting consists in selecting from the continuous series of motions of different energy content which accord with the classical mechanics a discrete series of "quantized" motions which fulfill certain special requirements. The criterion for this selection is known as the Wilson-Sommerfeld rule of quantization.

On the basis of the classical mechanics, if we have a steady state of internal motion, it is evident that the electrons and nuclei composing a molecule would be executing periodic or more generally conditionally periodic motions. In such a conditionally periodic motion there is a fundamental period associated with each degree of freedom which is the time necessary for the corresponding coordinate and momentum to return to a given starting place. If now we consider the coordinate q and momentum p for any particular degree of freedom, and take the value of the *phase integral* $\int pdq$ over a complete period, the *Wilson-Sommerfeld rule of quantization* states that this integral must be equal to some simple multiple of Planck's constant h, if the molecule is in a steady quantized state of motion. The rule can be stated in the form of the following equation

$$\oint p_i dq_i = n_i h \tag{159}$$

The circle on the integral sign indicates that the integration is to be taken over the complete period belonging to the degree of freedom in question (i.e. to the i'th degree of freedom).[5] And the ***quantum number*** n_i gives the particular multiple of h that the phase integral equals.

[5] In case there is indefiniteness as to the period over which the integration is to be taken, we may encounter the phenomenon of "weak" quantization. Under these circumstances there appears to be a tendency for the system to assume a state which corresponds to the quantization of the truly periodic motion with which the actual nearly periodic motion is in approximate agreement. See Ehrenfest and Tolman, *Phys. Rev.*, **24**, 287 (1924). The status of weak quantization in the newer dynamics is not yet apparent. It seems possible that the idea will no longer be valuable.

For different quantized states this quantum number will assume different values which can differ from each other, however, only by *integral* amounts. In addition it is found experimentally that the lowest possible value of the quantum number is either zero or one-half depending on the kind of degree of freedom involved. Hence for the different quantized states of a molecule, the quantum numbers for the various degrees of freedom present, all have values $n = 0, 1, 2, 3 \ldots$ or $n = 1/2, 3/2, 5/2 \ldots$.

We thus see that the Wilson-Sommerfeld rule of quantization selects out from the different classically allowed motions, which would present a continuous range of values for the different phase integrals $I_i = \oint p_i dq_i$ a discrete set of allowed quantized motions with definite values of the phase integrals and corresponding definite values of the energy content of the molecule.

96. The further development of quantum theory dynamics is apparently going to be in the direction of concentrating attention on the quantized states of motion and giving up any idea of the existence of the intermediate classically possible motions. The development will probably involve the abandonment of the Wilson-Sommerfeld rule of quantization. The idea of the existence of states of definite energy content, however, which is the result of inescapable experimental evidence, cannot be abandoned and this is the idea of importance for our further development of statistical mechanics.

97. Mechanism of Passage of Molecules from One Quantum State to Another.—Having established the existence and to some extent the character of the different quantum states, we are next concerned with the processes by which a molecule can pass from one quantum state to another. Experimentally we find that this can occur either by interaction with radiation or by collision with other molecules.

Thus a molecule can pass from a lower to a higher quantum state by the absorption of radiation and in the reverse direction by emission. Furthermore a molecule can collide with another molecule and pass from a lower to a higher quantum state, by drawing on the supply of kinetic energy which is available (called collision of the first kind) or by taking internal energy from the other molecule if the latter is itself in an excited state (called collision of the second kind). On the other hand, a molecule can pass from a higher to a lower quantum state as a result of collision, either by giving up its energy of excitation in the form of kinetic energy or by transferring at least a portion of this energy to be used in raising the molecule with which it has collided into a higher quantum state (both called collisions of the second kind).

The experimental facts demonstrating the passage of molecules from one quantum state to another by the absorption and emission of radiation and by collisions of the first and second kind are very satisfactory. Concerning the internal mechanism of these changes, however, we still have only partial knowledge. There are, however, two characteristics of these processes concerning which we have definite information.

98. In the first place when the passage from one state to another occurs through the emission or absorption of radiation, there is a definite relation between the energy lost or gained and the frequency of the radiation. This relation is known as *Bohr's frequency postulate* and can best be stated in the form of the equation

$$\Delta E = h\nu \tag{160}$$

where ΔE is the difference in energy between the initial and final states of the molecule, ν is the frequency of the absorbed or emitted light and h is Planck's constant.[6]

99. In addition to this knowledge as to the frequencies involved when the transfer between states takes place through interaction with radiation, we can assert in general, that when molecules pass from one quantum state to another by any process, the laws of the *conservation of energy and of linear and angular momentum* must hold. This conclusion is necessarily true, at least statistically, since otherwise these laws could not hold, as we know they do, for macroscopic systems composed of molecules. In addition the experiments of Compton and Simon[7] and of Bothe and Geiger[8] indicate at least in one case that the conservation laws hold as well for individual molecular processes.

The conservation laws for energy and momentum have proved valuable in furnishing simple quantum theory explanations of the Doppler effect and the Compton effect, while the conservation of angular mo-

[6] Equation (160) would indicate that the light emitted by a system of molecules should be absolutely unifrequentic provided the temperature is low enough so as to eliminate the Doppler effect and the molecules far enough apart to eliminate deviations in the energy of the molecules due to their mutual interactions. This perhaps would be the view advocated by the most extreme holders of the so-called light-quantum theory. It seems possible, however, as an alternative that spectral lines become sharper and sharper, the less "weak" the quantization (see preceding footnote), and are never absolutely unifrequentic. Equation (160), however, is entirely satisfactory for the very sharp lines that we encounter in the usual case of "strong" quantization.

[7] Compton and Simon, *Phys. Rev.*, **26**, 289 (1925).

[8] Bothe and Geiger, *Zeits. für Phys.*, **32**, 639 (1925).

mentum can be used to interpret the polarization of the radiation emitted when an atom falls from one quantum state to another.

100. Probability of Passage from One Quantum State to Another.—We must next inquire as to the frequency with which molecules pass from one quantum state to another, and since the internal mechanism of the transitions from state to state is largely unknown, we can merely ask as to the *probability* of transition under a given set of external conditions.

101. In the case of changes involving the interaction of radiation, we have definite evidence from absorption experiments that the probability of transition *from a lower to a higher quantum state* is directly proportional to the intensity of the radiation needed for the change. While for a transition *from a higher to a lower quantum state,* the measurements of Wien [9] on the decay in the intensity of the light coming from the excited atoms in a canal ray, have shown us that we must ascribe to excited atoms a definite probability in unit time of spontaneously emitting energy and dropping to a lower quantum state.

102. As the quantitative laws governing the probabilities of absorption and emission we shall accept the views presented by Einstein [10] in 1917. In accordance with these views, the chance per second that a molecule will pass from a lower state i to a higher state j by the absorption of radiation will be $B_{i\rightarrow j}u_\nu$ where $B_{i\rightarrow j}$ is a proportionality constant depending on the nature of the states, and u_ν is the radiation density at the frequency ν which is involved in the transfer. For the probability of a transition in the opposite direction from state j to state i, with the emission of radiation, we shall have the sum of two terms $A_{j\rightarrow i} + B_{j\rightarrow i}u_\nu$, where $A_{j\rightarrow i}$ is the chance per second of spontaneous emission, and $B_{j\rightarrow i}$ is a proportionality constant for induced emission under the action of radiation of the frequency ν. The implication of these laws will be more fully treated in Chapter 16, where theoretical relations connecting the three proportionality constants will be obtained.

103. The chance of spontaneous emission $A_{j\rightarrow i}$ will evidently determine the mean life of an excited molecule in the absence of collisions with other molecules and in a field free enough from radiation so that we can neglect the probability of the induced transition. If there is only a single lower state i to which molecules in the state j can jump, the mean life in the upper state will evidently be given by

[9] W. Wien, *Ann. der Physik*, **73**, 483 (1924) and earlier work.
[10] Einstein, *Phys. Zts.*, **18**, 121 (1917).

$$\tau = \frac{1}{A_{j \to i}} \tag{161}$$

and in case there are a number of lower states i, we must obviously write

$$\tau = \frac{1}{\sum_{i} A_{j \to i}} \tag{162}$$

Actual values of the mean life may vary all the way from the neighborhood of 10^{-7} seconds for a mercury atom excited to the $2p_2$ state, to the comparatively long life of an oscillating hydrogen chloride molecule and the practically indefinite life of an organic isomer.[11]

104. In interpreting this life time of excited atoms and molecules, we encounter some difficulty in distinguishing between the time during which the atom stays in its upper quantum state (Verweilzeit) and the time occupied in travelling from the upper to the lower state (Abklingungzeit). Indeed it is possible that our present concepts of time are inadequate for the treatment of interatomic phenomena. Nevertheless, we obtain for the present a very satisfactory representation of reality if we assume that the time taken by the atom in travelling from one state to another is negligibly short compared with the time spent in the different quantum states.

105. Turning now to the probability of transition from one state to another by collision with other molecules, we can merely say that the chance of transition will be proportional to the number of collisions occurring and to proportionality constants whose inner meaning is still to be elucidated, although certain theoretical relations connecting the probabilities of transition in reverse directions will be derived in Chapter 17.

On the experimental side the work of Glockler (see Chapter 17) indicates that the transition of an atom from a lower to a higher quantum state by electron impact, has an appreciable probability of occurrence, only when the electron has not more than a very slight excess over the necessary amount of kinetic energy to make the change possible at all. And with regard to the transfer of energy of excitation from one atom to another the work of Stuart (see Chapter 21) has shown that the probability for this to occur is a very specific property of the atoms involved.

106. Classical Mechanics as a Limiting Case.—In the method of presentation chosen for the purposes of this chapter, we have considered that the quantum theory mechanics arises from the classical mechanics

[11] See Tolman, "Duration of Molecules in Upper Quantum States," *Phys. Rev.*, **23**, 693 (1924), and Chapter 16 of this book.

by the introduction of extra restrictions, which were not appreciated until the world of atomic phenomena had been studied. For the final method of presentation which will ultimately be satisfactory, it seems possible that a more general dynamics will be invented of which the classical dynamics will be a limiting case. Whatever the future may bring, however, it is already evident that many dynamical phenomena—including the motions of molar masses whose study led to the classical mechanics and including also the translational motion of the molecules of a gas at ordinary temperatures, which has been of main interest for the preceding applications of statistical mechanics—will not be subject to quantum restrictions.

To show this let us return to the Wilson-Sommerfeld rule of quantization for the i'th degree of freedom of a system

$$\oint p_i dq_i = n_i h$$

Referring to equation (12) in Chapter 2 for the definition of p_i, we can rewrite this equation in the evident form

$$\int_0^\tau p_i \frac{dq_i}{dt} dt = \int_0^\tau \frac{\partial L}{\partial \dot{q}_i} \dot{q}_i dt = n_i h \qquad (163)$$

where L is the Lagrangian function for the system and τ is the period of the motion for the particular degree of freedom.

For an ordinary mechanical system, however, the Lagrangian function L is the difference between the kinetic and potential energies of the system, and the kinetic energy can be expressed as a sum of quadratic terms depending on the velocities, giving us [12]

$$L = T - V = \Sigma \tfrac{1}{2} m_i \dot{q}_i^2 - V \qquad (164)$$

Hence by substituting into equation (163) we can obtain

$$\int_0^\tau m_i \dot{q}_i^2 dt = n_i h \qquad \text{or} \qquad [\tfrac{1}{2} m_i \dot{q}_i^2]_{\text{av.}} = \frac{n_i h}{2\,\tau} \qquad (165)$$

Since the successive quantized states for i'th degree of freedom differ by changes of unity in the quantum number n_i, we see that for such a system the average kinetic energy associated with this degree of freedom will increase by very small steps whenever the period of the motion τ becomes long. In addition, of course, the kinetic energy could

[12] A similar treatment is also possible when cross products of the velocities appear in the expression for T.

be regarded as increasing practically continuously as compared with that already present, whenever we are in the region of large quantum numbers. Finally it should be noted that for all ordinary systems small changes in potential energy will be associated with small changes in kinetic energy.

Thus, since the energy levels will lie close together, we see that the classical dynamics can be regarded as applying to any degree of freedom with a long period of motion, and in general in the region of large quantum numbers. Hence we can neglect quantum restrictions in the case of translational motion, and in the case of periodic motion, provided the period is long or the number of quanta of energy already present is large.

107. The Correspondence Principle.—Although the postulates already introduced in the preceding sections contain all that we need in order to make our applications to statistical mechanics, nevertheless it would be unfortunate to proceed without also including *Bohr's correspondence principle,* a postulate which has been extraordinarily fruitful in the development of the quantum theory. This postulate demands a certain correspondence between the *rate* of emission of radiation by molecules as given by the quantum theory, and the *rate* of emission which would have been predicted on the basis of the older classical theory. The nature of the postulates can best be understood if we first derive what may be called the *correspondence theorem* which connects the *frequency* of radiation as given by the quantum theory with the *frequency* which would have been predicted by the classical theory. We shall then be in a better position to understand the postulated correspondence between the rates given by the two theories.

108. On the basis of the classical theory, a radiating molecule could be treated as a system performing a conditionally periodic motion. If such a system has s degrees of freedom, there will be present, in the motion of the system, s fundamental frequencies[18] $\omega_1 \ldots . \omega_s$ which are connected with the energy of the system E and the s phase integrals $I_1 \ldots . I_s$ by the equation

$$dE = \omega_1 dI_1 + \ldots . + \omega_s dI_s \tag{166}$$

the phase integrals being given by the equations

$$I_1 = \oint p_1 dq_1, \ldots \ldots, I_s = \oint p_s dq_s \tag{167}$$

[18] The development is given for non-degenerate systems in which the number of degrees of freedom is the same as the number of different fundamental periods. The modifications for degenerate systems will be obvious.

with the limits of integration taken around the complete period for the degree of freedom involved.

In accordance with the classical theory, the fundamental frequencies $\omega_1 \ldots \omega_s$ present in the motion of the molecule and the frequencies of their combination overtones $(\tau_1\omega_1 + \ldots + \tau_s\omega_s)$, the order numbers τ being integers, would be the frequencies of the emitted radiation. So that we could write for the frequency of the radiation

$$\nu_{\text{Classical}} = (\tau_1\omega_1 + \ldots + \tau_s\omega_s) \tag{168}$$

or substituting equation (166)

$$\nu_{\text{Classical}} = \left(\tau_1 \frac{dE}{dI_1} + \ldots + \tau_s \frac{dE}{dI_s}\right) \tag{169}$$

109. On the other hand on the basis of the quantum theory a molecule can also be treated as performing conditionally periodic motions in each of its quantum states, but instead of gradually radiating away its energy, it suddenly jumps to a new quantum state, the change in energy ΔE being accompanied by radiation of frequency ν in accordance with the equation

$$\Delta E = h\nu \tag{170}$$

and the change in quantum state being such that the phase integrals change by whole multiples of h in accordance with the equations

$$\Delta I_1 = \Delta n_1 h, \ldots, \Delta I_s = \Delta n_s h \tag{171}$$

where $\Delta n_1 \ldots \Delta n_s$ are integers giving the changes in quantum number that occur.

To correlate the two points of view, we may now rewrite equation (170) in the form

$$\nu_{\text{Quantum}} = \frac{\Delta E}{h} = \frac{1}{h}\left(\frac{\Delta E}{\Delta I_1}\Delta I_1 + \ldots + \frac{\Delta E}{\Delta I_s}\Delta I_s\right) \tag{172}$$

or substituting equations (171) we may write

$$\nu_{\text{Quantum}} = \left(\Delta n_1 \frac{\Delta E}{\Delta I_1} + \ldots + \Delta n_s \frac{\Delta E}{\Delta I_s}\right) \tag{173}$$

110. On comparing equations (169) and (173), we observe a great similarity, the integral order numbers and differential coefficients occurring in the classical theory expression being replaced by the integral changes of quantum number and difference coefficients in the quantum theory expression.

This result may be called the *correspondence theorem*. It indicates in the first place a correspondence between any given harmonic present in the classical motion with order numbers $\tau_1 \ldots \tau_s$ and the quantum transition which is brought about by changes in the quantum numbers of amount $\Delta n_1 = \tau_1 \ldots \Delta n_s = \tau_s$. And in the second place it shows at once that in the region of large quantum numbers, where we have a continuous series of quantum states and the difference coefficients and differential coefficients become identical, we can expect the quantum theory frequencies which are actually emitted to become identical with the frequencies calculated on the basis of the classical theory.

111. This correspondence between the *frequencies* calculated on the basis of the two theories has led to the Bohr *postulate* of a relation between the *intensities of radiation* which would be emitted on the basis of the two theories—a relation which would be exact in the region of large quantum numbers and give approximate information in the region of small quantum numbers.

112. On the basis of the classical theory it was possible to calculate the intensity of radiation corresponding to a given harmonic with the help of the Fourier analysis of the motion of the electric charges in the molecule. As the result of such an analysis the value of any component of the electric moment of the molecule, say for example the component parallel to the X-axis could be expressed by an equation of the form

$$\xi = \sum_{\tau_1} \ldots \sum_{\tau_s} D_{\tau_1 \ldots \tau_s} \cos 2\pi \left[(\tau_1 \omega_1 + \ldots + \tau_s \omega_s) t + \gamma_{\tau_1 \ldots \tau_s} \right] \quad (174)$$

where the quantities $D_{\tau_1 \ldots \tau_s}$ are the amplitudes corresponding to the different harmonics of frequencies $(\tau_1\omega_1 + \ldots + \tau_s\omega_s)$, the quantity $\gamma_{\tau_1 \ldots \tau_s}$ is a phase angle, and the summations are to be taken for all values of the integers $\tau_1 \ldots \tau_s$ from minus to plus infinity. And in the classical theory the rate of emission of any frequency $(\tau_1\omega_1 + \ldots + \tau_s\omega_s)$ would depend on the value of the corresponding amplitudes $D_{\tau_1 \ldots \tau_s}$ in accordance with the equation

$$\left(\frac{dR}{dt}\right)_{\text{Classical}} = \frac{16\pi^4}{3c^3} \sum_{x,\, y,\, z} \left[D^2_{\tau_1 \ldots \tau_s} (\tau_1\omega_1 + \ldots + \tau_s\omega_s)^4 \right] \quad (175)$$

where the summation sign indicates that we must sum for the amplitudes parallel to each of the three axes.

113. On the other hand, in accordance with the quantum theory, the rate of energy emission of a particular frequency per molecule in the upper state would evidently be

$$\left(\frac{dR}{dt}\right)_{\text{Quantum}} = A_{\tau_1 \ldots \tau_s} h\nu \quad (176)$$

where $A_{\tau_1 \ldots \tau_s}$ is to be taken as the value of Einstein's probability coefficient for the spontaneous transition which *corresponds* to the harmonic under consideration by resulting from changes in the quantum numbers $\Delta n_1 = \tau_1 \ldots . \Delta n_s = \tau_s$.

114. We now introduce the *correspondence principle* by postulating an approximate relation between the rates as given by equations (175) and (176) which can be written in the simplified form

$$Ah\nu \approx \frac{16\pi^4}{3c^3} \sum_{x, y, z} [D^2\omega^4] \qquad (177)$$

It is to be expected that this relation would hold exactly in the region of large quantum numbers and would give approximate information in the region of small quantum numbers.

The necessarily approximate nature of these latter results is indicated by the form of the relation itself. In the region of large quantum numbers where the quantized states lie close together the values of the amplitude D and the frequency ω which then equals ν will be the same in the initial and final quantum states and the method of substituting into (177) is obvious. On the other hand in the region of small quantum numbers, D and ω will have different values in the initial and final states and it is not evident which set we are to use, or whether some average of these quantities taken for the two states or taken over the intermediate classical states is to be employed. Apparently one of the greatest advantages that we may expect from the newer dynamics, which has been mentioned above, will be the removal of this difficulty.

115. In spite of these uncertainties as to the correct method of applying the correspondence principle, it has proved extremely useful. Especially in the hands of Kramers [14] it has given excellent predictions as to the *relative* intensities of related spectral lines, while with its help Tolman and Badger [15] have been able to account for the *absolute* intensities of pure rotation lines in the far infra red spectrum of hydrogen chloride.

This concludes our necessarily somewhat unsatisfactory account of quantum theory mechanics and we are now ready to discuss the modifications in statistical mechanics which must be introduced in order that we may consider quantum phenomena.

[14] Kramers, *Kgl. Danske Vidensk. Selsk. Shrifter*, **8**, III, 287 (1919).
[15] Tolman and Badger, *Phys. Rev.*, **27**, 383 (1926). This treatment will need modification on the basis of the new quantum dynamics.

Chapter 8

The Incorporation of Quantum Theory into Statistical Mechanics

116. Nature of the Modification.—In Section **104** of the preceding chapter, we introduced the new idea that molecules spend at least the major portion of their time in quantized states of internal motion, and have, if any, only a fleeting existence in a condition of transition from one state to another. It can at once be seen that this necessitates some modification in the older statistical mechanics which would ascribe no special probability to states of motion having a particular energy content.

It is evident that the general nature of the modification, which must be introduced into statistical mechanics, will be to assign in some way a preponderating γ-weight to regions in the γ-space which correspond to systems in which the molecules have the definite energy contents allowed by the quantum restrictions. This can be accomplished, moreover, by going at once to the phase space, belonging to each kind of molecule, μ-space, and assigning a larger μ weight to states of the molecule, having the desired energy contents, than would have been assigned on the basis of the classical theory.

117. There are two ways in which this modification might be introduced. In the first place leaving the dependence of energy on the coordinates and momenta unchanged, we could give up the conclusion, reached with the help of Liouville's theorem, that different regions in the phase space have a μ-weight equal to the volume $\int \ldots \int dq_1 \ldots dp_n$ involved, and assume rather that the μ-weight has been gathered in from intervening regions of the phase space and concentrated on those surfaces of the phase space which correspond to quantized motions. As a second possibility we might keep the distribution of μ-weight uniform throughout the phase space, and change the dependence of energy on the coordinates and momenta in such a way that a considerable volume in the phase space would all belong to states of the molecule with practically the same energy content. The two possibilities can be made clearer with the help of diagrams in which we plot energy E against phase integral $I = \oint pdq$.

118. Method of Concentrating μ-Weight.—The first method in which we concentrate μ-weight on particular boundaries in the phase space, leaving the dependence of energy on the coordinates and momenta unchanged, is illustrated by Figure 1.

The system considered is a non-harmonic oscillator of one degree of freedom, and the dotted curve gives the relation between energy E and phase integral I, which would accord with the classical theory.

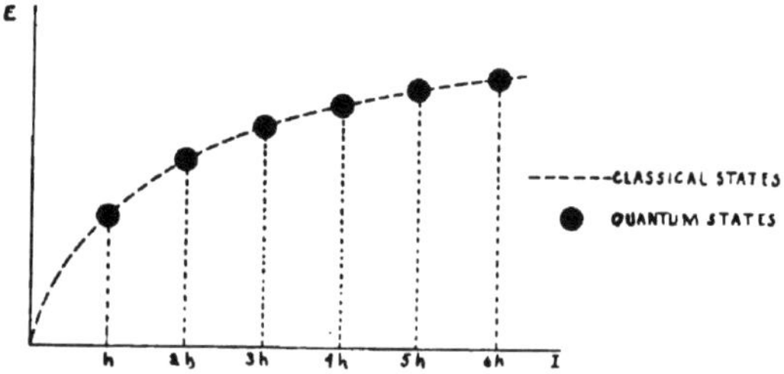

FIGURE 1.

The particular oscillator illustrated by the figure is one in which there is a deviation from Hooke's law, such that the elastic restoring forces get weaker as the amplitude increases. This evidently results in decreasing values of the frequency of oscillation ω as the energy increases. Hence, since energy and phase integral are connected with frequency, on the basis of the classical theory, by the equation

$$\frac{dE}{dI} = \omega \tag{178}$$

it is evident that the energy will increase less and less rapidly with the phase integral, as shown by the curve.

On the basis of the classical theory the oscillator could have the energy and phase integral corresponding to any point on the dotted curve. Indeed the a-priori probability that the oscillator would have a value of its phase integral between I and $I + dI$ would be directly proportional to the area $dI = dqdp$. Furthermore, in accordance with the classical theory, the oscillator, assuming the moving part to be charged, would be emitting radiation no matter where the representative point might be located on the curve, the frequencies of the radiation being the fundamental frequency ω and the harmonics $\tau\omega$.

To agree with the quantum theory, we might modify the above picture as follows. In the first place, since the oscillators are in general only to be found with values of the phase integral I equal to some multiple of h, we could gather together the a-priori probability previously spread uniformly along the I-axis and concentrate it at the points $I = nh$. This has been indicated in the figure by inserting heavy dots to correspond to the allowed values of phase integral and energy. In the second place, in spite of the vibrations which the oscillators are performing, we should have to assume that for some unknown reason they can remain for an appreciable length of time in the quantized states without emitting radiation, and then suddenly emit radiation with the quantum theory frequency

$$\nu_{\Delta n} = \frac{\Delta E}{h} = \Delta n \frac{\Delta E}{\Delta I} \tag{179}$$

instead of the classical frequency

$$\nu_\tau = \tau\omega = \tau \frac{dE}{dI} \tag{180}$$

119. Method of Changing the Relation Connecting Energy with the Coordinates and Momenta.—The other method of assigning greater μ-weight to the quantized states would be to leave the μ-weight uniformly distributed over the phase space, but change the form of the functional relation connecting the energy of the molecule with its coordinates and momenta. The method is illustrated by Figure 2.

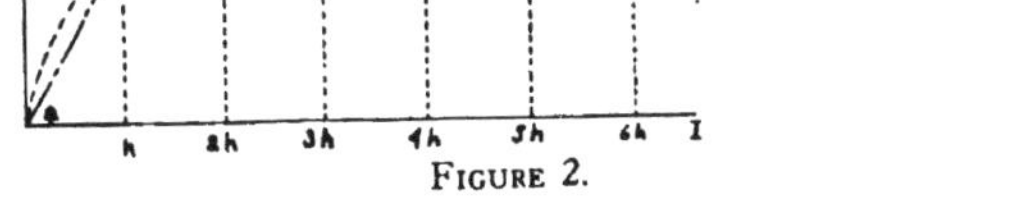

FIGURE 2.

The dotted curve again gives the relation between energy E and phase integral I, which on the basis of the classical theory would belong to a non-harmonic oscillator. The heavy line, however, giving a step-like rise of the energy, represents the changed relation between energy and phase integral which is proposed. It is evident that the molecule in question has a-priori probability of existence only at the energy levels corresponding to the horizontal portions of the line and these energy levels are chosen so as to agree with the accepted rules of quantization.

It may furthermore be noted, in accordance with the relation $\omega = dE/dI$, that the oscillator would have zero frequency in the quantized states and hence no longer vibrate, in agreement with the fact that it can remain in a quantized state without emission of radiation. This would also have the advantage, if it is such, of giving us the static atom so long desired by chemists.

To round out the picture we should also have to provide some way by which changes from one quantum state to another could occur from time to time in accordance with the laws of chance through the absorption or emission of radiation. This could perhaps be arranged by connecting the energy levels by paths which would give the molecule the desired behavior when it is interacting with radiation, and assuming that the molecule sometimes slips into one of these connecting paths. Two of these paths *ab* and *bc* have been indicated in the figure by dot and dash lines, and remarkably enough all the necessary paths can be correctly obtained by drawing straight lines connecting all the points on the classical curve which represent possible values of E and I.

The path *ab* would represent the motion of the oscillator when it is absorbing or emitting radiation in between the first two energy levels. It should be specially noticed that the frequency of the oscillation on this path as given by the classical laws of motion would be the actual frequency of the radiation known to be absorbed or emitted in the transition between these two levels. This is evident since we have from the figure

$$\omega_{\text{Classical}} = \frac{dE}{dI} = \frac{\Delta E}{h} = \nu_{\text{Quantum}} \tag{181}$$

The path *bc* relates to transitions between the levels $I = h$ and $I = 4h$, corresponding to a change in quantum number of 3. The fundamental frequency of the oscillation on this path would also be correctly given by the slope of the line, and the frequency of the radiation actually emitted or absorbed is known to be the third harmonic of this fundamental frequency, corresponding to the change of 3 in the quantum number.

The length of time which a molecule spends in the intermediate paths would be determined by the damping factor for the oscillation, and the time which it spends in the quantized states would depend on the chance of "slipping" into one of the connecting paths. These factors would have to be adjusted so as to agree with experiment.

120. Comparison of the Two Methods.—It will be seen that the two methods of assigning increased a-priori probability to the quantized states have much in common. The first method is rather less complicated and provides the language usually adopted.[1] The second method tries to retain more of the classical mechanics and to go more deeply into an explanation of the frequencies actually absorbed and emitted. The postulated connecting paths have some of the properties and advantages of the virtual oscillators introduced by Bohr, Kramers and Slater.[2] The method was first considered by the author a number of years ago and a model constructed illustrating these connecting paths between different energy levels for the case of a hydrogen atom.[3]

Of course neither method is entirely satisfactory, and we shall have to wait for a relatively complete solution of the problem of quantum mechanics before we can obtain a really acceptable picture of the state of affairs.

121. Result for Statistical Mechanics.—For our present purposes, it is only of interest to note that the result of both the above methods or of any others which are otherwise acceptable, will be the same as far as statistical mechanics is concerned. They agree in dividing up, between the different quantized states, the a-priori probability belonging to the whole phase space, and assigning a μ-weight of magnitude h to each quantized degree of freedom. This result can be incorporated in statistical mechanics by the introduction of the following postulate, which is expressed in a general enough form so as to include both "classical" and "quantized" degrees of freedom. (See Section **106** in the preceding chapter.)

Consider a molecule having n degrees of freedom with respect to which the motion may be considered as classical, and s degrees of freedom for which the motion is quantized. The μ-weight for any given state of this molecule in which the classical coordinates and momenta fall in the range $dq_1 \ldots dp_n$ and the quantized coordinates and momenta have the values corresponding to any given quantum state will then be

$$\mu = h^s dq_1 \ldots dp_n \tag{182}$$

This postulate is evidently directly required by the second method of assigning a-priori probabilities discussed above, and is in entire agreement with the first method. Certain special difficulties may arise as to the assignment of μ-weight to the lowest possible quantum state, but these need not detain us now. The postulate also obeys certain restrictions imposed by the second law of thermodynamics which have been dis-

[1] See for example Ehrenfest and Tolman, *Phys. Rev.*, **24**, 287 (1924).
[2] Bohr, Kramers and Slater, *Phil. Mag.*, 47, 785 (1924).
[3] See Tolman, *Phys. Rev.*, 11, 261 (1918); *J. Opt. Soc. Amer.*, **4**, 211 (1922).

cussed by Ehrenfest [4] and is the postulate which has always been at least implicitly chosen.[5]

122. Resulting Modification in the Maxwell-Boltzmann Distribution Law.—It is evident at once that this new postulate will lead to modifications in the Maxwell-Boltzmann distribution law in the case of degrees of freedom for which the motion is quantized. The nature of these modifications is evident, provided we accept the postulate.

For a molecule of n degrees of freedom, all of which are "classical," the Maxwell-Boltzmann distribution law can be written in accordance with equation (72) in Chapter 4 in the form

$$dN = \frac{Ne^{-\frac{\epsilon}{kT}} dq_1 \ldots . dp_n}{\int \ldots \int e^{-\frac{\epsilon}{kT}} dq_1 \ldots . dp_n} \tag{183}$$

where dN is the number of molecules having coordinates and momenta falling in the range $dq_1 \ldots . dp_n$ and the integrations in the denominator are to be taken over the whole phase space.

For a molecule of s degrees of freedom, all of which are quantized, in analogy with equation (183) we can then write in accordance with our postulate, for the number of molecules N_i falling in the i'th quantum state, the expression

$$N_i = \frac{Ne^{-\frac{\epsilon_i}{kT}} h^s}{\sum_j e^{-\frac{\epsilon_j}{kT}} h^s} = \frac{Ne^{-\frac{\epsilon_i}{kT}}}{\sum_j e^{-\frac{\epsilon_j}{kT}}} \tag{184}$$

where the summation in the denominator is taken over all possible quantum states.

Evidently for a molecule having n classical degrees of freedom and s quantized degrees of freedom we could use an expression of the form

$$dN_i = \frac{Ne^{-\frac{\epsilon_i}{kT}} e^{-\frac{\epsilon}{kT}} dq_1 \ldots . dp_n}{\sum_j e^{-\frac{\epsilon_j}{kT}} \int \ldots . \int e^{-\frac{\epsilon}{kT}} dq_1 \ldots . dp_n} \tag{185}$$

provided the energy increase could be divided into a part ϵ_i characteristic of the quantum state and a part ϵ dependent on the q's and p's.

Cases often arise, in connection with degenerate periodic motions,

[4] See Ehrenfest and Trkal, *Proc. Amst. Acad.*, **23**, 179 (1920).
[5] See Bohr, "On the Quantum Theory of Line Spectra," Copenhagen Academy, 1918, Part I, p. 25. For degenerate states, we must multiply by the number of individual non-degenerate states considered together in the single degenerate state.

in which several quantized states have the same energy content. When this happens it is often convenient to write equation (184) in the form

$$N_i = \frac{Np_i e^{-\frac{\epsilon_i}{kT}}}{\sum_j p_j e^{-\frac{\epsilon_j}{kT}}} \tag{186}$$

where the quantities p_i, etc., are the number of quantum states with the same energy content which have been lumped together. The quantities p_i are of course integers and are themselves called the a-priori probabilities for the collection of degenerate states which are considered as combined together into a single state.[6]

[6] See preceding footnote.

Chapter 9

The Specific Heat of Diatomic Gases

123. Rotational and Vibrational Specific Heat.—The first application which we shall make of our new form of the Maxwell-Boltzmann distribution law, modified so as to take account of the quantum theory, will be to investigate the specific heat of diatomic gases. For this purpose it will be a close approximation if we consider a diatomic molecule as a rigid dumb-bell like structure which can rotate about its center of gravity, and calculate the contribution to the specific heat made by the possibility thus provided of acquiring rotational energy. In the first analysis it will not be necessary to consider the vibrational energy of the atoms along the line joining their centers of gravity, since the constraints holding the atoms together prove to be so strong that appreciable vibrational energy usually does not enter until relatively high temperatures. It is also unnecessary to allow for changes in the distance apart of the two atoms due to centrifugal force, since the effects turn out to be inappreciable.

124. Quantization of a Simple Rotator.—The different quantized states of rotation for a dumb-bell like molecule are the same as for any simple rotator and can easily be determined from the Wilson-Sommerfeld rule of quantization as given by equation (159) in Chapter 7.

$$\oint p dq = nh \tag{187}$$

For describing the motion of any simple rotator we need only a single coordinate

$$q = \phi \tag{188}$$

which gives the angle through which the rotator has turned with respect to a fixed direction. The momentum p_ϕ corresponding to the coordinate ϕ can be obtained in accordance with equation (12) in Chapter 2, by differentiating the Lagrangian function (difference of kinetic and potential energies) by $\dot{\phi}$ giving us

$$p_\phi = \frac{\partial L}{\partial \dot{\phi}} = \frac{\partial}{\partial \dot{\phi}}(\tfrac{1}{2} J \dot{\phi}^2) = J\dot{\phi} \tag{189}$$

where J is the moment of inertia of the rotator.

Substituting equations (188) and (189) in (187), we obtain

$$\oint p dq = \oint p_\phi d\phi = \int_0^{2\pi} J\dot{\phi} d\phi = mh \qquad (190)$$

where in accordance with convention we have used m to indicate rotational quantum numbers. Since the angular velocity $\dot{\phi}$ is constant in a steady state of motion, the integration is easily performed and we obtain the well-known equation

$$p_\phi = J\dot{\phi} = \frac{mh}{2\pi} \qquad (191)$$

for the different values of the angular momentum of a rotator which are allowed by the quantum theory.

125. Application to a Diatomic Molecule.—For a rigid diatomic molecule, the energy of rotation will all be kinetic as given by the equation

$$E = \tfrac{1}{2} J\dot{\phi}^2 \qquad (192)$$

which on substitution of (191) gives

$$E = \frac{m^2h^2}{8\pi^2 J} \qquad (193)$$

as the possible values of the rotational energy.

In applying equation (193), it will be noted that for a rigid diatomic molecule the value of the moment of inertia J will be a constant quantity independent of the velocity of rotation and given by the equation

$$J = m_1 r_1^2 + m_2 r_2^2 \qquad (194)$$

where m_1 and m_2 are the masses of the nuclei of the two atoms and r_1 and r_2 their distances from the center of gravity. Introducing the condition which locates the center of gravity

$$m_1 r_1 = m_2 r_2 \qquad (195)$$

and using the symbol

$$\mu = \frac{m_1 m_2}{m_1 + m_2} \qquad (196)$$

for the mutual mass of the system, and

$$r = r_1 + r_2 \qquad (197)$$

for the total distance between the nuclei, equation (194) can then be rewritten when desired in the form

$$J = \frac{m_1 m_2}{m_1 + m_2}(r_1 + r_2)^2 = \mu r^2 \qquad (198)$$

126. Expression for Rotational Specific Heat.—Combining equation (193), for the different possible values of the rotational energy of our molecule, with the Maxwell-Boltzmann law, as given by equation (186) in the last chapter, for the number of molecules in the different quantum states, we evidently obtain for the rotational energy of the N diatomic molecules in one mol of a gas

$$E_{\text{Rot.}} = \frac{N\Sigma p_m e^{-\frac{m^2h^2}{8\pi^2JkT}} \frac{m^2h^2}{8\pi^2J}}{\Sigma p_m e^{-\frac{m^2h^2}{8\pi^2JkT}}} \qquad (199)$$

where the summations are taken for all values of m. By differentiating this expression with respect to temperature, we obtain what may be called the rotational specific heat

$$C_{\text{Rot.}} = \frac{dE_{\text{Rot}}}{dT}.$$

and the resulting expression can be written in the very simple form

$$\frac{C_{\text{Rot.}}}{R} = \sigma^2 \frac{d^2 \log Q}{d\sigma^2} \qquad (200)$$

provided we make the substitutions

$$\sigma = \frac{h^2}{8\pi^2 JkT} \qquad (201)$$

and [1]

$$Q = \sum_m p_m e^{-m^2\sigma} \qquad (202)$$

In order to determine the specific heat at any temperature with the help of equations (200), (201) and (202), we must substitute values

[1] The general quantity

$$Q = \sum_i p_i e^{-\frac{\epsilon_i}{kT}}$$

which often occurs in statistical mechanical calculations has been given the name "condition sum" (Zustandsumme). In the classical mechanics it would have the form

$$Q = \int \cdots \int e^{-\frac{\epsilon}{kT}} dq_1 \ldots dp_n$$

and is thus the reciprocal of the Maxwell-Boltzmann C

for the quantum numbers m and the a-priori probabilities p_m into the expression for the "condition sum" Q. The correct way of doing this is a little more uncertain than might at first thought have been supposed,—in the first place because fractional quantum numbers have to be used in connection with the rotation of molecules, and in the second place because the periodic motion involved is a degenerate one and it is not certain exactly how many completely quantized states there are corresponding to each energy level.

127. Assignment of Quantum Numbers.—A determination of the values which can be assumed by the rotational quantum number m can be obtained by comparing equation (193) for the different possible values of rotational energy with the results of spectrum analysis. Spectral lines corresponding solely to changes in the rotational energy of a molecule occur in the far infra red and have now been studied in the very simple case of rotating hydrogen chloride molecules.[2] The analysis of the series of lines definitely shows that, although the successive values of the rotational energy of a hydrogen chloride molecule can be obtained from equation (193) as would be expected by substituting successive values of m which increase by steps of unity, nevertheless the lowest value of m is not itself an integer but has the value one-half.

The full meaning of the occurrence of such half quantum numbers is not yet entirely evident. It seems possible, however, on the basis of the newer dynamics mentioned in Chapter 7, that this is fundamental. In any case the experimental facts indicate that in the evaluation of rotational specific heats we must substitute for the successive values of the rotational quantum number

$$m = \frac{1}{2}, \frac{3}{2}, \frac{5}{2}, \frac{7}{2}, \ldots\ldots \tag{203}$$

128. Assignment of A-priori Probabilities.—In assigning values to the a-priori probabilities in the expression (202) for the condition sum Q, we must take account of the fact that our system is a degenerate one, since there is only a single fundamental period corresponding to the two rotational degrees of freedom. As a result of this degeneration we have obtained only a single independent quantum condition as expressed for example by the equation

$$p_\phi = \frac{mh}{2\pi} \tag{204}$$

for the possible values of the total angular momentum, in spite of the fact that the system has two degrees of freedom which are performing

[2] Czerny, *Zeitschr. für Phys.*, 34, 227 (1925).

periodic motions. If the system had not been degenerate, we should have obtained two independent quantum conditions. On account of these two conditions, we should have then had to introduce another quantum number besides m, in quantizing the motion, and should have found that a given value of m could be associated with more than one value of the other quantum number. Hence from this point of view we may regard equation (204) as determining a group of quantum states rather than a single quantum state, and in accordance with the ideas of Bohr as given in our postulate in Chapter 8, we should assign to the whole group the weight $p_m h^2$ where p_m is the number of individual states belonging to the group.

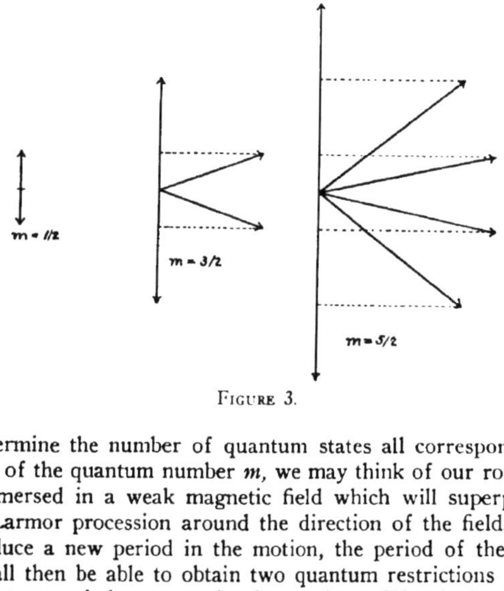

FIGURE 3.

To determine the number of quantum states all corresponding to the same value of the quantum number m, we may think of our rotating molecule as immersed in a weak magnetic field which will superpose on the motion a Larmor procession around the direction of the field. This will then introduce a new period in the motion, the period of the precession, and we shall then be able to obtain two quantum restrictions corresponding to the two periods present in the motion. We obtain in this way the number of separate quantum states which correspond, in this weak magnetic field, to a given value of m, and at the limit of zero magnetic field shall still assume the same effective number of quantum states.

As a result of such investigations, it has been found that the two quantum conditions prescribe not only that the total value of the angular momentum must increase by multiples of $h/2\pi$, but that the component parallel to the external field must also increase by multiples of $h/2\pi$. This result makes it at once possible to determine the number of quantum states corresponding to a given value of m as shown by the foregoing figures in which the vectors represent the possible values of the angular momentum. (The direction of the external field is taken as vertical.) On the basis of these results we may evidently put[8] for the a-priori probabilities of the degenerate states

$$p_m = 2m + 1 = 2,\ 4,\ 6,\ 8, \ldots\ldots \tag{205}$$

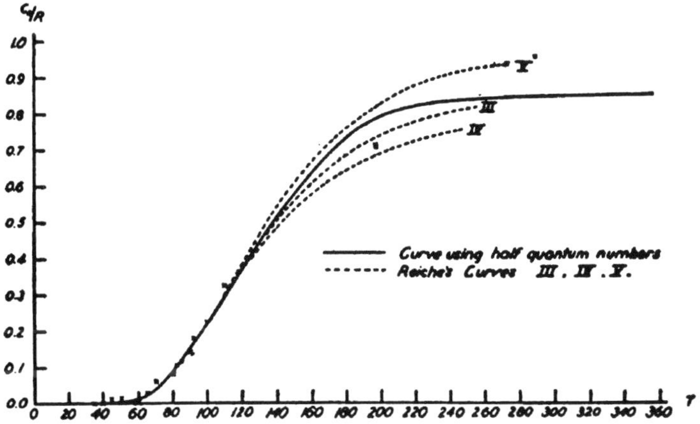

FIGURE 4.

129. Application to the Specific Heat of Hydrogen.—The foregoing considerations have been applied to an interpretation of the specific

[8] The above result has been obtained by taking the numbers 1/2, 3/2, 5/2 etc., not only as the possible values of the rotational (azimuthal) quantum number m which determines the total angular momentum, but also as the values for the equatorial quantum number r which determines the quantized component of angular momentum parallel to the field. The former choice is necessitated, as we have remarked, by Czerny's results on the far infra red spectra of hydrogen chloride. The second choice is more arbitrary. If we should use the values 0, 1, 2, 3, etc., for the equatorial quantum number r we should obtain the a-priori probabilities 1, 3, 5, etc., for the successive values of m. This choice is favored by Fowler, *Phil. Mag.*, **49**, 1272 (1925); Dieke, *Zeitschr. für Phys.*, **33**, 161 (1925), *Nature*, **115**, 875 (1925); Kemble, *Zeitschr. für Phys.*, **35**, 286 (1925), and Bourgin and Kemble, *Phys. Rev.*, **27**, 802 (1926), since it agrees with the summation rule of Dorgelo and Burger. On the other hand the work of Pauling, *Phys. Rev.*, **27**, 586 (1926), indicates that such an assignment of quantum numbers would lead to an impossible *negative* dielectric constant for hydrogen chloride gas.

The correct decision between the two possibilities is, however, quite possibly $p_m = 1, 3, 5$, etc., since Pauling's difficulty disappears in the new quantum dynamics. See Mersing and Pauli, *Phys. Zeits.*, **27**, 509 (1926), Kronig, *Proc. Nat. Acad.*, **12**, 488 (1926), and Van Vleck, *Nature*, **118**, 226 (1926). Nevertheless the method of substituting the a-priori probabilities into equation (202) would be the same with heat of hydrogen by substituting equations (203) and (205) into (202), and performing the indicated calculations. Assuming a value of the moment of inertia J which would give a good fit, the full line curve shown in Figure 4 was obtained.[4] The value of J from the curve was found to be 1.387×10^{-41} c.g.s. units. The value is of the right order of magnitude from other information as to the hydrogen molecule.

The failure of the curves at high temperatures is presumably due to the over simplification of our hypothetical model of the molecule. Calculations of Kemble and Van Vleck [5] indicate that at high temperatures the potential energy due to extension of the molecule by centrifugal force and due to the presence of vibrational energy must *both* be taken into account.

130. Application to the Entropy of Diatomic Gases.—Hydrogen is the only diatomic gas of low enough boiling point and low enough moment of inertia so that we can make any direct application of the quantum theory of rotational specific heat. The theory, however, finds an important application in calculating the entropy of diatomic gases at temperatures where the rotational specific heat has already reached the value R.

This application is based on the assumption that the entropy of *any* gas would be that given by the Sackur formula for the entropy of monatomic gases, *provided* we go down to temperatures low enough so that the specific heat has fallen to the value for monatomic gases.[6] If this assumption is correct, and it seems to be a necessary consequence of considerations which we shall later introduce, we can evidently calculate the entropy of any perfect gas by adding to the Sackur expression an additional term obtained by a suitable integration from the absolute zero to the temperature of interest. We obtain for the entropy of one mol of any perfect gas

$$S = \left[\frac{5}{2} R \log T - R \log P + R \log \left\{\frac{(2\pi m)^{3/2} k^{5/2}}{h^3}\right\} + \frac{5}{2} R\right] + \int_0^T \left(C_p - \frac{5}{2} R\right) \frac{dT}{T} \quad (206)$$

where the first term is the Sackur expression,[7] m being the mass of a

either choice, and experimental data are not yet sufficient to permit a decision on the basis of the difference in the predicted values of the specific heat.

The finally acceptable treatment of the quantum states of the dipole rotator and their a-priori probabilities will perhaps have to be based on the new dynamics.

[4] Tolman, *Phys. Rev.*, **22**, 470 (1923). The above treatment would have to be modified on the basis of the new quantum mechanics, but the correct modification is still uncertain. See Van Vleck, *Phys. Rev.*, **28**, 480 (1926).

[5] Kemble and Van Vleck, *Phys. Rev.*, **21**, 653 (1923).

[6] See Tolman, *J. Amer. Chem. Soc.*, **42**, 1185 (1920).

[7] For small modifications which must be made in this formula where the molecule has a symmetry number $\sigma \neq 1$, and the lowest state of rotation has an a-priori probability $p \neq 1$, see Chapters 12 and 13.

molecule, and the second term is the excess entropy due to the rise of specific heat above the value for monatomic gases.

For diatomic gases at sufficiently low temperatures the excess specific heat $(C_p - 5/2R)$ will evidently be the rotational specific heat, so that the second term in equation (206) can be written in the following form, by introducing equation (200) for the rotational specific heat,

$$S_{\text{Rot.}} = \int_0^T \left(C_p - \frac{5}{2}R\right)\frac{dT}{T} = \int_0^T C_{\text{Rot.}}\, d \log T$$

$$= \int_0^T R\sigma^2 \frac{d^2 \log Q}{d\sigma^2}\, d \log T \tag{207}$$

Introducing the relation between σ and T given by equation (201), we may transform the above expression as follows:

$$S_R = -\int_0^T R\sigma^2 \frac{d^2 \log Q}{d\sigma^2}\, d \log \sigma$$

$$= R \left| \log Q - \sigma \frac{d \log Q}{d\sigma} \right|_0^T \tag{208}$$

Since most expressions for the condition sum Q give simple expressions for log Q at high and low temperatures, equation (208) is often an easy one to apply. This method of treatment is due to Tolman and Badger.[8]

The most accurate test of equation (208) is afforded by the case of hydrogen chloride where we know from spectral data that the moment of inertia of the molecule is $J = 2.594 \times 10^{-40}$ gm. cm.^{2} and also know that half quantum numbers are to be used. Substituting in the foregoing equations we obtain for the theoretical value of the entropy of one mol of hydrogen chloride gas at 25° C. and 1 atm. pressure the value 43.2 cal./deg. as compared with the experimental value 43.3.

131. The above methods have many practical applications in connection with a number of other gases. Allowance must be made for the entrance of vibrational specific heat at the higher temperatures, and this explains the reason why diatomic molecules have specific heats above 5/2 R at higher temperatures. It may also be noted that the weaker the binding force between the atoms the lower the temperature will be at which vibrational energy will become appreciable. This is found to connect satisfactorily the chemical facts that we know about diatomic molecules with their known specific heats.

[8] Tolman and Badger, *J. Amer. Chem. Soc.*, **45**, 2277 (1923).

Chapter 10

The Specific Heat of Solids

132. Nature of the Experimental Facts.—We shall next apply the Maxwell-Boltzmann distribution law, modified so as to take account of the quantum theory, to a determination of the specific heat of solids. On the basis of the classical theory we were led to the conclusion that a solid composed of N atoms would have at any given temperature the heat capacity $3Nk$, corresponding to an equipartition both of kinetic and potential energy among the $3N$ degrees of freedom. The actual experimental facts, however, show that although the heat capacity of many solids has approximately this value at room temperature, it falls below $3Nk$ at lower temperatures, approaching zero indeed at the absolute zero of temperature. On the basis of our new information as to the quantization of periodic motions, it is evident that the principal cause of this fall below the classical value is due to the fact that the atoms in a solid must themselves be performing periodic motions, and hence can only assume certain discrete amounts of energy as allowed by the quantum theory, rather than the continuous succession of energies allowed by the classical theory. We shall now show how to introduce these quantum restrictions, following substantially the method of Debye.

133. Modes of Vibration of an Elastic Solid.—Let us consider the heat capacity of an isotropic solid composed of N equivalent atoms. Since the thermal energy of such a solid will be due to the periodic oscillation of the atoms composing it, we may regard the solid as a whole as a conditionally periodic system of $3N$ degrees of freedom.

In describing the periodic motions of this system it will be most advantageous to make use of results obtained from the theory of elasticity as to the periodic motions which can be present in an elastic solid. In accordance with elastic theory, the internal motions of an elastic solid can be analyzed into longitudinal and transverse wave motions travelling with the velocities

$$c_L = \sqrt{\frac{\lambda + 2\mu}{\varrho}} \quad \text{and} \quad c_T = \sqrt{\frac{\mu}{\varrho}} \tag{209}$$

where λ and μ are the elastic constants of the solid and ϱ its density. In the case of a body of finite size these wave motions will be reflected at the boundaries and will give rise to sets of standing waves of different frequencies. Each possible way in which such a standing wave can be formed is called a mode of vibration of the solid, and it can be shown that any steady state of internal motion for the solid can be described by specifying the amplitude and phase for each mode of vibration.

The number of modes of vibration lying in a given frequency range will evidently depend on the volume of the solid V, the velocities of the longitudinal and transverse wave motions, c_L and c_T and the frequency range ν to $\nu + d\nu$ under consideration, and it can be shown that the actual number of modes of vibration of the two kinds together are given by the equation

$$dZ = \frac{4\pi V \nu^2 d\nu}{c_L^3} + \frac{8\pi V \nu^2 d\nu}{c_T^3} \tag{210}$$

the first term giving the number of longitudinal modes of vibration and the second, the number of transverse modes, which is twice as great owing to the possibility of polarization for transverse waves. The complete derivation of equation (210) is somewhat complicated, but except for the numerical factors 4π and 8π it is evident that the expressions for the two numbers are a necessary consequence of the principle of dimensional homogeneity, and we may accept the complete result as "manna" from the mathematician without more ado.[1]

134. Modes of Vibration in an Actual Solid of 3*N* Degrees of Freedom.—The theory which leads to equation (210) is based on the assumption of a continuous solid, corresponding to an infinite number of degrees of freedom, and gives, as will be noted from the equation, an infinite number of modes of vibration provided we place no upper limit on the frequencies considered. Our actual solid, however, has only $3N$ degrees of freedom and hence its internal motion would be completely described by specifying the amplitudes and phases for $3N$ modes of vibration instead of an infinite number. This leads quite naturally to the introduction of the assumption of Debye that the actual internal energy of our solid can be satisfactorily accounted for if we assume it to be resident in $3N$ modes of vibration whose frequencies are distributed in accordance with equation (210), running from zero frequency to a maximum frequency ν_m, determined by the requirement that the total number of modes of vibration is to be $3N$.

This hypothesis is amply justified by the results of its application. As an a-priori justification, moreover, it should be noted that the dis-

[1] See Jeans, "Dynamical Theory of Gases," Second Edition, "Chapter XVII."

tribution of modes of vibration with frequencies low enough so that the wave lengths are long compared with the interatomic distances will obviously be practically the same as predicted for a continuous solid. And indeed, it turns out that even the maximum frequency usually corresponds to a wave length which is fairly long compared with interatomic distances. Hence the success of a theory built on these lines does not seem surprising.

The value of the maximum frequency ν_m can easily be determined with the help of equation (210), since the total number of modes of vibration is to be $3N$. We obtain

$$3N = \int dZ = \int_0^{\nu_m} 8\pi V\nu^2 \left(\frac{1}{2c_L^3} + \frac{1}{c_T^3}\right) d\nu = \frac{8\pi V\nu_m^3}{3}\left(\frac{1}{2c_L^3} + \frac{1}{c_T^3}\right) \tag{211}$$

or solving for the maximum frequency

$$\nu_m = \left\{\frac{9N}{8\pi V} \frac{2c_L^3 c_T^3}{2c_L^3 + c_T^3}\right\}^{1/3} \tag{212}$$

135. Quantization of a Harmonic Oscillator.—In the foregoing treatment we have given the number of modes of vibration for a solid which lie in any given frequency range ν to $\nu + d\nu$. Our next interest will be to determine the average energy associated with a mode of vibration of a given frequency. We shall then be in a position to integrate the energy for all the frequencies and hence determine the total energy content and heat capacity of our solid.

To carry out this program we must first obtain the quantum restrictions for the oscillating motion of a mode of vibration. The energy for a mode of vibration or for any harmonic oscillator of one degree of freedom is given in terms of a single coordinate q and momentum p by the equation

$$E = \tfrac{1}{2} aq^2 + \tfrac{1}{2} bp^2 \tag{213}$$

where a and b are constants, and the quantum condition will be given by the Wilson-Sommerfeld equation

$$\oint pdq = nh \tag{214}$$

In accordance with the equations of motion in the Hamiltonian form, however, we may write for dq

$$dq = \dot{q}dt = \frac{\partial E}{\partial p} dt = bpdt \tag{215}$$

Substituting in (214), we obtain

$$\oint bp^2 dt = nh \tag{216}$$

The integral, however, is to be taken around the complete period of the motion τ, so that equation (216) can be rewritten in the form

$$[bp^2]_{av.}\, \tau = nh$$

The period τ, however, is the reciprocal of the frequency ν and the average value of bp^2 is evidently twice the average potential energy of the oscillator, which from the equality of the average kinetic and potential energies of the oscillator, is equal to the total energy. We thus obtain

$$E = nh\nu \tag{217}$$

for the possible values of the energy of any quantized harmonic oscillator.

136. Average Energy of a Harmonic Oscillator.—If we have a system containing a large number of modes of vibration, or other harmonic oscillators, of frequency ν, we may determine their average energy with the help of the Maxwell-Boltzmann distribution law.[2] For the number of oscillators in a given quantum state, having the quantum number n, we may write

$$N_n = \frac{Ne^{-\frac{\epsilon_n}{kT}}}{\sum\limits_n e^{-\frac{\epsilon_n}{kT}}} \tag{218}$$

and hence for the average energy

$$\varepsilon_{av.} = \frac{\sum\limits_n e^{-\frac{\epsilon_n}{kT}}\, \varepsilon_n}{\sum\limits_n e^{-\frac{\epsilon_n}{kT}}} \tag{219}$$

Substituting the value $nh\nu$ for the energy in terms of the quantum number n as given by equation (217), and writing out the summations we obtain[3]

[2] It should be noted in accordance with Chapter 3, Section 30, and Chapter 4, that the Maxwell-Boltzmann distribution law applies just as reasonably to oscillators or modes of vibration as to those components of a system which are more commonly called molecules.

[3] This substitution assumes that the quantum number n has the successive values 0, 1, 2, 3 . . . . The newer dynamics mentioned in Chapter 7 would give the values 1/2, 3/2, 5/2 . . . .

If the half quantum numbers were employed the average energy would come out

$$\varepsilon_{av.} = \frac{h\nu}{e^{\frac{h\nu}{kT}} - 1} + \frac{h\nu}{2}$$

thus differing from (221) by a constant additive term which need not concern us.

$$\varepsilon_{\text{av.}} = \frac{h\nu\left(e^{-\frac{h\nu}{kT}} + 2e^{-\frac{2h\nu}{kT}} + 3e^{-\frac{3h\nu}{kT}} + \ldots\right)}{\left(1 + e^{-\frac{h\nu}{kT}} + e^{-\frac{2h\nu}{kT}} + e^{-\frac{3h\nu}{kT}} + \ldots\right)} \tag{220}$$

and by division this is easily shown to reduce to

$$\varepsilon_{\text{av.}} = \frac{h\nu}{e^{\frac{h\nu}{kT}} - 1} \tag{221}$$

which is a general expression for the average energy of harmonic oscillators of frequency ν at the temperature T.

137. For a system containing N harmonic oscillators the total energy would evidently be

$$E = \frac{Nh\nu}{e^{\frac{h\nu}{kT}} - 1} \tag{222}$$

and the heat capacity of such a system would be

$$C_v = \frac{dE}{dT} = \frac{Nke^{\frac{h\nu}{kT}}\left(\frac{h\nu}{kT}\right)^2}{\left(e^{\frac{h\nu}{kT}} - 1\right)^2} \tag{223}$$

If the atoms of a solid had only a single frequency, instead of the range of frequencies given by the Debye theory, the specific heat of a solid would evidently be three times the value given by equation (223). This was the original proposal of Einstein which led to the more complete treatment of Debye.

138. Energy Content and Specific Heat of a Simple Solid.—We may now combine equation (210) for the number of modes of vibration in a given frequency range with equation (221) for their average energy, and by integrating over all possible values of the frequency from zero to the maximum ν_m, obtain the following expression for the total energy of the solid

$$E = \int_0^{\nu_m} 8\pi hV\left(\frac{1}{2c_L^3} + \frac{1}{c_T^3}\right)\frac{\nu^3}{e^{\frac{h\nu}{kT}} - 1}\,d\nu \tag{224}$$

This expression can be simplified if we introduce equation (212) for the maximum frequency

$$\nu_m = \left\{\frac{9N}{8\pi V}\,\frac{2c_L^3 c_T^3}{2c_L^3 + c_T^3}\right\}^{1/3}$$

and substitute the symbols

$$\frac{h\nu_m}{k} = \Theta \quad \text{and} \quad \frac{h\nu}{kT} = x \tag{225}$$

Assuming Θ which depends on the elastic properties of the body as independent of the temperature, we obtain for the energy content and heat capacity

$$E = 9NkT\left(\frac{T}{\Theta}\right)^3 \int_0^{\frac{\Theta}{T}} \frac{x^3}{e^x - 1}\, dx \tag{226}$$

and

$$C_v = 9Nk\left\{4\left(\frac{T}{\Theta}\right)^3 \int_0^{\frac{\Theta}{T}} \frac{x^3}{e^x - 1}\, dx - \frac{\Theta}{T}\frac{1}{e^{\frac{\Theta}{T}} - 1}\right\} \tag{227}$$

139. It should be noted in accordance with these equations that the energy content and heat capacity of the simple solids under consideration are functions of a single parameter Θ, which is dependent on the elastic properties of the solid. This parameter has the dimensions of temperature and is sometimes called the characteristic temperature for the substance. For purposes of representing the experimental facts the value of Θ is of course best taken from actual measurements on specific heats. It is, however, also calculable, from the elastic properties of the solid, and in accordance with equations (212) and (225) has the value

$$\Theta = \left\{\frac{9Nh^3}{8\pi Vk^3}\,\frac{2c_L^3 c_T^3}{2c_L^3 + c_T^3}\right\}^{1/3} \tag{228}$$

in terms of the velocities of the two kinds of wave motions. These velocities, however, are themselves dependent on the quantities usually measured, density ϱ, compressibility $\varkappa$, and Poisson's ratio σ in such a way that the above value for Θ may be reexpressed in the form

$$\Theta = \frac{3.62 \times 10^{-3}}{\varkappa^{1/2}\varrho^{1/6}M^{1/3}\left\{\left(\frac{1+\sigma}{3(1-\sigma)}\right)^{3/2} + 2\left(\frac{2}{3}\,\frac{1+\sigma}{1-2\sigma}\right)^{3/2}\right\}^{1/3}} \tag{229}$$

where M is the molecular weight, and the other quantities are in c.g.s. units.

140. Approximations at Low and High Temperatures.—The general evaluation of the definite integral occurring in equations (226) and (227) is not possible in algebraic form, and we may first consider the simple approximations which hold at low and high temperatures.

When T is small the upper limit of the definite integral may evidently be taken as infinity instead of Θ/T. With this limit the integral is known to have the value $\pi^4/15$, and equations (226), and (227) assume the form

$$E = \frac{3\pi^4 NkT^4}{5\Theta^3} \tag{230}$$

and

$$C_v = \frac{12\pi^4 NkT^3}{5\Theta^3} \tag{231}$$

It is of special interest to note that at low temperatures the energy content and specific heat thus become proportional respectively to the fourth and third powers of the temperature. This is not only in accordance with the actual experimental facts, but should receive special attention, since the result differs from that obtained in earlier less complete theories of the specific heat of solids. The result, however, is in agreement with considerations based on the principle of similitude.[4]

Attention is also called to the fact that the approximations valid at low temperatures are in reality independent of the assumption that the solid has a discontinuous atomic structure. This will be seen by substituting into equations (230) and (231) the value given for Θ by equation (228) and noting that the number of atoms N disappears from the expressions. The possibility of treating the system at low temperatures as a continuous solid is due to the fact that all the periodic motions, which are of low enough frequency to have appreciable energy at low temperatures, have a wave length so long that the discontinuities do not enter into consideration.

141. At high temperatures, noting the definition of x given by equation (225), we can introduce the approximation

$$e^x = 1 + x \tag{232}$$

Equations (226) and (227) then reduce to the classical expressions in accordance with the law of Dulong and Petit

$$E = 3RT \tag{233}$$

$$C_v = 3R \tag{234}$$

a result which is of course to be expected.

142. Values at Intermediate Temperatures.—At intermediate temperatures, values for the integral occurring in equations (226) and (227) have been computed by methods of approximation. We give below a table[5] in which the heat capacity per mol in accordance with the Debye

[4] See Tolman, *Phys. Rev.*, 4, 145 (1914).

[5] Taken in part from Lewis and Randall, "Thermodynamics," McGraw-Hill, 1923.

theory is given as a function of T/Θ. The table also includes values which will be found useful for the total energy, E_v and entropy S_v which the solid acquires on heating up at constant volume from the absolute zero. These quantities are defined by the equations

$$\frac{E_v}{\Theta} = \int_0^T C_v \frac{dT}{\Theta} \tag{235}$$

and

$$S_v = \int_0^T C_v \frac{dT}{T} \tag{236}$$

TABLE I

T/Θ	$\log_{10} T/\Theta$	C_v	E_v/Θ	S_v
.251	–0.600	0.11	0.01	0.04
.316	–0.500	0.21	0.02	0.07
.398	–0.400	0.43	0.04	0.15
.501	–0.300	0.84	0.11	0.30
.631	–0.200	1.43	0.26	0.53
.794	–0.100	2.19	0.57	0.98
1.000	0.000	2.98	1.09	1.62
1.259	0.100	3.78	1.96	2.41
1.585	0.200	4.43	3.36	3.37
1.995	0.300	4.91	5.32	4.42
2.51	0.400	5.26	7.83	5.59
3.16	0.500	5.49	11.37	6.82
3.98	0.600	5.65	15.94	8.08
5.01	0.700	5.72	21.77	9.41
6.31	0.800	5.76	29.28	10.74
7.94	0.900	5.78	38.66	12.07
10.00	1.000	5.79	50.57	13.40
12.59	1.100	5.80	65.58	14.74
15.85	1.200	5.81	84.50	16.08

143. Comparison with Experimental Data.—We may now consider the correspondence between the Debye theory and the experimental facts. In actually making use of the Debye formulæ for energy content and specific heat, it is customary to employ values of the characteristic temperature Θ for different solids which have been obtained from the specific heat data themselves, rather than calculated from the elastic properties of the material. This of course leads to a better fit between experimental data and observations than would otherwise be obtained and increases the usefulness of the Debye formula as a practical method of correlating specific heat data. The improvement in fit thus obtained may be partially due to uncertainty as the proper values of the elastic constants. In addition, however, the evil effects of basing the derivation on an assumed homogeneity and isotropy of the solids may be partially eliminated by using values of Θ which give the best fit to the curve.

Using the values of Θ which do give the best fit, it is found that there are a considerable number of substances having specific heat curves which agree almost exactly with the Debye formula. Thus for aluminum, copper, lead, diamond, silver, mercury, thallium, and zinc the agreement is practically within the small experimental error in the measurements of specific heat.[6]

In the case of solids which do obey the Debye equation, it should especially be noted that a single measurement of specific heat at a temperature not too far removed from that where C_v has fallen to half the classical value will be sufficient to determine the whole specific heat curve, as well as the entropy and heat content at all temperatures. This is a fact of extreme practical importance in calculations of the free energy of solid substances.

144. Comparison of Thermal and Elastic Values of Θ.—Not only does the Debye theory give a very precise reproduction of the specific heat curve for a considerable number of solids when the value of Θ is chosen so as to give the best fit, but the values of Θ thus obtained are in very reasonable agreement with those calculated from the elastic properties, in accordance with equation (229). This is shown by the following table.[7]

TABLE II

Solid	Θ Thermal	Θ Elastic
Aluminum	396	402
Copper	309	332
Silver	215	214
Lead	95	73

In passing, attention is called to the fact that the light and resilient metals have large values of Θ and the dense and yielding ones have small values. This is a qualitative expression of the facts expressed by equation (229). Since large values of Θ mean that the temperature has to become high before the specific heat is freed from quantum restrictions, this agrees with the high frequencies of oscillation that we should expect

[6] As a matter of fact the values of C_v in Table I are those obtained by Lewis and Gibson, *J. Amer. Chem. Soc.*, **39**, 2554, 1917, from a curve representing actual experimental data. For other tables see Nernst, "Grundlagen des neuen Warmesatzes," Second Edition, Knapp, Halle, 1924.

[7] Taken from Born, "Atomtheorie des festen Zustandes," Teubner, Leipzig, 1923. The elastic constants used in calculating the values of Θ were obtained by Grüneisen (*Ann. d. Phys.*, **22**, 838, 1907; **25**, 845, 1908), and are the values at room temperature. At lower temperatures quite different values are obtained. It is believed, however, that these changes are due to the amorphous substance between the small crystals composing the metals rather than to a variation in the behavior in the crystals themselves. See "Die Theorie der Strahlung und der Quanten," 1, Solvay-Kongress, 1911 (Halle, 1914); Appendix by Eucken, p. 387.

for a light and resilient substance, and the high temperatures that must be attained before light and resilient crystals like the diamond obey the law of Dulong and Petit.

145. Extensions of Debye Theory.—The chief approximations in the Debye theory have arisen from the treatment of a solid as a continuous, homogeneous, isotropic structure rather than as the actual space lattice which we believe exists. Treatments based more closely on the theory of space lattices have been developed by Born, Kármán and Thirring.[8] Nevertheless, such treatments are complicated and although of great theoretical importance, the result for practical purposes is largely merely to alter the nature of the approximations which must be introduced before we can apply the conclusions to actual experimental results. There are two cases, however, in which extensions of the Debye theory can be made which are quite simple and of considerable practical importance.

146. In the case of solid compounds, Nernst has suggested the possibility of dividing the internal energy into two portions corresponding to the vibration of the compound molecule as a whole, and the mutual vibration of the atoms inside the molecule. For the energy belonging to the molecules as a whole we can use the Debye expression as given by equation (224) based on the presence of all possible frequencies from zero to the maximum ν_m, and for the energy of the atoms inside the molecule can use the Einstein expression as given by equation (222) substituting the actual frequencies of oscillation of the atoms in the molecule. We can thus obtain for the total energy an expression of the form

$$E = \frac{9Nh}{\nu_m^3}\int_0^{\nu_m} \frac{\nu^3 d\nu}{e^{\frac{h\nu}{kT}} - 1} + N\sum_i \frac{h\nu_i}{e^{\frac{h\nu_i}{kT}} - 1} \tag{237}$$

where the summation Σ in the last term is to be carried out for all the internal vibrations present inside the molecule. For a diatomic molecule this reduces to a single term.

The treatment is of course again only a partial approximation, since the absolute division of the internal energy into the two kinds is an over simplification, and the X-ray analysis of the crystalline substances to which the formula may be successfully applied usually do not give evidence of separate molecules inside the crystal. Nevertheless, as a matter of fact, Nernst has shown that equation (237) gives an excellent representation of the energy content for a number of substances. In applying the equa-

[8] See Born, previous reference.

tion to a number of crystalline salts, moreover, he has obtained good agreement using a value of ν_m calculated from the elastic properties of the crystal, and a value of ν_i which is the frequency of the infra red absorption band (Reststrahlen) as measured by Rubens for the salt. This latter identification seems reasonable, since we should expect such optical effects to result from the relative vibration of the ions.

The following table taken from Nernst [9] shows the close agreement between observed and calculated values of the heat capacity per mol of potassium chloride.

TABLE III

HEAT CAPACITY PER MOL FOR KCl

T degrees absolute	Debye term C_v	Einstein term C_v	Correction term $C_p - C_v$	Calculated C_p	Observed C_p
22.8	1.04	0.046	....	1.086	1.16
26.9	1.48	0.13	....	1.61	1.52
30.1	1.87	0.25	....	2.12	1.96
33.7	2.25	0.43	....	2.68	2.50
48.3	3.52	1.43	....	4.95	5.70
57.6	4.06	2.13	0.02	6.21	6.12
70.0	4.57	2.89	0.04	7.50	7.58
86.0	4.97	3.66	0.06	8.79	8.72
235	5.81	5.55	0.32	11.68	11.78
416	5.91	5.83	0.68	12.42	12.72
550	5.93	5.87	0.90	12.70	13.18

147. Another simple and important extension of the Debye theory is due to the work of Lewis, Eastman and Rodebush [10] on the specific heat of strongly electropositive metals, in which even at low temperatures the specific heat rises higher than would accord with the Debye formula and indeed at moderate temperatures may exceed the classical values of Dulong and Petit. The explanation of this departure is of course that the "free" electrons which account for the high conductivity of such strongly metallic substances have acquired a very considerable thermal energy even at moderate temperatures. Further quantitative theoretical study of the form of these specific heat curves would be desirable.

[9] Wolfskehl Kongress Göttingen, 1914 (Leipzig-Berlin, 1914), 4, Vortrag, W. Nernst, Kinetische Theorie Fester Körper, p. 81.

[10] Lewis, Eastman and Rodebush, *Proc. Nat. Acad. Sci.*, **4**, 25 (1918).

Chapter 11.

The Distribution of Radiation at Thermal Equilibrium.

148. Method of Treatment.—We shall next apply the Maxwell-Boltzmann law modified to allow for quantum theory to a determination of the distribution with respect to frequency of the radiation present in an enclosure or hohlraum which has come to temperature equilibrium with the walls. This problem has in addition to its great theoretical and practical importance, the historical interest of having led to the first introduction of the quantum theory by Planck, in his derivation of the famous *Planck radiation law.*

For the purpose of this chapter, we shall regard radiation after it has once been emitted as a wave motion propagated in accordance with the classical laws of the Maxwell electromagnetic theory of light, and hence shall be able to assign the energy present in a hohlraum to the different modes of electromagnetic vibration which are present. The quantum theory in this treatment will be introduced by quantizing the motion of these modes of vibration, as in the derivation of equation (217) in the last chapter, and thus permitting them to assume only such quantities of energy as are multiples of $h\nu$. It will be found that this treatment leads to the Planck radiation law.

In a later chapter we shall present another mode of treatment due to Einstein, which is based on considerations as to the mechanism of the absorption and emission of radiation by the material of the walls which are in thermal equilibrium with the radiation in the hohlraum. In this later treatment it will not be necessary to assume that the radiation consists of wave motions, and indeed the treatment on the whole adopts at least a friendly attitude towards the extreme position of the existence of light corpuscles. The quantum theory in this later mode of approach will be introduced by assuming that the emission and absorption of radiation result from changes in the quantum states of atoms or molecules in thermal equilibrium with the radiation. The final result will be the Planck radiation law just as in the treatment we are now going to undertake.

149. Modes of Vibration in the Hohlraum.—For the number of modes of electromagnetic vibration in a hohlraum of volume V, with frequencies between ν and $\nu + d\nu$ we may write the expression

$$dZ = \frac{8\pi V}{c^3} \nu^2 d\nu \tag{238}$$

where c is the velocity of light. The expression is analogous to that given in the last chapter for the number of modes of transverse vibration in an elastic solid and the derivation is similar. Except for the magnitude of the numerical constant 8π, the equation is evidently completely derivable from the principle of dimensional homogeneity.

150. Average Energy of a Mode of Vibration.—Each mode of vibration is a sub-system of one degree of freedom, its state being completely describable by the specification of the amplitude and phase of the vibration, or by the specification of the value of a single coordinate and the corresponding momentum. Since the coordinate performs a harmonic oscillation, the motion can be quantized in the same way as for any harmonic oscillator and the mode of vibration can only assume values of the energy which are multiples of $h\nu$, in accordance with the equation

$$\varepsilon = nh\nu \tag{239}$$

derived in the last chapter. Furthermore the average energy for a mode of vibration will be

$$\varepsilon_{\text{av.}} = \frac{h\nu}{e^{\frac{h\nu}{kT}} - 1} \tag{240}$$

as given by equation (221) in the last chapter.[1]

151. Planck Distribution Law.—Combining equations (238) and (240) for the number of modes of vibration in a given frequency range and their average energy and dividing by the volume V, we obtain

$$du = \frac{8\pi h\nu^3}{c^3} \frac{1}{e^{\frac{h\nu}{kT}} - 1} d\nu \tag{241}$$

for the density of energy in the frequency range $d\nu$ when the hohlraum has come to thermal equilibrium.

This is the Planck radiation law which agrees with the experimental facts within the limits of observational error.

In the region of low frequencies and at high temperatures, the quantity

[1] For the result obtained using half quantum numbers, see footnote to Section **136**.

$e^{\frac{h\nu}{kT}}$ can be approximated by the first two terms of its expansion and equation (241) rewritten in the form

$$du = \frac{8\pi h\nu^2}{c^3}\, kTd\nu \tag{242}$$

which is the classical Rayleigh-Jeans formula deducible from the principle of equipartition of energy.

In the region of high frequencies, however, equation (241) can evidently be rewritten in the form

$$du = \frac{8\pi h\nu^2}{c^3}\, e^{-\frac{h\nu}{kT}}\, d\nu \tag{243}$$

which is the so-called Wien equation.

152. Total Energy in the Hohlraum.—To obtain the total density of energy in the hohlraum, we may integrate equation (241) for all frequencies from zero to infinity and obtain

$$u = \int_0^\infty \frac{8\pi h}{c^3} \frac{\nu^3}{e^{\frac{h\nu}{kT}} - 1}\, d\nu = \frac{8\pi}{c^3} \frac{(kT)^4}{h^3} \int_0^\infty \frac{x^3}{e^x - 1}\, dx$$

$$= \frac{8\pi^5 k^4}{15c^3h^3}\, T^4 \tag{244}$$

which is the familiar Stefan-Boltzmann law for the total density of energy in the hohlraum. This form of this law is of course derivable from thermodynamics but the magnitude of Stefan's constant

$$a = \frac{8\pi^5 k^4}{15c^3h^3} \tag{245}$$

was not theoretically determinable until the introduction of the quantum theory. The value given by equation (245) is in agreement with the experimental results, within the observational error.

Equations (241) and (244) have a wide range of theoretical application and in addition well-known practical applications in the measurement of high temperatures.

Chapter 12

General Determination of the Most Probable State of a System

153. Introduction.—In the foregoing applications of statistical mechanics, we first derived the Maxwell-Boltzmann law for the distribution of atoms and molecules in different states and then applied this law to the problem of interest. We are next going to consider two problems—the vapor pressure of solids and liquids at low temperatures, and chemical equilibria in gases at high temperatures—in which the energy connected with an atom or molecule is very dependent on the presence of neighboring atoms or molecules with which it may be chemically combined or associated in the solid or liquid form. This dependence of the energy in such a far reaching way on the coordinates of more than one atom or molecule, makes it difficult to solve the problems by applying the Maxwell-Boltzmann distribution law, since, as we saw in the original derivation, the quantity ε occurring in the Maxwell-Boltzmann expression is the rate of change in the energy of the system per molecule introduced into a given region of the phase space when the system is in its state of maximum probability, and in the case of the systems now to be considered we could no longer make the simplification of taking ε as the individual energy of a single molecule.

This makes it desirable for our present purposes to return to the use of the same kind of considerations as to the probabilities of different states which were employed in deriving the Maxwell-Boltzmann distribution law itself. For this purpose, however, we must first reexamine some of our fundamental methods since we must now take account of the quantum theory and allow for the possibility of quantized states in a way not contemplated in the original derivation of the Maxwell-Boltzmann law.

154. The Microcanonical Ensemble.—In Chapter 3 we made a study of the behavior of ensembles of systems of identical structure, distributed uniformly in the QP phase space or γ-space with total energies between E and $E + dE$. Such collections were called microcanonical ensembles, and prove to be very useful in predicting the behavior of single systems of the same structure as those in the ensemble. In the first place, since the infinitesimal dE may be chosen as small as desired, all the systems of the ensemble may be regarded as having substantially the same energy E. In the second place it was shown with the help of Liouville's theorem that the systems composing a microcanonical ensemble will move in such a way as to maintain the uniform distribution of their representative points in the γ-space. Hence if we accept some form of the ergodic hypothesis the following propositions will hold:

1. Setting up a microcanonical ensemble of systems with energies between E and $E + dE$ and then selecting a system from the ensemble at random is equivalent to starting off a single system with the energy E and then taking it in the phase it happens to be in at some random time after starting.

2. The probability that a system will be found at any instant in a given state will be proportional to the γ-weight of the state, that is to the volume in the phase space belonging to those systems in the corresponding microcanonical ensemble which are in the state in question.

These propositions make it possible, as we have already seen, to use microcanonical ensembles in predicting the most probable states of systems.

155. Fully Excited, Partially Excited, and Frozen-In Degrees of Freedom.—The above considerations as to the relation between probability and volume in the phase space apply in the first instance to systems that obey the classical laws of dynamics which permit the representative point for a system to assume a continuous range of positions in the γ-space. Under these circumstances the volume in the phase-space or γ-weight *belonging* to a given state of the system is evidently the total volume which can actually be occupied by the representative point.

In determining the γ-weight, however, for systems where some of the degrees of freedom may be affected by quantum restrictions, we find it desirable to distinguish three different types of degrees of freedom:— 1. Fully excited degrees of freedom such that the corresponding coordinates and momenta can be regarded as governed by the classical laws of motion. 2. Partially excited degrees of freedom such that the corresponding coordinates and momenta can only assume values belonging to particular quantized states. 3. Frozen-in degrees of freedom such that the corresponding coordinates and momenta are in the lowest possible quantum state.

In the calculation of the γ-weight for a complex system of the above nature, we assign to the fully excited degrees of freedom the continuous hyper-volume which can be occupied by the phase point of the system when the system is in the state of interest. In the case of the other degrees of freedom, however, which are affected by quantum restrictions, we must assign, in accordance with the considerations of Chapter 8, to each quantum hyper-surface the volume in the γ-spaces which lies between it and the next higher quantum hyper-surface, thus introducing a factor h for each non-degenerate degree of freedom involved.

156. The Possibility of Gas Degeneration.—In applying the methods of this chapter to gases, we shall assume that the three translational degrees of freedom for each molecule of the gas can be treated as fully excited and not subject to quantum restrictions. This agrees with the considerations of Chapter 7, Section **106**, where it was shown that quantum restrictions became negligible when the periods of motion are long or many quanta of energy are already present.

The possibility certainly arises, however, that at high concentrations when the space available for the motion of the molecules becomes small, and at low temperatures where the average energy of the molecules is small, that quantum restrictions will also affect the translational motion of the molecules. If this should occur we should have a phenomenon which has been given, in advance of its elucidation and discovery, the name of *gas degeneration*.

The problem of possible gas degeneration has already been the subject of much speculation, but unequivocally acceptable results have not yet been obtained owing to doubts as to the proper method of quantizing the translational motions of molecules. In any case it can be remarked that all the tentative results which have so far been obtained agree in making appreciable degeneration occur only as the result of extreme conditions which are not met in ordinary experimental work.

Hence we shall have no hesitation in treating gases in the proposed manner and accepting the results as valid for ordinary purposes. Thus for example, the Sackur-Tetrode formula for the entropy of gases, which will be derived in the next chapter, can be regarded as valid for all the purposes to which it is usually put.

157. Method of Determining the γ-Weight.—We now have the principles which are to be used in computing the γ-weight of a state of our system. In the actual determination of γ-weight we find it convenient to proceed as follows.[1]

Consider a system composed, for example, of $X + Y + Z \ldots$ atoms of different kinds and assume that three degrees of freedom are sufficient for each atom.[2] Then choose some particular exemplar of the state of interest, with the representative point correspondingly located in the γ-space of $6\,(X + Y + Z + \ldots)$ dimensions.

[1] The method is that of Ehrenfest and Trkal, *Proc. Amsterdam Acad.*, **23**, 162 (1920).

[2] Ultimately perhaps 3 degrees of freedom for each proton and electron.

a. Generation of Region (γ_1).—First now, without making any changes in the combination of the atoms into molecules or other aggregates, move the representative point through all possible positions of the γ-space, lying in the energy range E to $E + dE$, which correspond to the state in question. This motion of the point will give the molecules all positions, orientations, and velocities consistent with the state of interest, leaving the combination of the atoms unchanged, and will describe a region in the γ-space, of a given shape and volume, say γ_1. The region will of course be partly generated by continuous motions for fully excited degrees of freedom and partly arise from the assignments of volume due to quantized degrees of freedom, but this need cause no confusion.[8]

b. Permutation of Atoms.—Second, having thus generated one region in the γ-space, of weight γ_1, which corresponds to the state of interest, we may evidently now obtain further such regions by permutations of similar atoms with each other. This will lead to further regions of the same volume and shape as before, but differently located owing to changes in axes accompanying the permutation of atoms. The total γ-weight of the state will then evidently be the product of the weight γ_1 for one of these regions by the total number of such regions obtained by the permutations.

To calculate the number of these regions, we note that the total number of possible permutations of similar atoms will evidently be $(\underline{|X}\ \underline{|Y}\ \underline{|Z} \ldots .)$ where $X, Y, Z \ldots .$ are the numbers of atoms of each of the various kinds. The result of some of these possible permutations, however, would evidently be the same as already reached by the motion of the representative point in generating the region of weight γ_1. Thus if all the atoms of two similar molecules are interchanged with each other, we reach a new phase which can also be attained by a motion of the representative point without leaving the original region of weight γ_1. Furthermore, molecules may contain similar atoms symmetrically placed so that a rotation of the molecule will lead to the same result as a permutation of atoms. Such permutations may be called *internal* and evidently must not be counted in getting the total γ-weight.

c. Expression for γ-weight.—We may now evidently write as an expression for the total γ-weight

$$\gamma = \frac{\underline{|X}\ \underline{|Y}\ \underline{|Z} \ldots\ldots}{\underline{|N_1}\ \underline{|N_2}\ \underline{|N_3} \ldots\ldots \sigma_{1_{N1}}\, \sigma_{2_{N2}}\, \sigma_{3_{N3}} \ldots .}\, \gamma_1 \qquad (246)$$

where γ_1 is the weight obtained by motions of the representative point which do not involve permutation of the atoms, and the rest of the ex-

[8] In general we shall assign the weight h to each quantized non-degenerate degree of freedom, whether partially excited or frozen-in. See, however, the remarks in the last two paragraphs at the end of the next chapter.

pression is the number of new similar regions which can be obtained by the permutation of atoms, $X, Y, Z \ldots$ being the number of atoms of each of the different kinds, $N_1, N_2, N_3 \ldots$ the number of molecules of each of the different kinds, and $\sigma_1, \sigma_2, \sigma_3 \ldots$ the symmetry factors for the different kinds of molecules which give the number of equivalent rotational orientations of the molecule in question.

158. Volume of Hypersphere and Hyperellipsoid.—In calculating the weight of a region generated by the motion of a representative point for our system, it often becomes necessary to know the value of certain definite integrals which may be regarded as the volumes of a hypersphere or hyperellipsoid in a space of n dimensions.[4]

Consider the hyper-volume in a space of n dimensions included inside a surface which is defined by the equation

$$x_1^2 + x_2^2 + \ldots + x_{n-1}^2 + x_n^2 = r^2 \tag{247}$$

This surface may be regarded as a hypersphere with the volume defined by the equation

$$V = \int\int \ldots \int\int dx_1 dx_2 \ldots dx_{n-1} dx_n \tag{248}$$

where the integration is to be taken for all values of $x_1 \ldots x_n$ such that

$$x_1^2 + x_2^2 + \ldots x_{n-1}^2 + x_n^2 < r^2 \tag{249}$$

Introducing limits that correspond to integration over positive values of the variable and allowing for this by multiplying by the proper power of two, we may rewrite equation (248) in the form

$$V = 2^n \int_0^r \int_0^{\sqrt{r^2 - x_n^2}} \ldots \int_0^{\sqrt{r^2 - x_3^2 - \ldots - x_n^2}} \int_0^{\sqrt{r^2 - x_2^2 - \ldots - x_n^2}} dx_1 dx_2 \ldots dx_{n-1} dx_n \tag{250}$$

The evaluation of this integral can be obtained in the following simple manner. Let us denote the volume of a sphere of n dimensions and radius r by the symbol $V_n(r)$. Then it is evident from equation (250) that we may connect the volume of a sphere of n dimensions with that of one of $n - 1$ dimensions and radius $\sqrt{r^2 - x^2}$ by the relation

$$V_n(r) = 2\int_0^r V_{n-1}(\sqrt{r^2 - x^2})\,dx \tag{251}$$

or substituting for simplification

$$x = r \sin\phi \tag{252}$$

[4] See Schoute, "Mehrdimensionale Geometrie," Part II, page 288, Sammlung Schubert, Leipzig, 1905.

we may write

$$V_n(r) = 2r\int_0^{\frac{\pi}{2}} V_{n-1}(r\cos\phi)\ \cos\phi\ d\phi \tag{253}$$

Making use of our knowledge of the volume of spheres of low dimensions, we see, moreover, that the solution of equation (253) must be of the form

$$V_n(r) = a_n r^n \tag{254}$$

where a_n is a function merely of n, and by comparison with (253) can write

$$a_n = 2a_{n-1}\int_0^{\frac{\pi}{2}} (\cos\phi)^n d\phi \tag{255}$$

The integral, however, has known values,[5] which permit us to write equation (255) in the forms

$$\left.\begin{aligned} a_n &= 2\,\frac{2.4.6\,.\,.\,.\,.\,.\,(n-1)}{1.3.5\,.\,.\,.\,.\,.\,.\,.\;n}\,a_{n-1} \qquad (n \text{ is odd}) \\ a_n &= 2\,\frac{1.3.5\,.\,.\,.\,.\,.\,(n-1)}{2.4.6\,.\,.\,.\,.\,.\,.\,.\;n}\,\frac{\pi}{2}\,a_{n-1} \qquad (n \text{ is even}) \end{aligned}\right\} \tag{256}$$

or in general

$$a_n = \frac{2\pi}{n}\,a_{n-2} \tag{257}$$

Since a_1 and a_2, however, have the evident values of 2 and π respectively, we may at once write for the volume of our hypersphere the expressions

$$\left.\begin{aligned} V_n(r) &= \frac{2^{\frac{n+1}{2}}\,\pi^{\frac{n-1}{2}}}{1.3.5\,.\,.\,.\,.\,.\;n}\,r^n \qquad (n \text{ is odd}) \\ V_n(r) &= \frac{(2\pi)^{\frac{n}{2}}}{2.4.6\,.\,.\,.\,.\,.\;n}\,r^n \qquad (n \text{ is even}) \end{aligned}\right\} \tag{258}$$

or in terms of the Gamma function the volume of an n-dimensional hypersphere of radius r can be expressed in the general form

$$V_n(r) = \frac{\left\{\Gamma\left(\frac{1}{2}\right)\right\}^n}{\Gamma\left(\frac{n}{2}+1\right)}\,r^n = \frac{\pi^{\frac{n}{2}}}{\Gamma\left(\frac{n}{2}+1\right)}\,r^n \tag{259}$$

[5] See B. O. Pierce, "Short Table of Integrals," No. 483.

159. Just as we have regarded the surface bounded by the equation

$$x_1^2 + \ldots\ldots + x_n^2 = r^2$$

as a hypersphere we may regard the surface bounded by the equation

$$\frac{x_1^2}{a_1^2} + \frac{x_2^2}{a_2^2} + \ldots\ldots + \frac{x_{n-1}^2}{a_{n-1}^2} + \frac{x_n^2}{a_n^2} = b^2 \tag{260}$$

as a hyperellipsoid with semi-axes a_1b, a_2b, etc.

The volume of the hyperellipsoid can easily be found by making the substitutions

$$z_1 = \frac{x_1}{a_1}, \; z_2 = \frac{x_2}{a_2} \ldots\ldots \tag{261}$$

to have the value:

$$V = \frac{\pi^{\frac{n}{2}}}{\Gamma\left(\frac{n}{2}+1\right)} b^n a_1 a_2 \ldots\ldots a_n \tag{262}$$

These formulæ will be needed for our treatment of the vapor pressure of crystals and glasses, and of chemical equilibria in gases.

Chapter 13

The Vapor Pressure of Crystals and Glasses at Very Low Temperatures

160. Introduction.—We shall first apply the considerations of the preceding chapter on the general determination of the most probable state of a system, to investigate the most probable distribution of molecules between a condensed phase and a vapor phase at very low temperatures. We shall take the temperature so low that all the degrees of freedom of the atoms are frozen in except the three degrees of translational freedom for the vapor molecules. This means that the condensed phase, crystal or glass, regarded as a quantum system will be in the lowest possible quantum state, and the molecules of the vapor will behave like those of a monatomic gas.

The investigation is of interest because of the bearing of the results on the *third law of thermodynamics*, which has been much employed in the calculation of chemical equilibria and the calculation of the free energy values of chemical substances. We shall find that a given substance must be regarded as having the same entropy when present at the absolute zero in different *crystalline* forms. On the other hand a substance at the absolute zero in the form of a *supercooled liquid or glass* may have a greater entropy than in the crystalline form in agreement with the ideas of Lewis and Gibson. We shall also find that the entropy of a monatomic gas or of a polyatomic gas at temperatures low enough so that it behaves as a monatomic gas, will be given by the Sackur formula with the Tetrode value for the constant.

161. The γ-weight of Vapor and Crystal at Very Low Temperature.—Consider a system composed of $X + Y + Z \ldots.$ atoms of different sorts contained in a volume V, and combined into similar molecules of mass M, and a symmetry number σ. Let N of these molecules be in the form of a vapor, and the remainder N', in the form of a perfect crystal. And let the temperature be so low that all the degrees of freedom except the $3N$ translational degrees of freedom for the vapor molecules are frozen in, these latter, however, being regarded as fully excited.

In accordance with the considerations of the preceding chapter it is evident that we may write for the γ-weight of this state of the system the expression

$$\gamma = \frac{\underline{|X}\ \underline{|Y}\ \underline{|Z}\ldots}{\underline{|N}\ \sigma^N} h^{3(X+Y+Z+\ldots-N)} \gamma_N \qquad (263)$$

where the last factor γ_N is the weight of the $6N$ dimensional surface in the γ-space that would be generated by moving our representative point so that the N vapor molecules assume all positions and velocities consonant with the volume of the vessel and energy available, the factor $h^{3(X+Y+Z+\ldots-N)}$ arises from the weight that must be assigned to the frozen in degrees of freedom, and the factor $(\underline{|X}\ \underline{|Y}\ \underline{|Z}\ldots)/\underline{|N}\ \sigma^N$ is the number of permutations of atoms that will lead to the generation of similar surfaces of weight γ_N.

162. The value of the weight γ_N which arises from the possibility of assigning different positions and velocities to the vapor molecules will evidently be given by the multiple integral

$$\gamma_N = \int\ldots\int dx_1dy_1dz_1\ldots dx_Ndy_Ndz_N\, M^{3N}d\dot{x}_1d\dot{y}_1d\dot{z}_1\ldots d\dot{x}_Nd\dot{y}_Nd\dot{z}_N \qquad (264)$$

where the limits of integration must be taken consonant with the volume and energy available.

We may at once introduce the limits imposed on the positional coordinates by the fact that the total volume available is V, and hence can immediately rewrite equation (264) in the form

$$\gamma_N = V^N M^{3N} \int\ldots\int d\dot{x}_1d\dot{y}_1d\dot{z}_1\ldots d\dot{x}_Nd\dot{y}_Nd\dot{z}_N \qquad (265)$$

To obtain the limits to be imposed on the velocities we may proceed as follows. Let K be the kinetic energy of the N vapor molecules, $N\varepsilon_{\text{pot.}}$ their potential energy and $N'\varepsilon'_{\text{pot.}}$ the potential energy of the molecules of the crystal, so that the total energy E of the system is given by the equation

$$E = K + N\varepsilon_{\text{pot.}} + N'\varepsilon'_{\text{pot.}} \qquad (266)$$

Hence in a microcanonical ensemble of systems in which the total energy lies between E and $E + dE$, it is evident that the kinetic energy

$$K = E - N\varepsilon_{\text{pot.}} - N'\varepsilon'_{\text{pot.}} \qquad (267)$$

must lie in the range

$$dK = dE \qquad (268)$$

The kinetic energy, however, is connected with the velocities of the molecules by the equation

$$\dot{x}_1^2+\dot{y}_1^2+\dot{z}_1^2+\ldots.+\dot{x}_N^2+\dot{y}_N^2+\dot{z}_N^2=\frac{2K}{M} \tag{269}$$

and this is the equation of a $3N$ dimensional hypersphere of radius $(2K/M)^{\frac{1}{2}}$ which in accordance with equation (259) of the last chapter has the volume

$$V_{3N}\left\{\left(\frac{2K}{M}\right)^{\frac{1}{2}}\right\}=\frac{\pi^{\frac{3N}{2}}}{\Gamma\left(\frac{3N}{2}+1\right)}\left(\frac{2K}{M}\right)^{\frac{3N}{2}} \tag{270}$$

Evidently then the representative point for the velocities must lie in the spherical shell, between the radii $\sqrt{2K/M}$ and $\sqrt{2(K+dK)/M}$ and this volume is seen by differentiating the expression in equation (270) to have the value

$$\int\ldots.\int d\dot{x}_1 d\dot{y}_1 d\dot{z}_1\ldots.d\dot{x}_N d\dot{y}_N d\dot{z}_N=\frac{3N\pi^{\frac{3N}{2}}}{2\Gamma\left(\frac{3N}{2}+1\right)}\left(\frac{2K}{M}\right)^{\frac{3N}{2}}K^{-1}dE. \tag{271}$$

Substituting in equation (265) and then substituting the value thus obtained for γ_N in equation (263), we obtain as our final expression for the γ-weight of the system of crystal and vapor the result

$$\gamma=\frac{\lfloor X\ \lfloor Y\ \lfloor Z\ldots.}{\lfloor N\ \sigma^N}\,h^{3(X+Y+Z+\ldots.-N)}\,V^N\,\frac{3N}{2}\,\frac{(2\pi MK)^{\frac{3N}{2}}}{\Gamma\left(\frac{3N}{2}+1\right)}K^{-1}dE \tag{272}$$

163. Most Probable Distribution between Crystal and Vapor.— The above expression for the γ-weight is obviously dependent on the number of molecules N in the vapor, and hence we can determine the most probable distribution of molecules between crystal and vapor by putting the variation of γ equal to zero. Using for greater convenience the logarithm of γ, introducing the Stirling approximation for factorials

$$\lfloor N=\sqrt{2\pi N}\left(\frac{N}{e}\right)^N \tag{273}$$

putting

$$\Gamma\left(\frac{3N}{2}+1\right)=\lfloor\left(\frac{3N}{2}\right) \tag{274}$$

and omitting terms which are negligible when N is large, we easily obtain from equation (272) the expression,

$$\delta \log \gamma = \delta N \left\{ -\log N - \log \sigma - 3 \log h + \log V + \frac{3}{2} \log 2\pi MK - \frac{3}{2} \log \frac{3N}{2} + \frac{3N}{2} \frac{d \log K}{dN} \right\} = 0 \quad (275)$$

as the condition for the most probable distribution of molecules between crystal and vapor.

164. Vapor Pressure of the Crystal.—Introducing the relation

$$\delta K = (\varepsilon'_{\text{pot.}} - \varepsilon_{\text{pot.}}) \delta N \quad (276)$$

which is an obvious result of equation (267), giving the kinetic energy of the vapor molecules the value of which it has at the low concentrations involved

$$K = \frac{3}{2} NkT \quad (277)$$

and putting for the pressure of the vapor

$$P = \frac{N}{V} kT \quad (278)$$

the condition for maximum probability easily reduces to

$$\log P = -\frac{\varepsilon_{\text{pot.}} - \varepsilon'_{\text{pot.}}}{kT} + \frac{5}{2} \log kT + \frac{3}{2} \log \frac{2\pi M}{h^2} - \log \sigma \quad (279)$$

which is the desired expression for the vapor pressure of a crystal at low enough temperatures so that the vapor acts as a monatomic gas with its $3N$ translational degrees of freedom fully excited, and all other degrees of freedom frozen in, in the lowest quantum state.

165. Vapor Pressure of a Glass.—Similar methods to the above can be used to obtain an expression for the vapor pressure of a glass or supercooled liquid at very low temperatures. In the case of a glass, however, an exceedingly important difference arises when we consider the number of ways of arranging the molecules which make up the condensed phase.

In a perfect crystal having a given size, shape, and orientation each molecule occupies a definite equilibrium position about which it will oscillate when the temperature is raised and the total number of different exemplars of the crystal which can be obtained will be found by permutations of the atoms which do not involve changes in equilibrium positions. In the case of a supercooled glass, however, especially one composed of complicated molecules, there may be a number of ways of fitting the molecules together and obtaining an exemplar of the glass, and we must take account of this in determining the γ-weight for a given distribution of molecules between the condensed and vapor phases.

In determining the number of such arrangements of the molecules of the glass which involve a shift in the positions of the molecules, we must proceed, however, with care. We are applying our considerations to a system which has all of its degrees of freedom frozen in, and are assigning in any case a weight equal to the area $h = \int p dq$ to each degree of freedom. Hence a shift in the position of a molecule which does not move it out of a region in the phase space already allotted to it as belonging to the lowest quantum state must not be counted. To determine the correct number of shifts, let us consider the properties of the glass when the degrees of freedom which locate the positions of the molecules have been excited to the next to the lowest quantum state. Since in accordance with the considerations of Chapter 8 we should assign to the lowest or frozen-in quantum state all the γ-weight corresponding to systems in between the lowest and next to the lowest states, it is evident that rearrangements, involving shifts in the equilibrium positions of the molecules, when the system is thus excited, should all of them be counted as giving new exemplars of the glass. It is evident that the number of such exemplars of the glass will depend on the number of molecules of N' in the glass so that we may represent the number of exemplars of glass by the symbol $\phi(N')$.

166. We can now obtain an equation for the vapor pressure of a glass in the same way that we derived the vapor pressure of a crystal. The expression for the γ-weight of a given distribution of molecules between the condensed and vapor phases will evidently have the same form as equation (272) except that it will be multiplied by the number of exemplars of glass $\phi(N')$, and this leads to a vapor pressure formula exactly similar to (279) except that it contains the additional term $\frac{d \log \phi(N')}{dN'}$ as shown below.

$$\log P = -\frac{\varepsilon_{\text{pot.}} - \varepsilon'_{\text{pot.}}}{kT} + \frac{5}{2}\log kT + \frac{3}{2}\log\frac{2\pi M}{h^2} - \log o - \frac{d \log \phi(N')}{dN'} \qquad (280)$$

The nature of the function $\phi(N')$ can now, however, easily be determined. Since we assume that we have taken a large enough sample of glass so that the vapor pressure is not appreciably dependent on the number of molecules N' in the glass, it is evident that the last term in equation (280) must be a constant. Hence we may write

$$\frac{d \log \phi(N')}{dN'} = \log a \tag{281}$$

where a is a constant and by integration obtain

$$\log \phi(N') = N' \log a + \text{const.} \qquad \phi(N') = ca^{N'} \tag{282}$$

where c is a constant. Furthermore, substituting (281) in (280) we obtain

$$\log P = -\frac{\varepsilon_{\text{pot.}} - \varepsilon'_{\text{pot.}}}{kT} + \frac{5}{2} \log kT + \frac{3}{2} \log \frac{2\pi M}{h^2} - \log \sigma - \log a \tag{283}$$

167. This development, however, also provides some information as to the magnitude that the quantity a can be expected to assume. Equation (282) gives us an expression for the number of ways we can rearrange the N' molecules of a glass and still obtain an exemplar of the glass. If we increase the number of molecules by δ, the number of rearrangements becomes

$$\begin{aligned} \phi(N' + \delta) + \delta) &= ca^{N' + \delta} \\ &= a^{\delta} \phi(N^1) \end{aligned} \tag{284}$$

so that a^{δ} is the factor by which the number of arrangements is increased when we increase the number of molecules by δ. We see then that a may be regarded as the average number of ways in which a molecule can be rearranged in the glass, excluding rearrangements which can be obtained by permutations of the atoms and excluding rearrangements which would disappear when the positional degrees of freedom for the molecules of the glass have been excited to the next to lowest quantum state.

We see that on this basis a will presumably be a small number which increases with the complexity of the molecule. This conclusion will be of interest in connection with a consideration of the entropy of liquids.

168. Applications in Thermodynamics.—The above conclusions as to the vapor pressure of crystals and glasses at very low temperatures are of great interest in connection with the postulates made by the so-called third law of thermodynamics as to the entropy of substances in the neighborhood of the absolute zero. Indeed they permit us to obtain a very precise notion of the implications of the third law.

a. Entropy of Different Crystalline Forms of the Same Substance.—In the first place it can be shown from our formula for the vapor pressure of crystals, equation (279), that the different crystalline forms of the same substance will have the same entropy at very low temperatures. To prove this consider two crystalline forms 1 and 2 of the same substance and transform one mol of the former into the latter by a reversible evaporation, followed by a reversible change in pressure and a reversible condensation at the new vapor pressure. The entropy change accompanying this process will evidently be

$$\Delta S = \frac{\Delta H_1}{T} + R \log \frac{P_1}{P_2} - \frac{\Delta H_2}{T} \tag{285}$$

where ΔH_1 and ΔH_2 are the reversible heats of evaporation of the two crystals and P_1 and P_2 are their vapor pressures. We may evidently substitute, however,

$$\begin{aligned} \frac{\Delta H_1}{T} - \frac{\Delta H_2}{T} &= N \frac{\left(\varepsilon - \varepsilon_1' + \frac{5}{2}kT\right)}{T} - N \frac{\left(\varepsilon - \varepsilon_2' + \frac{5}{2}kT\right)}{T} \\ &= \frac{N(\varepsilon_2' - \varepsilon_1')}{T} \end{aligned} \tag{286}$$

where N is Avogadro's number and ε_1' and ε_2' are the potential energies of the molecules in the two crystalline forms, and ε in the gaseous form, and it will be seen from equation (279) that this quantity will just cancel the term containing the ratio of the vapor pressures $R \log (P_1/P_2)$ leading to the desired result.

$$\Delta S = 0 \tag{287}$$

b. Entropy of Crystalline Compounds and Elements.—It is also easy to see that there will be no change in entropy at very low temperatures if we transform a crystalline compound into its elements provided they are also in the crystalline form. We can think of the transformation as carried out by a reversible evaporation of the compound, followed by a reversible change in the pressure of the vapor of the compound to that pressure which it has in equilibrium in the vapors from the crystals of the elements, followed by a reversible condensation into the crystals of the element. It is evident that a formula can be derived for the vapor pressure of the compound in equilibrium with the vapors from the crystals of its elements by exactly the same methods as equation (279) for the vapor pressure over a single crystal, and indeed the formula will have the same form as (279) except that the quantity $\varepsilon'_{\text{pot.}}$ will be the sum of the potential energies in the crystals of the elements of the atoms formed from one molecule of the compound. Using this result we can obviously proceed as in the last paragraph and show here too that

$$\Delta S = 0 \tag{288}$$

for the transformation.

c. Entropy of Glasses at Low Temperatures.—It is also important to note that our two formulæ (279) and (283) for the vapor pressures of crystals and glasses lead at once by the method used above to the conclusion that the entropy of a supercooled glass at very low temperatures is greater than that of the same material in the form of a pure crystal by an amount

$$\Delta S = R \log a \tag{289}$$

Since we have seen above that a is apparently a small number increasing with the complexity of the molecule, we can form at once an idea as to the order of magnitude of the differences we can expect between the entropy of crystals and glasses. These results are in agreement with the more qualitative ideas of Lewis and Gibson [1] and the experimental work of Gibson and Giauque [2] on the entropy of glasses and crystals of glycerine.[3]

d. The Entropy of Monatomic Gases.—Equation (279) for the vapor pressure of crystals at low temperatures provides the most satisfactory method of deriving the Sackur-Tetrode equation for the entropy of monatomic gases. The change in entropy in changing one mol of substance at low temperature from crystal to vapor at pressure P will evidently be

$$\Delta S = \frac{\Delta H}{T} + R \log \frac{P_{\text{vap.}}}{P} \tag{290}$$

where ΔH is the reversible heat of evaporation and $P_{\text{vap.}}$ is the vapor pressure. Introducing the obvious relation

$$\Delta H = N\left(\varepsilon - \varepsilon' + \frac{5}{2}kT\right) \tag{291}$$

and substituting equation (279) for the vapor pressure we obtain

$$\Delta S = \frac{5}{2} R \log T - R \log P + \frac{3}{2} R \log M + R \log \frac{(2\pi)^{3/2} k^{5/2}}{h^3} + \frac{5}{2} R - R \log \sigma \tag{292}$$

If we take the entropy of the crystal itself as zero as is customary, the above expression is the value for the entropy of the gas itself. For gases which actually are monoatomic σ is unity and the expression reduces to that of Sackur with the Tetrode value for the constant. There seems, however, no reason why the Sackur-Tetrode equation is not also valid for gases in which σ is not unity provided we are at low enough temperatures so that rotations of the molecules are frozen in. (See Chapter 9, Section **130**.)

[1] Lewis and Gibson, *Jour. Amer. Chem. Soc.*, **42**, 1529 (1920).
[2] Gibson and Giauque, *Journ. Amer. Chem. Soc.*, **45**, 93 (1923).
[3] The above treatment of the vapor pressure and entropy of glasses is due to Pauling and Tolman, *Journ. Amer. Chem. Soc.*, **47**, 2148 (1925).

The foregoing results contain the full import of the third law of thermodynamics as far as concerns the usual calculations of entropy and free energy values.

The treatment given in this chapter has been based on the assumption that each frozen in degree of freedom contributes the factor h to the total γ-weight of a state of the system. The possibility has recently been suggested, however, that certain degrees of freedom which have been so treated may contribute a somewhat larger factor to the γ-weight. If this suspicion should prove to be true, the above formulae would be subject to small modifications.

For example, if the lowest rotational state of a vapor molecule should have the a-priori probability p_0, conditions in the crystal remaining as assumed, the Sackur Tetrode formula (292) would contain an additional term $+ R \log p_0$. Since p_0, if not unity, would presumably be a small integer, it is doubtful if our present entropy data are accurate enough to establish the presence of such a term.

Chapter 14

Gaseous Equilibria at Temperatures of Constant Specific Heat

169. Introduction.—The investigations of the last chapter have given us values for the entropy of crystals and gases at the absolute zero, and hence make possible the calculation of the entropy change accompanying a gaseous reaction at higher temperatures, by the consideration of a cycle in which we start with the initial substances at the absolute zero, vaporize, raise to the higher temperature, permit the reaction to occur, cool the product to the absolute zero, and then condense to crystalline form. Combining the entropy change, thus obtained for a gaseous reaction, with the heat change we can then calculate the free energy change and equilibrium constant of the reaction.

Such a thermodynamic method of calculation has the advantage of being familiar and of permitting the use of actual empirical rather than theoretical formulæ for the specific heats of the gases involved when the temperature is raised. Nevertheless, it is evident that the methods of Chapter 12 should make it possible to obtain a direct statistical mechanical calculation of gaseous equilibria by determining what state of chemical combination has the greatest probability or γ-weight. Such a statistical mechanical method is theoretically illuminating and often has practical advantages.

The treatment given below is substantially that of Ehrenfest and Trkal.[1]

170. The System to be Considered.—Let us consider a gaseous system of volume V and temperature T, composed of X, Y, Z, etc., atoms of different kinds, combined into N_1 $N_2 \ldots . N_j$ molecules of different kinds. And let us consider that the temperature T is such that all the degrees of freedom can be considered as either fully excited or fully frozen in. Thus a monatomic molecule will be taken as having 3 degrees of freedom, corresponding to its translational motion, a diatomic molecule as having 5 degrees of freedom corresponding to translation and rotation about two axes, and a polyatomic molecule as having 6 degrees

[1] Ehrenfest and Trkal, *Proc. Amsterdam Acad.*, **23**, 162 (1920).

of freedom corresponding to translation and rotation about all three axes. With these assumptions we shall evidently have to be in a temperature range of constant specific heats $C_v = 3/2R$, $5/2R$ and $3R$ corresponding respectively to monatomic, diatomic and polyatomic molecules. Vibrational degrees of freedom will be taken as completely frozen-in.

We shall use the symbols M_1, P_1, Q_1, $R_1, \ldots . M_j$, P_j, Q_j, R_j for the masses and moments of inertia of the different kinds of molecules. For monatomic molecules all three moments of inertia P, Q, R will be equal to zero. For diatomic molecules we shall have $P = Q$, $R = 0$; and for polyatomic molecules all three moments will in general be different.

For the total number of degrees of freedom we shall evidently have $3(X + Y + Z + \ldots .)$. Some of these degrees of freedom will be fully excited and some frozen-in. For the number of *fully excited* degrees of freedom we shall use the symbol F and shall write

$$F = \sum_i N_i f_i \qquad f_i = 3, 5 \text{ or } 6. \tag{293}$$

where N_i is the number of molecules of the i'th kind and f_i has one of the values indicated depending on whether the molecule is monatomic, diatomic or polyatomic. For the number degrees of freedom which are frozen in, we shall evidently have $3(X + Y + Z + \ldots .) - F$.

For the symmetry factors for the different kinds of molecules we shall use the symbols o_1, $o_2, \ldots . o_j$.

171. Probability of a Given State of Combination.—In accordance with the principles of Chapter 12, we can now evidently write for the γ-weight of a given state of chemical combination of the atoms in the gas

$$\gamma = \frac{\lfloor X \, \lfloor Y \, \lfloor Z \ldots . .}{\lfloor N_1 \, \lfloor N_2 \ldots . \lfloor N_j \; o_1^{N_1} \, o_2^{N_2} \ldots o_j^{N_j}} h^{3(X + Y + Z + \ldots) - F} \gamma_1 \tag{294}$$

where γ_1 is the area of the hyper-surface in the γ-space corresponding to fully excited degrees of freedom over which the representative point for the system can be moved, without changing the state of chemical combination and without permuting the atoms.

172. The value of γ_1 can, however, easily be seen to be given by the multiple integral for the coordinates and momenta $q_1 \ldots . p_F$ for the fully excited degrees of freedom

$$\gamma_1 = \int \ldots . \int dq_1 \ldots . dp_F \tag{295}$$

provided we use the proper limits of integration. In the case of the three positional coordinates for each molecule we can evidently integrate over the whole volume V and this will introduce the factor $V^{(N_1 + N_2 + \ldots + N_j)}$ In the case of the two angular coordinates, which determine the direction of the axis of a diatomic molecule, we must integrate over the surface of the whole solid angle 4π and shall hence obtain the further factor $(4\pi)^{N_d}$ where N_d is the number of diatomic molecules and, in the case of the three angular coordinates necessary to determine the orientation of the axes of the polyatomic molecules, will evidently obtain on integration the factor $(8\pi^2)^{N_p}$ when N_p is the number of polyatomic molecules.

Hence integrating first of all the coordinates, we can rewrite equation (295) in the form

$$\gamma_1 = V^{(N_1 + N_2 + \cdots + N_j)} (4\pi)^{N_d} (8\pi^2)^{N_p} \int \ldots \int dp_1 \ldots dp_F \quad (296)$$

We must next inquire into the limits of integration for the momenta $p_1 \ldots p_F$. The values of these momenta will evidently determine the kinetic energy of the system in accordance with the equation

$$K = \frac{p_1^2}{2A_1} + \frac{p_2^2}{2A_2} + \ldots + \frac{p^2_F}{2A_F} \quad (297)$$

where the constants $A_1\, A_2 \ldots A_f$ are evidently the masses and moments of inertia $M_1P_1Q_1R_1 \ldots M_jP_jQ_jR_j$, and the F-dimensional volume in which the phase-point would have to be for any value of the kinetic energy less than K would evidently by a hyper-ellipsoid, having a volume in accordance with Section **159** of Chapter 12, of the amount

$$V_F = \frac{\sqrt{(2\pi K)^F}}{\Gamma\left(\frac{F}{2}+1\right)} \sqrt{A_1A_2 \ldots A_F} \quad (298)$$

We are interested, however, in a microcanonical ensemble of systems which have total energies between E and $E + dE$, and since the potential energy of the systems is *definitely* fixed by the state of chemical combination, it is evident that the kinetic energy must be in the range

$$dK = dE \quad (299)$$

Hence by differentiating equation (298) with respect to K we obtain for the volume of the ellipsoidal shell of interest

$$dV_F = \frac{F}{2} \frac{\sqrt{(2\pi K)^F}}{\Gamma\left(\frac{F}{2}+1\right)} \sqrt{A_1A_2 \ldots A_F}\; K^{-1}dE \quad (300)$$

Substituting in (296), we then obtain

$$\gamma_1 = V^{\Sigma N_i} (4\pi)^{N_d} (8\pi^2)^{N_p} \frac{F}{2} \frac{\sqrt{(2\pi K)^F}}{\Gamma\left(\frac{F}{2}+1\right)} \sqrt{A_1 A_2 \ldots A_F}\ K^{-1} dE \qquad (301)$$

This expression can, however, be somewhat simplified if we note that the constants $A_1 \ldots . A_f$ are evidently in accordance with equation (297) the masses and moments of inertia of the molecules. The mass for each molecule enters three times into the product corresponding to three components of momentum and each moment of inertia enters once. Hence if we put

$$\alpha_i \begin{cases} = \sqrt{M_i^3} & \text{(for monatomic molecules)} \\ = 4\pi \sqrt{M_i^3 P_i^2} & \text{(for diatomic molecules)} \\ = 8\pi^2 \sqrt{M_i^3 P_i Q_i R_i} & \text{(for triatomic molecules)} \end{cases} \qquad (302)$$

We can rewrite equation (301) in the form

$$\gamma_1 = V^{\Sigma N_i} \frac{F}{2} \frac{\sqrt{(2\pi K)^F}}{\Gamma\left(\frac{F}{2}+1\right)} \left(\prod_i \alpha_i^{N_i}\right) K^{-1} dE \qquad (303)$$

where Π_i indicates a product taken for all the kinds of molecules present.

173. Substituting now in equation (294), and taking for convenience the logarithm, we obtain as an expression for the total γ-weight of a given state of chemical combination

$$\begin{aligned} \log \gamma = {} & \log\ (\underline{|X}\ \underline{|Y}\ \underline{|Z} \ldots) h^{3(X+Y+Z+\ldots)}\, dE \\ & - \sum_i \log \underline{|N_i} - \sum_i N_i \log \sigma_i - F \log h \\ & + \sum_i N_i \log V + \log \frac{F}{2} + \frac{F}{2} \log 2\pi K \\ & - \log \Gamma\left(\frac{F}{2}+1\right) + \sum_i N_i \log \alpha_i - \log K \end{aligned} \qquad (304)$$

where for convenience constant quantities have been collected in the first term.

This expression, however, can be written in a simpler form, if we denote the constant first term by the symbol I, introduce equation (293) for the number of fully excited degrees of freedom

$$F = \sum_i N_i f_i$$

and introduce the approximations

$$\log \underline{|N} = N\,(\log N - 1)$$
$$\log \Gamma\left(\frac{F}{2}+1\right) = \log \underline{|F/2} = \frac{F}{2}\left(\log\frac{F}{2} - 1\right) \tag{305}$$

We obtain

$$\begin{aligned}\log\gamma = I &- \sum_i N_i\,(\log N_i - 1) - \sum_i N_i \log \sigma_i - \sum_i N_i f_i \log h\\ &+ \sum_i N_i \log V + \log\frac{F}{2} + \sum_i N_i f_i \log\sqrt{2\pi K}\\ &- \frac{F}{2}\left(\log\frac{F}{2} - 1\right) + \sum_i N_i \log a_i - \log K\end{aligned}$$

or collecting terms and introducing a new quantity α_i, which is a property of molecules of the i'th kind, defined by the equation

$$\log \alpha_i = -\log\sigma_i - f_i \log h + f_i \log\sqrt{2\pi} + \log a_i$$

or (306)

$$\alpha_i = \frac{a_i}{\sigma_i}\left(\frac{\sqrt{2\pi}}{h}\right)^{f_i}$$

We can write as our *final expression for the total γ-weight of any given state of chemical combination*

$$\begin{aligned}\log\gamma = I &- \sum_i N_i(\log N_i - 1) - \frac{F}{2}\left(\log\frac{F}{2} - 1\right)\\ &+ \sum_i N_i\left(\log V + \frac{f_i}{2}\log K + \log\alpha_i\right)\end{aligned} \tag{307}$$

where the two terms $\log F/2$ and $-\log K$ have been dropped as negligible compared with $F/2 \log F/2$ and $F/2 \log K$.

174. Change in Probability with Chemical Reaction.—If a chemical reaction takes place between the molecules of the gas the probability of the state as given by equation (307) will in general be altered, since the number of molecules N_i in the different states will be altered. For the change in the log γ accompanying a reaction we can then write

$$\begin{aligned}\delta\log\gamma = &-\sum_i \delta N_i \log N_i - \frac{\delta F}{2}\log\frac{F}{2}\\ &+ \sum_i \delta N_i\left(\log V + \frac{f_i}{2}\log K + \log\alpha_i\right)\\ &+ \sum_i N_i \frac{f_i}{2}\frac{\delta K}{K}\end{aligned} \tag{308}$$

It is evident from equation (293), however, that we can put

$$\delta F = \sum_i \delta N_i f_i \tag{309}$$

Furthermore, if we denote the *potential energy* of a molecule of the i'th kind by ε_i we can write

$$K = E - \sum_i N_i \varepsilon_i \tag{310}$$

and put

$$\delta K = -\sum_i \delta N_i \varepsilon_i \tag{311}$$

Finally let us specify the chemical reaction, which we are considering as producing the change in composition, by stating that ν_i is the number of molecules of the ith kind that are produced when the reaction as written takes place. Thus if we write the reaction in the familiar form $aA + bB + \ldots = cC + dD + \ldots$
the quantities ν_i are the numbers $a, b, \ldots . c, d \ldots .$, taken with the proper sum, which give the number of molecules entering into the chemical reaction. We may then write

$$\delta N_i = \nu_i \delta Q \tag{312}$$

if we denote by δQ the extent to which the reaction occurs. Substituting equations (309), (311) and (312) in (308), we obtain

$$\begin{aligned}\delta \log \gamma = \delta Q \Bigg\{ & -\sum_i \nu_i \log N_i - \sum_i \nu_i \frac{f_i}{2} \log \frac{F}{2} \\ & + \sum_i \nu_i \left(\log V + \frac{f_i}{2} \log K + \log a_i \right) \\ & - \left(\sum_i \frac{N_i f_i}{2} \right) \frac{\sum_i \nu_i \varepsilon_i}{K} \Bigg\}\end{aligned} \tag{313}$$

for the change in probability with chemical reaction.

175. Condition of Chemical Equilibrium.—The condition for chemical equilibrium with respect to the reaction in question is evidently obtained by setting expression (313) equal to zero, and thus securing the chemical state of maximum probability. Doing this, rearranging and noting $F = \sum N_i f_i$, we obtain

$$\sum_i \nu_i \log \frac{N_i}{V} = -\frac{F}{2K} \sum_i \nu_i \varepsilon_i + \sum_i \nu_i \frac{f_i}{2} \log \frac{2K}{F} + \sum_i \nu_i \log a_i \tag{314}$$

The quantity on the left hand side of this equation is, however, evidently equal to $\log K_c$, where K_c is the familiar equilibrium constant for the reaction, expressed with concentrations in numbers of molecules per unit volume. Furthermore we can evidently write from the principle of equipartition of energy which holds for the fully excited degrees of freedom

$$K = \frac{F}{2} kT \tag{315}$$

Substituting in equation (314), we obtain as our final expression for the ***equilibrium constant*** of the reaction

$$\log K_c = -\frac{1}{kT} \sum_i \nu_i \varepsilon_i + \sum_i \nu_i \frac{f_i}{2} \log kT + \sum_i \nu_i \log a_i \tag{316}$$

In applying this equation it should be noted that K_c is the equilibrium constant expressed with concentration in numbers of molecules per cubic centimeter, and that $\Sigma \nu_i \varepsilon_i$ is the change in potential energy and $kT \Sigma \nu_i f_i / 2$ is the change in kinetic energy when ν_i molecules react. Furthermore, in accordance with equations (302) and (306), it should be noted that the theoretical "chemical constants" have the values

$$a_i \begin{cases} = \sqrt{M_i^3} \left(\frac{\sqrt{2\pi}}{h}\right)^3 \\ = \frac{4\pi}{\sigma_i} \sqrt{M_i^3 P_i^2} \left(\frac{\sqrt{2\pi}}{h}\right)^5 \\ = \frac{8\pi^2}{\sigma_i} \sqrt{M_i^3 P_i Q_i R_i} \left(\frac{\sqrt{2\pi}}{h}\right)^6 \end{cases} \tag{317}$$

for monatomic, diatomic and polyatomic molecules respectively.

176. Some interesting consequences of equations (316) and (317) may be noticed. In the first place at low temperatures the preponderating influence of the first term throws the equilibrium in the direction of decreased potential energy, while at high temperatures the influence of the second term throws the equilibrium in the direction of increased number of fully excited degrees of freedom, that is in the direction of dissociation.

With regard to the "chemical constants" as given by equation (317), it should be noted of course that transformations must be made in order to obtain the usual values of the chemical constants used in thermodynamic calculations. It should also be noted that the words monatomic, diatomic, and polyatomic are used in connection with these constants in a purely conventional sense to indicate the number of excited degrees of freedom. Thus hydrogen (H_2) at low temperatures would be regarded as monatomic, and carbon dioxide ($O = C = O$) which is known from spectral data to have a linear structure would be regarded as diatomic.

The interesting effect of the symmetry factor σ_i in decreasing the tendency of a molecule to form should also be noted. Thus "other things being equal" there is twice the tendency of a molecule of the linear structure *A A B* to form as for one of the structure *A B A*. This can be seen qualitatively from the consideration that the first molecule could form by the two steps $A + AB$ *and* $AA + B$, but the second only by the step $AB + A$.

177. Application to Thermal Ionization.—As an important example of the application of these methods, we may calculate the degree of thermal ionization of a metal vapor into positive ions and electrons at high temperatures. This is a problem of great interest for astrophysics.[a]

Consider the vapor of a metal *A* which can exist in its *normal* unionized state or dissociated into a positive ion and electron in accordance with the reaction

$$A = A^+ + E^-$$

We can immediately apply equation (316) to this reaction. In doing so we may neglect the slight differences between the masses and moments of inertia of *A* and A^+ so that the chemical constants for these two constituents will balance out, except for the fact that it would seem reasonable to take the symmetry factor σ for the atom *A* as equal to the number of equivalent electrons held in the most lightly bound orbits, and hence susceptible to ionization. Applying equation (316) and noting equations (317), with the modification resulting from the introduction of a symmetry factor into the expression for a monatomic molecule, we evidently obtain

$$\log \frac{(A^+)(E^-)}{(A)} = -\frac{\varepsilon_\infty}{kT} + \frac{3}{2}\log kT + \log \frac{(2\pi m)^{3/2}\,\sigma}{h^3}$$

or

$$\frac{(A^+)(E^-)}{(A)} = \frac{(2\pi mkT)^{3/2}\,\sigma}{h^3}\, e^{-\frac{\varepsilon_\infty}{kT}} \qquad (318)$$

where the concentrations, in molecules per cubic centimeter, are indicated by enclosing the symbol for the atom in brackets, ε_∞ is the work of ionization, *m* is the mass of the electron and σ is the symmetry factor for the unionized atom.

We shall find a later application for this equation.

[a] See Fowler and Milne, *Monthly Notices R. A. S.*, **83**, 407 (1923).

Chapter 15

Some General Principles Regarding Molecular States and Molecular Processes

178. Introduction.—In our preceding statistical mechanical discussions we have been mainly interested in determining the numbers of molecules of a system in different states when the system as a whole is in a condition of equilibrium and have paid scant attention to the processes by which molecules change from one state to another. It is evident, however, that even in a condition of equilibrium, molecules will be continually changing from one state to another, the constancy of the quotas in the different possible states being maintained by some form of opposed processes which balance each other. Furthermore, in the case of systems which are not in a state of equilibrium, where gross physical chemical changes are taking place, it is evident that the progress of these changes is intrinsically dependent on the mechanisms by which molecules can change from one state to another. Hence it can easily be seen that a complete understanding of the processes, by which systems approach the state of equilibrium and maintain this state after it is reached, is dependent upon a knowledge of the nature of the individual mechanisms by which molecules change from one state to another.

The present chapter will be devoted to a somewhat abstract classification of molecular states and molecular processes and the discussion of some very general principles as to the nature of molecular processes. The next two chapters will investigate, from this special point of view of the mechanism of molecular processes, two important cases of equilibrium, and later chapters will use a knowledge of the laws governing these processes to investigate the rates of gross physical chemical change.

179. The Specification of Molecular States.—In most statistical mechanical considerations we treat systems which are dilute enough so that we can specify the states of the individual molecules composing the system without any ambiguity such as would be introduced by the continuous action of closely neighboring molecules.[1]

[1] The necessity of sufficient dilution does not thereby completely rule out the treatment of solids, liquids and compressed gases by the methods under discussion,

We shall denote a given state of a molecule by giving the values of the n generalized coordinates $(q_1 \ldots . q_n)$ and n momenta $(p_1 \ldots . p_n)$, which determine not only the location and motion of the center of gravity of the molecule, but also the positions and motions of its internal parts. And in order to assist in visualization and in providing a convenient quasi-geometrical language, we shall use our earlier device of representing the state of a molecule by the location of a representative point in a $2n$ dimensional phase-space (μ-space) corresponding to the $2n$ coordinates and momenta of the molecule.

In general the state of a molecule will be specified not by the precise location of a single point in the phase-space, but rather by a small region in the phase-space within which the representative point will lie. Thus in the case of a molecule having n "classical" degrees of freedom the state would be specified by the infinitesimal region $dq_1 \ldots . dp_n$ in which the representative point was located, while the state of a molecule with n "quantized" degrees of freedom would usually be specified by a finite region of magnitude h^n. In making these specifications of state, regions in the phase-space of just the same volume will be used for the different conditions of a molecule in order to preserve an equivalence between the probabilities of different possible molecular states.

180. The Classification of Molecular States.—We shall first find it convenient to introduce a rather abstract classification of molecular states into congruent, enantiomorphous and reverse states.

a. Congruent States.—Since a mere change in the spatial location or orientation of a molecule, leaving undisturbed its internal configuration and the magnitudes and relative directions of the velocities of its different parts, will not in general alter its energy content nor the nature of its behavior towards other molecules, which are suitably placed or radiation which is properly disposed, we shall often desire to group together such molecules and shall speak of them as being in congruent states. Indeed it is often customary to speak of molecules as being in the same state when more strictly they are in congruent states.

Our definition of congruence can be constructed from analogy with the use of the term in geometry. *Two states of a molecule will be called congruent, if the molecule can be changed from one state to the other, by a movement which does not alter the relative positions or velocities of its parts and does not change its relation to any external field of force in which it may be situated.* In carrying out such imaginary movements from one congruent state to another, the velocities of the different parts of the molecule and of its center of gravity can conveniently

since the state of interest of the molecule may be much more dependent on its internal configuration than on its proximity to other molecules.

be thought of as geometric vectors, rigidly attached to the molecule, so that they will partake in the translation and rotation of the molecule as a whole. Congruence for molecular physics thus means the same as congruence for geometry with the additional restrictions as to velocities and external field.

b. Enantiomorphous States.—It is also possible to generalize the idea of enantiomorphous states. *Two states of a molecule will be called enantiomorphous when one of the states is congruent with the mirror image of the other.* This idea of enantiomorphous states is of course a common one in organic chemistry and is here merely extended to include the idea of the reflection of the vectors giving the velocities of the parts of the molecule as well as those that determine their relative position. It is interesting to note that this addition leads to enantiomorphism in the case of molecules which do not contain any asymmetric atom.

c. Reverse States.—In addition to the ideas of congruent and enantiomorphous states adapted from geometry, we must add for the purposes of physics, the further idea of reverse states, which differ by a reversal in direction of the components of velocity. *Two states of a molecule will be called reverse, if the molecule can be changed from one of the states to a state congruent with the other by a reversal in direction of all components of velocity without change in magnitude or change in the relative positions of the constituents of the molecules.*

The possibility of the existence and stability of molecules in reverse states is guaranteed by the principle of dynamical reversibility to which we shall now turn our attention.

181. The Principle of Dynamical Reversibility.—The possibility of the "time reversal" of any dynamical motion is a well-known mechanical principle. The proof of this principle, that the reverse of any given motion of a system is also a possible motion of the system, can easily be obtained if we consider the equations of motion in the Lagrangian form (see Chapter 2, Section 13)

$$\frac{d}{dt}\frac{\partial L}{\partial \dot{q}_i} - \frac{\partial L}{\partial q_i} = Q_i \qquad i = 1, 2, 3 \dots n \tag{319}$$

Rewriting these equations so as to show the dependence of the Lagrangian function L and the external forces Q_i on the generalized coordinates $q_1 \dots q_n$ and generalized velocities $\frac{dq_1}{dt} \dots \frac{dq_n}{dt}$ we obtain

$$\frac{d}{dt}\frac{\partial}{\partial\left\{\frac{dq_i}{dt}\right\}} L\left(\dots q_i \dots \frac{dq_i}{dt} \dots\right) - \frac{\partial}{\partial q_i} L\left(\dots q_i \dots \frac{dq_i}{dt} \dots\right) = Q(\dots q_i \dots) \tag{320}$$

Since in general, however, L is quadratic in the velocity terms $\frac{dq_i}{dt}$, equation (320) can obviously be rewritten in the form

$$\frac{d}{d(-t)} \frac{\partial}{\partial\left\{\frac{dq_i}{d(-t)}\right\}} L\left(\ldots q_i \ldots \frac{dq_i}{d(-t)} \ldots\right) - \frac{\partial}{\partial q_i} L\left(\ldots q_i \ldots \frac{dq_i}{d(-t)} \ldots\right) = Q_i(\ldots q_i \ldots) \quad (321)$$

Hence if

$$q_i = \phi_i(t) \quad (322)$$

is a solution of the equations of motion, it is evident from the identity of the forms of the two equations (320) and (321) that

$$q_i = \phi_i(-t) \quad (323)$$

is also a possible solution. The result means that if at any given instant we reverse the direction of all the velocities of such a dynamical system, it will then retrace its former path, having at each instant on the reverse path the same position and reverse direction of its velocities as at the corresponding time in the original motion.

182. Limitations to the above "principle of reversibility" will obviously arise if the Lagrangian function L or the external forces Q_i should contain terms of the first or other odd degree in the velocities. An interesting example of this occurs in the case of the motion of a charged particle in an external magnetic field. Here a reversal in the direction of that component of velocity which is perpendicular to the field, will lead to a reversal in the direction of the force on the particle and thus exclude the possibility of the reverse motion. Thus, as has been pointed out in this connection by Einstein and Ehrenfest, if we consider a Bohr hydrogen atom in an external magnetic field with the electron rotating in a plane perpendicular to the field, the time reversal of this motion will not be possible since the magnetic force will, for one direction of rotation be acting towards the nucleus, and for the other direction of rotation away from it. Hence one of the orbits will have a different velocity of rotation than the other.

Special attention, however, should be called to the possibility and indeed perhaps propriety of including in the time reversal, a reversal in the direction of the electric current which produces the external magnetic field. Under these conditions the complete time reversal of the original motion can then be secured.

183. For our present purposes, the thing of interest, arising from the above considerations, is to note that in the absence of an external field, the reverse of any state of a molecule is also a dynamically possible state. In addition it is easy to see that if one of the states is quantized, then the other will also be quantized with the same quantum numbers, since if for the one state we have

$$\oint p_i dq_i = \int_0^\tau p_i \dot{q}_i dt = n_i h,$$

then for the other state we shall have

$$\oint p_i' dq_i' = \int_0^{\tau'} p_i' \dot{q}_i' dt' = \int_0^{-\tau} (-p_i)\ (-\dot{q}_i) d(-t) = \int_0^{\tau} p_i \dot{q}_i dt = n_i h.$$

We may conclude that the reverse of any state of a molecule can also exist, and indeed in a system which has come to equilibrium we shall expect an equal number of molecules in each of any pair of reverse states.

184. On the Congruence of Reverse States.—The nature and behavior of reverse states is going to be very important for our later work. In particular we must consider whether or not such states are congruent, and may say in advance that reverse states are apparently sometimes congruent and sometimes not congruent with each other, depending on the complexity of the molecule involved. In addition in many cases, although perhaps not necessarily in all, a reverse state which is not congruent with the original state may become so at a later time through elastic collisions or through the rotations which are taking place in the molecule.

185. As a first example let us take a simple spherical electron moving with a definite velocity in a given direction. Here evidently the reverse state, obtained by reversing the direction of motion, will be entirely congruent with the original state, owing to the symmetry of the electron itself.

186. Proceeding to a more complicated case where the molecule does not have spherical symmetry, let us next consider for example a chloroform molecule, CH_3Cl, with the chlorine atom and the three hydrogen atoms arranged approximately at the corners of a tetrahedron, and the whole molecule moving say in the direction of the chlorine atom. Here it is evident, even neglecting the effect of the reversal of the motion of the electrons inside the atoms, that the reverse state will not be congruent with the original state, since in one case the molecule as a whole will be moving in the direction of the chlorine atom and in the other case in the opposite direction. Moreover, this reversal might have immediate physical consequences, since a collision with the chlorine atom ahead could certainly produce different results from one with the hydrogen atoms in the lead. Nevertheless, it must be noted that through elastic collisions with other molecules, or by rotation around an axis perpendicular to the line joining the chlorine atom to the center of gravity of the molecule, the new molecule can assume a state congruent with the original one. Since numerous collisions always do take place in the systems of interest, and since in addition the occurrence of half rotational quantum numbers makes it possible that molecules are necessarily rotating even at the lowest temperatures, we must hence conclude that the lack of instantaneous congruence after reversal of the direction of motion will have no effect on the long run behavior of the chloroform molecule. In general we conclude that any lack of congruence in a reversed molecule, due to the change in the direction of the motion as a whole, will have in the long run no effect on the physical behavior of the molecule.

187. Turning our attention next to the effect of the reversal of the internal motions of the electrons inside of molecules, let us first take a simple Bohr hydrogen atom with the electron in a circular orbit, the atom as a whole having a velocity, for example, in the direction from the nucleus to the instantaneous position of the electron. Consider now the reverse state, obtained by reversing the direction of rotation of the electron and reversing the direction of motion of the atom as a whole. By rotating this reverse atom through 180° around an axis lying in the plane of the orbit perpendicular to the direction of motion of the atom, it is evident that we give this new atom the same direction of motion as a whole as the original atom and also give its electron the same direction of rotation in the orbit. The two atoms, however, are *not* congruent, since the original atom is moving in the direction of a line from nucleus to electron, and the new atom in the direction of a line from electron to nucleus. Here too at first sight this lack of congruence might seem to be of physical importance, since a collision with the nucleus in advance might have very different consequences from one with the electron in advance. It must be noted, however, that the new atom will evidently change to a state congruent with the original state of our first atom, provided only we wait for a half period of rotation to complete itself. Hence, if we consider the long run behavior of the two atoms, the lack of instantaneous congruence is again of no physical significance. Indeed it might be more in accord with the spirit of modern quantum theory to disregard any mention of instantaneous positions in an orbit, since the transfer of our ordinary concepts of space and time to describe the internal behavior of an atom may not be justified.

188. Although in the case of the simple Bohr hydrogen atom, the reverse of any state is congruent with the original state, at least after a proper time interval, it is more difficult to be sure as to what would be true in this respect in the case of more complicated atoms. Consider, for example, an atom with a number of electrons rotating in approximately plane elliptical orbits, and for simplicity take the atom as a whole at rest. It is quite evident that the immediate result of reversing the direction of motion of all the electrons in such an atom will not be to produce a congruent state, since if we choose any pair of orbits in the reversed atom and turn the atom over through 180° around the line of intersection of the planes of these two orbits, we shall indeed rectify the directions of rotation in these two orbits, but shall in general put the planes of the other orbits in positions which are not parallel with their old positions in the original atom, and in addition shall have the further difficulty that electrons in the original atom which are moving towards perihelion will in the reversed atom be moving away from perihelion and vice versa. For an atom with a number of electron orbits it is hence evident that at the best the congruence of the reverse state will only be achieved after a suitable time interval.

To investigate if the changes in state occurring in an atom might lead at a later time to such congruence, we may recall that the interpretation of spectral levels and their fine structure in accordance with the Bohr correspondence principle,[2] leads to the belief that the electrons in a general atom can be regarded as moving in approximately plane elliptical orbits, with a precession of the perihelion in the plane of the orbit and in addition a precession of the plane of the orbit itself. Hence, if this be the case, considering any single given orbit in the reversed atom, it is evident that in the course of time it will adjust both the direction of its perihelion and direction of rotation to that of the corresponding orbit in the original state. This means of course that in the absence of any relation between the periods of the precessions for different orbits, the reversed atom would in course of time come into a state congruent with the original state of the first atom. It seems possible, however, that there could be an interlocking between different orbits, which would introduce relations between the periods of precession, which would prevent the reversed atom from ever assuming through such precessions a state congruent with the original. Such interlocking might also prevent the occurrence of the necessary changes through ordinary elastic collision,

[2] Bohr, *Ann. der Physik*, 71, 248 (1923).

especially since the inner X-ray levels would hardly be affected by ordinary collisions.

189. It must apparently be concluded that, in the present state of atomic theory,[*] we cannot assert that the reversed state of an atom will, through ordinary collision or its own internal motions, always necessarily become congruent with the original state. We have of course in this matter the slight experimental evidence that such a difference between reversed atoms has never been detected, but the separation and detection of the two forms if they do exist would be an almost insuperable problem. Under the circumstances, even though it will complicate our later considerations, we shall proceed as much as possible on the basis that the reverse state of a molecule is not necessarily equivalent to the original state.

190. Changes of Molecular State through Collision.—In general any far-reaching change in the state of a molecule involves a change in its energy content which can of course only result through collision with other molecules or interaction with radiation. We shall treat the case of collision first.

a. Definition of Collision and Constellation.—Following the procedure of Boltzmann, it will be convenient to consider that a *collision* occurs when the centers of gravity of the two molecules involved approach within a certain set distance b, which is chosen large enough so that the mutual interaction of the molecules when they are farther apart than this may be regarded as negligible. At any instant during the course of the collision we shall designate the state of the two molecules as a *constellation*.

b. Classification of Constellations.—We shall use the terms *congruent*, *enantiomorphous* and *reverse* to classify different constellations in the same sense as for single molecules. In addition in the case of constellations we shall need two further categories.

We shall use the term *critical constellation* to designate the state of the system of two molecules at the beginning or at the end of the collision when the distance apart of the two centers of gravity assumes exactly the value b. Further two constellations will be designated as *corresponding constellations* when they differ only in an interchange in the positions of the centers of gravity of the two molecules, leaving unchanged all the velocities and the values of those coordinates which determine the internal configurations of the molecules.

It will be convenient to designate different constellations by symbols of the form (a, b) where the letters inside the bracket refer to the states

[*] The later development of atomic theory may well make use of the new mechanics of Heisenberg, Born and others rather than the quasi-classical mechanics hitherto used. See Chapter 7.

of the two molecules involved. The corresponding constellation to (a, b) will be designated by the symbol (b, a) and the reverse constellation by the symbol $(-a, -b)$.

If (a, b) designates a critical constellation occurring at the beginning of a collision, it is evident that the corresponding constellation (b, a) and the reverse constellation $(-a, -b)$ are both of them critical constellations with which a collision might end. Furthermore it should be noted that we can obtain the totality of all the final critical constellations with which a collision can end, if we consider the totality of all initial constellations and then construct either all the corresponding constellations or all the reverse constellations.

c. Classification of Collisions.—The different possible kinds of collision which can occur can be succinctly described by specifying the critical constellations with which the collision begins and ends, and we shall denote these different kinds of collision by symbols of the form $\begin{pmatrix} a, & b \\ c, & d \end{pmatrix}$ where (a, b) is the initial and (c, d) the final critical constellation.

In classifying different kinds of collisions we shall use the terms *congruent, enantiophorous,* and *reverse* in the evident sense. We shall also introduce the further terms *corresponding collision* and *inverse collision.*

Considering any given original collision $\begin{pmatrix} a, & b \\ c, & d \end{pmatrix}$, the *corresponding collision* will be defined as one which starts with the initial constellation (d, c) which corresponds to the final constellation (c, d) resulting in the first collision, and can be derived from (c, d) by interposing the positions of the centers of gravity of the two molecules. Thus $\begin{pmatrix} d, & c \\ x, & y \end{pmatrix}$ is the corresponding collision to $\begin{pmatrix} a, & b \\ c, & d \end{pmatrix}$.

Considering again some given original collision $\begin{pmatrix} a, & b \\ c, & d \end{pmatrix}$, we shall often be specifically interested in methods of getting the two molecules back into states congruent with their original states a and b. Such a collision will be called the *inverse collision* to the original collision and could be designated by the symbol $\begin{pmatrix} c', & d' \\ a', & b' \end{pmatrix}$ if desired, where the primes indicate the congruence rather than identity of states involved in the two collisions.

It is evident that for any given collision $\begin{pmatrix} a, & b \\ c, & d \end{pmatrix}$ there always exist congruent collisions $\begin{pmatrix} a', & b' \\ c', & d' \end{pmatrix}$ and the corresponding collision $\begin{pmatrix} d, & c \\ x, & y \end{pmatrix}$.

The existence of the reverse collision $\begin{pmatrix} -c, & -d \\ -a, & -b \end{pmatrix}$ and the inverse collision $\begin{pmatrix} c', & d' \\ a', & b' \end{pmatrix}$ must be given more special consideration.

191. Changes of Molecular State through Interaction with Radiation.—Changes in the state of a molecule may occur not only by collision with other molecules but also by the absorption or emission of radiation. A complete and correct description of the mechanism of the absorptive and emissive processes cannot now be given, and we shall have to wait for such an account until the science of physics has solved the difficult and baffling problem of the interrelation between the quantum theory and the wave theory of light. For our present purposes, however, we can schematically represent the absorptive process by which an atom or molecule A is raised from a given quantum state A_i to a higher state A_j, by the expression

$$h\nu_{ij} + A_i \longrightarrow A_j \tag{324}$$

where ν is the frequency of the radiation which leads to the absorption, $h\nu$ is the energy increase, and the subscrift ij is intended to indicate that the active radiation has not only the right frequency, but all other characteristics such as proper polarization, phase, spatial distribution and so forth which may be necessary for the phenomenon of absorption.[a]

If we turn now to the emissive process by which the molecule can drop from quantum state j to state i, it is evident that the radiation accompanying the emission will have the same frequency as that previously active in the absorption process, but in our present state of partial knowledge it cannot definitely be stated that all the properties, in particular the spatial distribution of the two radiations, are the same. For this reason it will be safest to designate the emission process by the formula

$$A_j \longrightarrow A_i + h\nu_{ji} \tag{325}$$

where the transposition of the subscripts from the order ij to ji indicates that we are not asserting an identity in the radiation active in emission and absorption except as to frequency.

[a] The more general case of interaction between radiation and molecules, in which the radiation is not completely absorbed, as, for example, in the Compton effect, could be designated by an expression of the form

$$h\nu_{ij} + A_i \rightarrow A_j + h\nu_{ji} \tag{324'}$$

In addition cases of a combined collisional and radiational nature must sometimes be considered, for example the capture of an electron by an ionized atom

$$Na^+ + E^- \rightarrow Na + h\nu \tag{324''}$$

Just as in the case of collisions we shall wish to speak of congruent, enantiomorphous, reverse, corresponding, and inverse interactions with radiation. In analogy with our previous usage, it is evident that the emissive process given by expression (325) would be the *corresponding* process to the absorption given by (324). The *reverse* process to (324) could be designated by the formula

$$A_{-j} \longrightarrow A_{-i} + h\nu_{-i,-j} \tag{326}$$

and the *inverse* process, if it exists, by the formula

$$A_j \longrightarrow A_i + h\nu_{ij} \tag{327}$$

where the change in the order of the subscripts from that in expression (325) indicates the complete identity of the emitted radiation with that absorbed in process (324).

192. On the Occurrence of Congruent Processes.—In the foregoing sections we have described both in the case of collisional and radiational interactions what we mean by congruent, enantiomorphous, reverse, corresponding, and inverse processes, and must now consider the possibility and frequency of occurrence of such processes.

It is at once evident that there is no limitation to the occurrence of molecular processes which are congruent with one another, since by a mere transformation of coordinates the equations describing two congruent processes assume the same form. Furthermore, since we have adopted the convention of always using the same volume of phase space (μ-space) $dq_1 \ldots . dp_n$ or h^n in specifying the state of a molecule, it is evident that in a large system which has come to equilibrium the probability for the occurrence of any molecular process is the same as for any process congruent with it. For this reason we often speak of congruent processes as being the same process.

193. On the Occurrence of Enantiomorphous Processes.—Except in the field of organic chemistry we are not usually specially interested in enantiomorphous states and enantiomorphous processes. It is interesting to note, however, that in a system which has come to complete "racemic" equilibrium we may also expect the same probability of occurrence for enantiomorphous processes.

194. On the Occurrence of Reverse Processes.—The actual possibility for the occurrence of reverse collisions and reverse absorptions and emissions is made certain by the principle of dynamical reversibility which we have already discussed in an earlier section of this chapter, since in the absence of an external magnetic field, this principle guarantees the possibility of the time reversal of any internal molecular process which may take place.

Furthermore, in a system containing a large number of molecules of each of the kinds involved, we must evidently expect, when equilibrium has been attained, that the rate at which any molecular process is taking place is equal to the rate at which the reverse process is taking place. Thus at equilibrium the collision $\begin{pmatrix} a, & b \\ c, & d \end{pmatrix}$ will occur with the same frequency as the reverse collision $\begin{pmatrix} -c, & -d \\ -a, & -b \end{pmatrix}$. And the absorption $h\nu_{ij} + A_i = A_j$ will occur with the same frequency as the emission $A_{-j} = A_{-i} + h\nu_{-i-j}$.

It should be especially noted, however, that the *reverse* of a given process does not leave the molecules involved in the state which they had at the start of the original process, but rather in the reverse of that state, a fact of considerable significance in connection with our later consideration of the mechanisms by which a system in equilibrium always maintains a constant number of molecules in any given state.

195. On the Occurrence of Corresponding Processes.—The possibility of the occurrence of corresponding processes is also evident. If we consider the collision $\begin{pmatrix} a, & b \\ c, & d \end{pmatrix}$, the corresponding collision, say $\begin{pmatrix} d, & c \\ e, & f \end{pmatrix}$, must certainly be possible, since it will occur by definition when molecules in the states d and c approach each other at the proper angle within the distance b, where appreciable interaction can begin. Also if we consider that $h\nu_{ij} + A_i \rightarrow A_j$ is a possible absorption, the occurrence of the corresponding emission $A_j \rightarrow A_i + h\nu_{ji}$ is guaranteed by the correspondence principle, since on the basis of the classical theory the possibility of absorption and emission go hand in hand.

196. In the case of corresponding collisions, a special theorem of interest and importance for our later work has been demonstrated by Boltzmann, which we may call *the theorem of the cycle of corresponding collisions* and shall now present.

Consider the series of collisions

$$\begin{pmatrix} 2, 1 \\ 3, 4 \end{pmatrix} \begin{pmatrix} 4, 3 \\ 5, 6 \end{pmatrix} \begin{pmatrix} 6, 5 \\ 7, 8 \end{pmatrix} \cdots\cdots \begin{pmatrix} k-3, k-4 \\ k-2, k-1 \end{pmatrix} \begin{pmatrix} k-1, k-2 \\ k, k+1 \end{pmatrix} \begin{pmatrix} k+1, k \\ k+2, k+3 \end{pmatrix} \cdots$$

where the integers denote states of the molecules, and each successive collision corresponds to the preceding one, that is starts with the two molecules in the states in which they were left by the preceding collision, but with their centers of gravity transposed.

Let us now define the different states of our molecules 1, 2, 3, 4, 5, 6, etc., by assigning to each state a specified region in the phase space (μ-space), which has a volume small enough so that we can regard the physical behavior of the molecule as determined by the location of the specified region, but a volume which is nevertheless finite. This procedure, which introduces no theoretical incorrectness, has the advantage of giving us only a finite number of possible states to consider.

Returning now to the series of collisions given above, it is evident, since we have only a finite number of possible states, that by extending the series to the right we must at last come to a collision in which the final states of the two molecules are the same as the initial states for some earlier collision, that is to a collision *to* which some earlier collision corresponds. We shall now prove that this earlier collision which corresponds to the last collision of the series must as a matter of fact be the first collision of the series $\begin{pmatrix} 2, & 1 \\ 3, & 4 \end{pmatrix}$.

To demonstrate this, let us assume that $\begin{pmatrix} k+1, & k \\ k+2, & k+3 \end{pmatrix}$ is the last collision of the series and that the earlier collision corresponding to it is not the first collision of the series but, for example, a later one $\begin{pmatrix} 6, & 5 \\ 7, & 8 \end{pmatrix}$. We have then from this correspondence that the collision $\begin{pmatrix} k+1, & k \\ k+2, & k+3 \end{pmatrix}$ must be the collision $\begin{pmatrix} k+1, & k \\ 5, & 6 \end{pmatrix}$ or, since the end states are identical, this collision assumed as the last must actually be the collision $\begin{pmatrix} 4, & 3 \\ 5, & 6 \end{pmatrix}$ and preceding it in the series the collision $\begin{pmatrix} k-1, & k-2 \\ k, & k+1 \end{pmatrix}$ must be $\begin{pmatrix} 2, & 1 \\ 3, & 4 \end{pmatrix}$ and the true last member of the series be $\begin{pmatrix} k-3, & k-4 \\ 2, & 1 \end{pmatrix}$ as was to be demonstrated.

This theorem, that the series of corresponding collisions between two molecules form a closed cycle, will be of use to us when we later come to the demonstration of Boltzmann's famous H-theorem.

197. It would be of interest to inquire whether a similar theorem holds for corresponding processes in the case of the absorption and emission of radiation. Here, however, we shall be on difficult ground until physics has solved the problem of the nature of radiation. Nevertheless, if we take the position of the extreme light quantum theory and assume that the nature of a quantum of light is uniquely determined by the initial and final states of the atom which emits or absorbs it, we can conclude that the quantum $h\nu_{ij}$ absorbed by the process

$$h\nu_{ij} + A_i \longrightarrow A_j$$

is identical with the quantum $h\nu_{ji}$ given out by the emission

$$A_j \longrightarrow A_i + h\nu_{ji}$$

and hence that these two steps alone would then form a cycle of corresponding processes.

198. There is a further theorem with respect to corresponding collisions which we must now deduce. This may be called the *theorem of equal probability coefficients for corresponding collisions.* The proof may be obtained as follows.

To specify *completely* the state of a molecule we must give the region in the phase space (μ-space) $\delta q_1\delta q_2\delta q_3\,\delta q_4 \ldots . \delta q_n\delta p_1 \ldots . \delta p_n$ in which its representative point falls. The first three coordinate ranges may be taken as determining the location of the center of gravity of the molecule and the remaining ranges as determining the values of its internal coordinates and its external and internal momenta. In considering different kinds of collisions, however, we are usually not interested in the particular location in space where the collision occurs, but desire to know merely the values of the internal coordinates and the external and internal momenta for a molecule that is entering a collision. Hence for such purposes we shall specify the state of a molecule by giving *merely* the range

$$\delta\omega = \delta q_4 \ldots \delta q_n \delta p_1 \ldots \delta p_n \tag{328}$$

in which the internal coordinates and the external and internal momenta lie. Furthermore we shall choose equal values of $\delta\omega$ for specifying different possible states of a molecule.

Consider now a system containing N_1, N_2, N_3 and N_4 molecules in the states given by the ranges $\delta\omega_1$, $\delta\omega_2$, $\delta\omega_3$ and $\delta\omega_4$, and let the molecules in the different states be uniformly distributed throughout the total volume V of the system. Furthermore let the states be so chosen that we can have a collision $\begin{pmatrix}2, & 1\\ 3, & 4\end{pmatrix}$ between molecules in states 1 and 2 leading to molecules in states 3 and 4. Since the molecules in the different states are uniformly distributed throughout the system, the chance of such a collision will be proportional to the numbers of molecules in the two states, so that we may write for the probable number of such collisions occurring per second the expression

$$Z^{2,\,1}_{3,\,4} = C^{2,\,1}_{3,\,4}\,N_1N_2 \tag{329}$$

where $C^{2,\ 1}_{3,\ 4}$ is a probability coefficient. Similarly for the number of corresponding collisions of the type $\begin{pmatrix} 4, & 3 \\ 5, & 6 \end{pmatrix}$ we may write

$$Z^{4,\ 3}_{5,\ 6} = C^{4,\ 3}_{5,\ 6} N_3 N_4 \tag{330}$$

where $C^{4,\ 3}_{5,\ 6}$ is another such coefficient. With the help of Liouville's theorem we shall now show, however, that these two coefficients are equal.

Let us follow the course of the collision $\begin{pmatrix} 2, & 1 \\ 3, & 4 \end{pmatrix}$ by the motion of a representative point in a combined phase-space for the two molecules, of such a nature that the position of the representative point will give us the location of the center of gravity of the first molecule *relative* to that of the second and give us the ranges $\delta\omega_1$ and $\delta\omega_2$ for the internal coordinates and external and internal momenta of the two molecules.

At the start of the collision when the distance apart of the two centers of gravity is b, let the center of gravity of the first molecule lie in a range $b^2\delta\lambda\delta s$, where δs is a distance laid off along the line connecting the centers of gravity of the two molecules and $\delta\lambda$ is a solid angle drawn from the center of gravity of the second molecule, the quantity $b^2\delta\lambda\delta s$ thus being a small volume. Similarly at the close of the collision, when the distance apart of the two centers again becomes b, let the center of gravity of the first molecule lie in a similar range $b^2\delta\Lambda\delta S$. Neglecting then the position of the center of gravity of the second molecule, which would merely tell us the particular location in space where the collision takes place, our representative point for the two molecules will lie in the region $\delta\omega_1\delta\omega_2 b^2\delta\lambda\delta s$ at the start of the collision and in the region $\delta\omega_3\delta\omega_4 b^2\delta\Lambda\delta S$ at the end of the collision. In accordance with Liouville's theorem, however, in the form giving the conservation of extension in phase as expressed by equation (31) in Chapter 3, we must have an equality between these two phase ranges

$$\delta\omega_1\delta\omega_2 b^2\delta\lambda\ \delta s = \delta\omega_3\delta\omega_4 b^2\delta\Lambda\ \delta S \tag{331}$$

We have, however, agreed to use equal regions $\delta\omega$ in specifying states of the molecules so that we obtain by cancellation

$$b^2\delta\lambda\ \delta s = b^2\delta\Lambda\ \delta S \tag{332}$$

Finally since the choice of δs and δS is still open to our disposal, provided we do not impose values on $\delta\lambda$ and $\delta\Lambda$, let us take them as equal to the changes in the distance between centers that will occur in a small time interval δt, respectively at the start and close of the collision, in accordance with the equations

$$\delta s = k\delta t \quad \text{and} \quad \delta S = K\delta t \tag{333}$$

where k and K are the components of the relative velocity of the two molecules, parallel to the line of centers, at the start and close of the collision. Substituting in (332) we obtain

$$b^2\delta\lambda\ k\delta t = b^2\delta\Lambda\ K\delta t \tag{334}$$

We are now ready to consider the probable frequency of the collisions in which we are interested. For the probable number of collisions of type $\begin{pmatrix}2, 1\\3, 4\end{pmatrix}$ occurring per second we shall evidently obtain

$$Z_{3,\ 4}^{2,\ 1} = \frac{b^2\delta\lambda\ k}{V} N_1 N_2 \tag{335}$$

since the chance that a collision will occur during any second is proportional to the chance that a molecule in the first state has its center gravity at the beginning of that time in the range $b^2\delta\lambda\ k$, and this in turn will be proportional to the ratio between this volume and the total volume V.

To determine the probable number of corresponding collisions $\begin{pmatrix}4, 3\\5, 6\end{pmatrix}$ we must merely note that the component of relative velocity K with which the molecules part after the collision $\begin{pmatrix}2, 1\\3, 4\end{pmatrix}$ is, by the definition of corresponding collisions, the same as that with which they approach in the corresponding collision $\begin{pmatrix}4, 3\\5, 6\end{pmatrix}$ and hence may write for the probable number occurring per second the expression

$$Z_{5,\ 6}^{4,\ 3} = \frac{b^2\delta\Lambda\ K}{V} N_3 N_4 \tag{336}$$

Comparing with our original equations (329) and (330), however, we see that we have thus obtained values for the probability coefficients of collision, giving us the identifications

$$C_{3,\ 4}^{2,\ 1} = \frac{b^2\delta\lambda\ k}{V} \quad \text{and} \quad C_{5,\ 6}^{4,\ 3} = \frac{b^2\delta\Lambda\ K}{V} \tag{337}$$

and from equation (334) we obtain the equality of these coefficients

$$C_{3,\ 4}^{2,\ 1} = C_{5,\ 6}^{4,\ 3} \tag{338}$$

We have thus demonstrated the equality of the coefficients which determine the probable rates at which corresponding collisions are taking place. This theorem will also be of the greatest value in our derivation of Boltzmann's H-theorem.

199. On the Occurrence of Inverse Processes.—If we consider any process in which molecules (including quanta if necessary) change from one set of states to another, we have defined the inverse of this process as one which starts with *all* the molecules involved in states congruent with those in which they have been left by the first process, and returns them to their original states.

The question of the possibility and frequency of occurrence of inverse processes is one of considerable interest, since it has sometimes been assumed that corresponding to any given molecular process there always exists an inverse process, and that, in a system which has come to equilibrium, the constant quotas of the molecules in the different states is assured by an equality between the rates with which every process and its inverse takes place. Of late years this postulate has often been used as a starting point for theoretical investigations. We shall find, however, that in the simple and general form in which the above principle is usually stated it certainly cannot be universally true, and we shall need in general to proceed carefully in deducing the consequences which have usually been based on this postulate.

200. The true state of affairs can most easily be understood by considering the relation between inverse processes and reverse processes. In the *inverse* of a given process the molecules must start in states congruent with those in which they have been left by the first process and return to their original states. In the *reverse* of a given process, however, the molecules start in states which are the reverse of those occurring at the end of the first process and change to states which are the reverse of the original ones. Hence, whenever the molecules involved are such that any state and its reverse are congruent, the reverse process will itself be an inverse process, and since the reverse of any process is always possible, we actually shall have in such cases for every molecular process an inverse process.

In general, however, we have found that the reverse of a given state is not necessarily congruent with the state itself, and hence cannot conclude in this way that an inverse does exist for every process.

201. As an example of a process in which every molecular state is congruent with its reverse, we may consider the collision of solid elastic spheres. On account of their symmetry a reversal in the direction of the motion of such spheres will obviously lead to a congruent state. Hence the reverse of any given collision will also be an inverse collision and will return the spheres to states congruent with their original ones, a result which can easily be verified in detail, by applying to such collisions the laws of the conservation of energy and momentum.

202. When we consider, however, the collisions of less symmetrical bodies we no longer find that reverse and inverse processes are the same. To make this evident by a simple example, let us consider the somewhat artificial case of a collision between a sphere and a wedge-shaped solid such as shown in Figure 5, where the sphere and wedge are taken as having the same mass, and the point of impact of the sphere on the

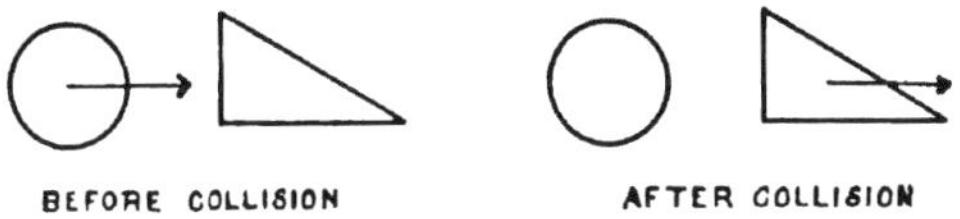

FIGURE 5.

wedge is taken so that the collision will result in a complete transfer of energy from the sphere to the wedge. It is immediately evident that no single collision is possible which will return the sphere and the wedge to states congruent with their original motions. Such a return to the original states can only be secured by a series of collisions, such as the cycle of corresponding collisions first considered by Boltzmann and discussed above in Section **196.**

203. Although the particular illustration given above was a trivial one, chosen for its simplicity, we find it necessary to conclude in general that there may be molecular processes for which no inverse exists at all, and hence in our further development shall avoid making use at all of the concept of inverse processes.[5]

204. The Principle of Microscopic Reversibility.—In our section on the occurrence of reverse processes, we came to the conclusion that in a system not subjected to an external field of the magnetic type, which would introduce odd powers of the velocities into the Lagrangian equations, the reverse of any molecular process is also a possible process. Furthermore, when such a system comes to a state of equilibrium it is evident that the rate at which any molecular process is taking place is equal to the rate at which the reverse process is taking place. We shall call this conclusion the *principle of microscopic reversibility.*

In more specific language, if i, j, k, l, etc., are states of molecules in the system and $-i$, $-j$, $-k$, $-l$, etc., the reverse of these states, this principle requires, when equilibrium has been reached, that the number of processes per second in which molecules change from i to j and k to l, etc., must be equal to the number of reverse processes in which

[5] Compare Lorentz, *Situngsber. d. Akad. d. Wiss. zu Wien,* 2 abt., **95,** 115 (1887).

molecules change from $-j$ to $-i$ and $-l$ to $-k$, etc. As already noted, this principle as just stated in its correct form must be carefully distinguished from the incorrect statement that the number of processes characterized by i to j, k to l, etc., is necessarily equal to the number j to i, l to k, etc., since the inverse processes may not even exist.

There are a number of corollaries to this principle which it will be well for us to consider.

(a) If we denote the rate at equilibrium of *any particular process* and its reverse, in which a molecule changes from one set of congruent states i to another set j, by symbols of the form $Z_{i\to j}$ and $Z_{-j\to -i}$ we may evidently write the equality

$$Z_{i\to j} = Z_{-j\to -i} \tag{339}$$

or summing up for *all processes* in which a molecule changes from set j to i, may write

$$N_{i\to j} = \Sigma Z_{i\to j} = \Sigma Z_{-j\to -i} = N_{-j\to -i} \tag{340}$$

Similarly we shall have

$$\begin{aligned} N_{-i\to -j} &= N_{j\to i} \\ N_{i\to -j} &= N_{j\to -i} \\ N_{-i\to j} &= N_{-j\to i} \end{aligned} \tag{341}$$

and if we desire to group together each state with its reverse, we may combine equation (340) with equations (341), and obtain

$$N_{\pm i\to \pm j} = N_{\pm j\to \pm i} \tag{342}$$

where the symbol $N_{\pm i\to \pm j}$ denotes the total rate at equilibrium at which molecules in the set of congruent states i or its reverse $-i$ are changing to molecules in the set of states j or its reverse $-j$.

(b) For those particular cases where each of the states of the molecule and its reverse are congruent, equations (339) and (340) reduce to the special form

$$Z_{i\to j} = Z_{j\to i} \tag{343}$$

and

$$N_{i\to j} = N_{j\to i} \tag{344}$$

(c) If $Z_{i\to j}$, the rate at equilibrium for a particular process by which molecules change from the set of congruent states i to the set j, happens to to be equal to the rate $Z_{-i\to -j}$ for an analogous process for the reverse states, we shall then have by the principle of microscopic reversibility

$$\begin{aligned} Z_{i\to j} &= Z_{-j\to -i} \\ Z_{-i\to -j} &= Z_{j\to i} \end{aligned} \tag{345}$$

which combined with our special assumption of the equality

$$Z_{i \to j} = Z_{-i \to -j} \tag{346}$$

will also give us for this special case the result

$$Z_{i \to j} = Z_{j \to i} \tag{347}$$

Such a condition of affairs in which this special result might be expected to hold would be if the processes i to j and $-i$ to $-j$ consisted in the emission of radiation, since in accordance with the correspondence principle we should then expect these processes i to j and $-i$ to $-j$ to occur with the same frequency.

In following chapters we shall find many occasions to apply the principle of microscopic reversibility and its corollaries.

Chapter 16

The Absorption and Emission of Radiation

205. Introduction.—In an earlier chapter we have deduced the statistical equilibrium between radiation and matter, by treating the hohlraum as containing modes of electromagnetic vibration of different frequencies, in analogy to the modes of vibration of an elastic solid, and then assigning to each mode of vibration the energy that it should have on the basis of the Maxwell-Boltzmann distribution law as modified to take account of quantum phenomena. In this way we have obtained a satisfactory derivation of the Planck radiation law for the distribution of energy between the different frequencies.

The above method of derivation gave us, however, no insight into the mechanism by which the continual absorption and emission of radiation by the walls of the container actually does maintain the equilibrium distribution of energy among the different frequencies. In addition the increasingly apparent possibility that radiational energy is actually corpuscular in nature makes the treatment of the hohlraum as a system of modes of electromagnetic vibration seem somewhat artificial. For these reasons a derivation of the Planck radiation law, by a method which is based on considerations of the probability of the absorption and emission by atoms of radiation of different frequencies, would contain elements of superiority. The considerations of the last chapter now make it possible to present such a derivation of the Planck radiation law, a derivation first due to Einstein. Furthermore the considerations now to be presented have the additional advantage of permitting a calculation of the mean life of molecules in different quantum states from a knowledge of the intensities of spectral lines.

206. Einstein's Coefficients of Absorption, Emission and Induced Emission.—Consider an atom or molecule capable of existing in different quantum states, and consider two of these states i and j, of such a nature that the atom can pass from i to j by the absorption of a quantum of radiation of magnitude $h\nu$, and pass in the reverse direction by the emission of a quantum of the same magnitude.

207. If we should have an atom of this kind present in the state j in free space, there would evidently be a certain probability in unit time of a spontaneous passage to the lower state i by the emission of a quantum of radiation. We shall call this quantity the *probability of spontaneous emission* and designate it by the symbol $A_{j\to i}$. It is the first of Einstein's three coefficients.

208. If, instead of being in free space our atom should be surrounded by radiation, it is evident from the correspondence principle that the chance of emitting radiation would be different, since on the basis of the classical theory the rate of energy emission of an isolated atom is different from the rate of emission of an atom whose moving electrons are interacting with the varying electromagnetic field present in a radiation field. Hence, to obtain the total probability of emission we must include a term depending on the density of radiation, and shall take the *total probability of emission* in unit time as given by the expression $A_{j\to i} + u_\nu B_{j\to i}$ where $u_\nu = (du/d\nu)_{\nu=\nu_{ij}}$ is the density of radiational energy per unit frequency interval[1] at the frequency ν_{ij} involved in the transition, and $B_{j\to i}$ is a second constant coefficient dependent on the nature of the states i and j. We shall call this quantity $B_{j\to i}$ the *coefficient of induced emission.*

209. Einstein's procedure in placing the total probability of emission equal to the sum of two terms of the form given above seems a natural one, but of course contains a certain element of seemingly arbitrary assumption which will only be justified by the later results. In thinking about the process of emission, it is perhaps unwise to consider spontaneous emission and induced emission as two different kinds of process. The quantities $A_{j\to i}$ and $u_\nu B_{j\to i}$ have the same physical dimensions, and if we are willing for the moment to use the language of virtual radiation fields, we may regard both kinds of emission as induced,—$A_{j\to i}$ being the probability of induced emission due to the virtual field of the atom itself, and $u_\nu B_{j\to i}$ that due to the virtual field of the surroundings.[2]

It must also be noted in connection with the nature of emitted light that on the basis of the correspondence principle we should expect radiation induced by an external field to be coherent with the radiation associated with that field.

210. Turning now to the probability of absorption, it is evident that this will also be dependent on the density of radiation of the frequency involved, and we shall take this probability as equal to $u_\nu B_{i\to j}$ where the

[1] In what follows we shall use the term *radiation density* to signify density of energy per unit volume per unit frequency interval.

[2] Note the conflicting view presented by G. N. Lewis, *Proc. Nat. Acad. Sci.*, **11**, 179 (1925). Compare footnote to Section **213**.

quantity $B_{i \to j}$ is a third constant dependent on the nature of the states i and j which we shall call the *coefficient of absorption.*

211. The Maintenance of Equilibrium by Absorption and Emission.—Consider now a system of atoms and radiation in equilibrium and let N_i and N_j, N_{-i} and N_{-j} be the numbers of atoms in the system in the states i and j and their reverse states $-i$ and $-j$. Then in accordance with the principle of microscopic reversibility, we may equate the number of changes per second from state i to j occurring through absorption with the number of changes occurring through emission between the reverse states $-j$ and $-i$, giving us the equality

$$N_i u_\nu B_{i \to j} = N_{-j}(A_{-j \to -i} + u_\nu B_{-j \to -i}) \tag{348}$$

In accordance with the principle of correspondence, however, we may expect the emission of radiation by atoms in states j and $-j$ to have the same probability of occurrence,[3] and hence in agreement with Corollary c of the principle of microscopic reversibility as given in the last chapter we may write

$$N_i u_\nu B_{i \to j} = N_j(A_{j \to i} + u_\nu B_{j \to i}) \tag{349}$$

as one of the conditions which must hold at equilibrium.

If we have a state of equilibrium, however, we must have the additional condition that the numbers of molecules in the two states i and j shall stand in the ratio prescribed by the Maxwell-Boltzmann distribution law, giving us

$$\frac{N_i}{N_j} = \frac{p_i}{p_j} e^{\frac{h\nu}{kT}} \tag{350}$$

where p_i and p_j, the a-priori probabilities of the two states, are unity in case the systems are non-degenerate, and $h\nu$ is evidently the proper value to use for the energy difference of the two states.

Hence, if equilibrium is to be maintained, the interaction of radiation and matter must be such that equations (349) and (350) both hold.

212. Derivation of the Planck Distribution Law.—Substituting equation (350) in (349) and solving for the radiation density u_ν, we easily obtain

$$u_\nu = \frac{A_{j \to i}/B_{j \to i}}{\dfrac{N_i}{N_j}\dfrac{B_{i \to j}}{B_{j \to i}} - 1} = \frac{A_{j \to i}/B_{j \to i}}{\dfrac{p_i}{p_j}\dfrac{B_{i \to j}}{B_{j \to i}} e^{\frac{h\nu}{kT}} - 1} \tag{351}$$

The apparently incongruous factor $p_i B_{i \to j}/p_j B_{j \to i}$ can easily be elimi-

[3] Note that at equilibrium $N_{+j} = N_{-j}$, and with the nomenclature used we may regard u_ν as equal to $u_{+\nu} + u_{-\nu}$, where $u_{+\nu} = u_{-\nu}$.

nated from this equation, since for an infinite value of the temperature T, the value of the radiation density u_ν must itself go to infinity, and this will only be true if the factor in question is equal to unity, giving us the equation

$$\frac{p_i B_{i\to j}}{p_j B_{j\to i}} = 1 \tag{352}$$

so that we may now rewrite (351) in the form

$$u_\nu = \frac{A_{j\to i}}{B_{j\to i}} \frac{1}{e^{\frac{h\nu}{kT}} - 1} \tag{353}$$

Finally the equation can be changed into the usual expression for the Planck distribution law, if we recall that in accordance with the Wien displacement law u_ν must in any case have the form $u_\nu = \nu^3\phi(T/\nu)$, permitting us to write

$$u_\nu = \frac{8\pi h\nu^3}{c^3} \frac{1}{e^{\frac{h\nu}{kT}} - 1} \tag{354}$$

where $8\pi h/c^3$ may be regarded as an experimental value of the constant to be introduced, or may be deduced from the hypothesis that the classical Rayleigh-Jeans formula, based on the equipartition of energy

$$u_\nu = \frac{8\pi\nu^2}{c^3} kT \tag{355}$$

must hold when the temperature T is high and the frequency ν small.

213. Relation between Einstein's Coefficients.—The foregoing treatment provides us not only with a derivation of the Planck distribution law, but also with a set of relations between the important quantities which we have called Einstein's coefficients of absorption and emission.

In accordance with equation (352), we may write

$$B_{j\to i} = \frac{p_i}{p_j} B_{i\to j} \tag{356}$$

and combining with equations (353) and (354) obtain

$$A_{j\to i} = \frac{8\pi h\nu^3}{c^3} B_{j\to i} = \frac{p_i}{p_j}\frac{8\pi h\nu^3}{c^3} B_{i\to j} \tag{357}$$ [4]

[4] Substituting equation (357) in our previous expression given in Section 208 for the total probability of emission, we obtain for this quantity the expression $\left(\frac{8\pi h\nu^3}{c^3} + u_\nu\right) B_{j\to i}$. Noting that $\frac{8\pi h\nu^3}{c^3}$ has the same physical dimensions as u_ν, we see that by regarding $\frac{8\pi h\nu^3}{c^3}$ as a virtual radiation density from the atom itself, we can consider all emission as induced, as already remarked.

We thus see that the three coefficients are so related that the value of any two can be calculated from the third. This is of importance in considering the intensities of absorption and emission lines.

214. Interpretation of the Width of Spectral Lines.—When we come to the application of the foregoing considerations to the interpretation of the intensities of absorption or emission of spectral lines, we are immediately met by the experimental fact that spectral lines actually have in general a small but definite width, the intensity usually falling off rapidly on both sides of an approximately central maximum.

The quantum theory explanation for this fact lies in the hypothesis that, in an actual gas, the molecules in a given quantum state (i.e. with given values of the quantum numbers) do not all have exactly the same energy content. This arises from the fact that molecules in a given quantum state have different velocities, and are differently affected by the stray fields from neighboring molecules. Hence when a transition takes place from one given set of quantum numbers to another, the energy emitted or absorbed may not be exactly the same as it would be if the molecules were stationary and in free space, and hence the magnitude of the quantum $h\nu$ will also vary slightly for a given change in quantum numbers. We have indeed experimental evidence that such causes would affect the frequency of the radiation, since the quantum theory interpretation of the Doppler effect shows the influence of kinetic energy on the frequency of spectral lines and the experimental results and theoretical considerations of the Stark effect illustrate the influence that is to be expected from stray electric fields.

In addition to these known causes for the broadening of spectral lines, it is also possible that there is an inherent weakness in the quantization of molecular states, and consequent lack of sharpness in the energy levels, which would give some width to the spectral lines even from a set of stationary molecules in free space. At the present time, however, we have no direct experimental evidence as to what this "natural" width of spectral lines would be, if it really exists. In the cases of usual interest, however, it would probably be small compared with the moderate widening produced by the Doppler effect and the usually much larger widening produced by the fields from neighboring molecules.

It is evident that these factors operative in disturbing the energy levels of molecules and thus affecting the width of spectral lines might also have an effect on the probability of transition from one level to another. Thus the probability that an isolated molecule surrounded by radiation of unit density at the frequency necessary for interaction would absorb a quantum of radiation might be different from the probability of absorption for a molecule also under the influence of radiation of the right frequency and unit density but affected by the fields from neighboring molecules. We have indeed some experimental evidence that this might be a possibility, since the experiments of Füchtbauer, Joos and Dinkelacker [5] show that in the case of a given amount of mercury vapor in the normal $1S$ state, the total number of quanta absorbed by transition to the $2p_2$ state (λ 2537), drops by about 50 per cent when nitrogen at a pressure of 50 atmospheres is added to the previously dilute mercury vapor, the intensity of light being kept the same.[6] Hence, since the total intensity of a line is affected by the presence of neighboring molecules, it seems possible that the relative intensities within the width of an absorption line might also be due, not only to the relative numbers of molecules having the requisite energy levels for the frequency in question, but to a variation in the probability of absorption as we pass across the line. We have, however, no direct experimental evidence on this point, and consideration of the question makes it seem probable that such a possible variation in probability of absorption as we pass across a line is of secondary importance in accounting for the intensity distribution within the line as compared with the variation in numbers of molecules having the requisite energy levels. Hence, in the interests of simplicity, at the present stage of experimental and theoretical development, we shall feel justified in treating Einstein's coefficients $A_{j \to i}$, $B_{j \to i}$, and $B_{i \to j}$ as constant quantities applying at any point within a spectral line.[7]

215. Relation between the Probability of Absorption $B_{i \to j}$ and the Ordinary Coefficient of Absorption.—We are now ready to interpret the intensity of absorption lines with the help of the foregoing considerations, and indeed to use the experimental measurements of intensities for calculating the values of Einstein's coefficients.

Consider an ordinary experiment on the intensity of absorption lines, in which a beam of light of total intensity I as measured in ergs per square centimeter per second falls on a layer of the absorbing medium of thickness dx and let us define the intensity of light I_ν for a given frequency by the expression

$$I_\nu = \frac{\partial I}{\partial \nu} \tag{358}$$

Since the energy is travelling with the velocity c it is evident that the radiation density u_ν will be related to the intensity I_ν by the equation

[5] Füchtbauer, Joos and Dinkelacker, *Ann. der Physik*, **71**, 204 (1923).
[6] More familiar deviations from Beer's law of absorption should also be noted.
[7] For treatments where this constancy is not assumed, see Pauli, *Zeitschr. für Phys.*, **18**, 272 (1923); Einstein and Ehrenfest, *Zeitschr. für Phys.*, **19**, 301 (1923); Tolman and Badger, *Phys. Rev.*, **27**, 383 (1926).

$$u_\nu = \frac{I_\nu}{c} \tag{359}$$

Let us now consider the diminution in the intensity of this light, on passage through the layer of thickness dx, due to the absorption of light by molecules passing from a lower state i to a higher state j.

If N_i is the total number of molecules per cubic centimeter in the lower quantum state, then the number in the layer of unit cross section and thickness dx will be $N_i\,dx$. In accordance with our discussion of the width of spectral lines, however, we shall only be interested for the moment in those molecules which are not only in the lower quantum state, but are so affected by their velocity and by the fields from neighboring molecules as to be capable of picking up energy in the range ν to $\nu + \delta\nu$, and can take for this number of molecules the expression $\frac{\partial N_i}{\partial \nu}\delta\nu dx$. Hence the rate at which energy in the frequency range ν to $\nu + \delta\nu$ is being removed from the beam in the layer dx will evidently be

$$h\nu\left(\frac{\partial N_i}{\partial \nu}\,\delta\nu dx\right) u_\nu B_{i\rightarrow j} \tag{360}$$

On the other hand, by induced emission, radiation coherent with the oncoming beam will evidently be returned at the rate

$$h\nu\left(\frac{\partial N_j}{\partial \nu}\,\delta\nu dx\right) u_\nu B_{j\rightarrow i} \tag{361}$$

The difference of these two terms for energy removed and returned per second will evidently give the decrease in the energy passing in one second through the layer, so that combining (360) and (361) and substituting (359), we can write

$$-d(I_\nu\delta\nu) = \frac{h\nu}{c} I_\nu\left(\frac{\partial N_i}{\partial \nu} B_{i\rightarrow j} - \frac{\partial N_j}{\partial \nu} B_{j\rightarrow i}\right)\delta\nu dx \tag{362}$$

or transposing and eliminating,

$$-\frac{1}{I_\nu}\frac{dI_\nu}{dx} = \frac{h\nu}{c}\left(\frac{\partial N_i}{\partial \nu} B_{i\rightarrow j} - \frac{\partial N_j}{\partial \nu} B_{j\rightarrow i}\right) \tag{363}$$

This logarithmic decrement of the intensity with distance is, however, the quantity usually defined as the absorption coefficient a_ν for light of a given frequency, and is the quantity ordinarily determined in absorption experiments. So that we now have the useful equation

$$a_\nu = \frac{h\nu}{c}\left(\frac{\partial N_i}{\partial \nu} B_{i\rightarrow j} - \frac{\partial N_j}{\partial \nu} B_{j\rightarrow i}\right) \tag{364}$$

connecting Einstein's coefficients with the experimentally measured quantity a_ν.

216. Calculation of Einstein's Coefficients from Absorption Measurements.—We shall now be able to calculate the value of Einstein's coefficients from intensity measurements. If we plot the measured values of a_ν as a function of ν across the whole width of an absorption line, and integrate under the curve, we evidently obtain in accordance with equation (364)

$$\int a_\nu d\nu = \int \frac{h\nu}{c}\left(\frac{\partial N_i}{\partial\nu} B_{i\rightarrow j} - \frac{\partial N_j}{\partial\nu} B_{j\rightarrow i}\right) d\nu \qquad (365)$$

Since in most cases ν varies but slightly in crossing the effective width of the line, and since in accordance with Section **214** we shall take $B_{i\rightarrow j}$ and $B_{j\rightarrow i}$ as constants, we can rewrite this in the form

$$\int a_\nu d\nu = \frac{h\nu}{c}\left[B_{i\rightarrow j}\int \frac{\partial N_i}{\partial\nu} d\nu - B_{j\rightarrow i}\int \frac{\partial N_j}{\partial\nu} d\nu\right]$$

$$= \frac{h\nu}{c}\left[B_{i\rightarrow j}N_i - B_{j\rightarrow i}N_j\right] \qquad (366)$$

Since in general the intensity of the light used in absorption experiments is not enough to cause any appreciable change in the number of molecules in different quantum states from that existing at equilibrium, we can substitute for the ratio of N_i to N_j that given by the Maxwell-Boltzmann distribution law as expressed by equation (350). Furthermore, we can substitute for $B_{j\rightarrow i}$ in terms of $B_{i\rightarrow j}$ in accordance with equation (356). Doing this and solving for the thing of interest, the coefficient of absorption $B_{i\rightarrow j}$, we obtain

$$B_{i\rightarrow j} = \frac{c}{h\nu N_i\left(1 - e^{-\frac{h\nu}{kT}}\right)}\int a_\nu\, d\nu \qquad (367)$$

At ordinary temperatures, even at frequencies as low as in the near infra red, the value of ν over T is large enough, so that this equation can often be simplified to the form

$$B_{i\rightarrow j} = \frac{c}{h\nu N_i}\int a_\nu d\nu \qquad (368)$$

The possibility thus provided of calculating the experimental value of Einstein's coefficient $B_{i\rightarrow j}$ from measurements of the intensity of absorption lines is of great value in the interpretation of atomic behavior. Knowing the value of $B_{i\rightarrow j}$, the other two coefficients $B_{j\rightarrow i}$ and $A_{j\rightarrow i}$ can of course at once be calculated from equations (356) and (357).

217. Mean Life of Molecules in Activated States.—Since $A_{j\to i}$ is the specific rate for the spontaneous transition of molecules by emission of radiation from state j to state i, it is of course easy to determine from this quantity the mean duration of molecules in state j before jumping to state i. For the number of molecules leaving state j in any time interval dt we have

$$-dN_j = N_j A_{j\to i}\, dt \tag{369}$$

and integrating obtain for the number present at any time t

$$N_j = (N_j)_0 e^{-A_{j\to i}t} \tag{370}$$

where $(N_j)_0$ is the number present at the time zero. Hence for the number having a life between t and $t+dt$ we may write $(N_j)_0 e^{-A_{j\to i}t} A_{j\to i} dt$ and for the mean life τ obtain

$$\tau = \frac{1}{(N_j)_0}\int_0^\infty (N_j)_0 e^{-A_{j\to i}t} A_{j\to i} t\, dt = \frac{1}{A_{j\to i}} \tag{371}$$

The reciprocal of Einstein's coefficient $A_{j\to i}$ is then the mean life of the molecule in state j, provided, however, of course that this transition from state j to state i is the only one that can occur. If the molecule can leave state j by a number of different processes, perhaps of a collisional as well as a radiational nature, then the actual mean life will be correspondingly shortened. For example, if there are a series of lower states i to which a spontaneous transition can take place we shall have

$$\tau = \frac{1}{\sum_i A_{j\to i}} \tag{372}$$

218. Examples of the Calculation of Einstein's Coefficients.—With the help of equations (367) and (368), values of Einstein's coefficients have been calculated for a considerable number of atomic and molecular transitions.[8]

As a good example of the method we may consider the calculation of the value of $B_{i\to j}$ for the transition of mercury from the normal $1S$ to the $2p_2$ state, corresponding to the line λ 2537, since the absorption data of Füchtbauer, Joos and Dinkelacker[9] on this line are the best that we now have available. The frequency of this line is high enough compared with the temperature at which the experiments were made

[8] See Tolman, *Phys. Rev.*, **23**, 693 (1924); Tolman and Badger, *ibid.*, **27**, 383 (1926).

[9] Füchtbauer, Joos and Dinkelacker, *Ann. der Physik.*, **71**, 204 (1923).

so that we may neglect the second term in the denominator of equation (367) and use equation (368)

$$B_{i \to j} = \frac{c}{h\nu N_i}\int a_\nu d\nu \qquad (368)$$

for our calculations. For ν we take the frequency corresponding to 2537 Å. For N_i we may take the total number of atoms of mercury per cubic centimeter, since the first energy level of mercury above the normal $1S$ state is so high that there are practically no atoms in any but the normal state at the temperatures employed. In the experiments of Füchtbauer, Joos and Dinkelacker this number could be calculated from the known vapor pressure of mercury, since liquid mercury at a definite temperature was present in the apparatus. The value of $\int a_\nu d\nu$ was determined by these authors by graphical integrations under the absorption curves obtained in runs in which the width of the line was artificially increased by the addition of hydrogen, nitrogen, argon, or carbon dioxide at various pressures. The final value was determined by extrapolation to zero pressure, and is quite possibly correct within ten per cent.

Substituting the values thus obtained into the equation, we obtain for $B_{i \to j}$ the value 2.88×10^{18}, everything being expressed in c.g.s. units. Taking the ratio of p_i to p_j as $1:3$ [10] we can then obtain from equation (357) for the rate of spontaneous decay of mercury in the $2p_2$ state, $A_{j \to i}$, the value 9.68×10^6. From this the mean life τ of mercury atoms in the $2p_2$ state comes out, in accordance with equation (371) 1.03×10^{-7} seconds.

219. This latter value has special interest as it has been checked, since its first calculation,[11] by two independent methods.

Thus Turner [12] has calculated the mean life τ from measurements of Cario and Franck, on the rate of production of activated hydrogen by collisions with mercury vapor which is activated by λ 2537. It is evident that this rate will be dependent on the mean life τ for the excited atoms, since in order to activate hydrogen they must collide with a hydrogen molecule before losing their excitation by spontaneous emission. By comparing the rates of reaction at different pressure of hydrogen Turner was thus able to determine that the mean life τ must be in the neighborhood of 10^{-7} seconds.

[10] According to Pauli, *Zeitschrift für Physik*, **31**, 765 (1925), the number of stationary states corresponding to the quantum number j which gives the total angular momentum of the atom is $2j + 1$. The $1S$ level has the value $j = 0$ and the $2p_2$ level the value $j = 1$.

[11] Communicated to the National Academy, Jan. 17, 1924.

[12] Turner, *Phys. Rev.*, **23**, 464 (1924).

Determinations of the mean life of atoms in excited states can also be obtained by observing the rate of decay in the intensity of the light emitted by canal rays, as the excited atoms causing these rays travel back of the slit through the cathode. Measurements of this kind were first made by Wien and gave indeed the earliest determinations that we have for values of τ. In the case of mercury excited to the $2p_2$ state, Wien[13] obtains in this manner the value $\tau = 0.98 \times 10^{-7}$ seconds in complete agreement with that calculated above.

220. As another example of the calculation of Einstein's coefficient, we may consider the transition of hydrogen chloride molecules from one rotational state to another. These transitions give rise to lines in the far infra red, the absorption for which have recently been measured by Czerny.[14] A high degree of accuracy can not be claimed for the results, owing to the extreme difficulty of working in the far infra red. The values of Einstein's coefficients that can be calculated, however, are of special interest because of the simplicity of the model, that of a rigid dipole rotator, which can be used for their interpretation.

For these calculations the energy levels involved are both so low that there are an appreciable number of molecules in both states i and j, and the reinforcement of the primary beam in the absorption measurements by induced emission is quite appreciable. Hence equation (367) in its complete form

$$B_{i \to j} = \frac{c}{h\nu N_i\left(1 - e^{-\frac{h\nu}{kT}}\right)} \int a_\nu d\nu \qquad (367)$$

must be used.

Czerny's data permits the application of this equation to lines Nos. 8, 9, and 10 in the pure rotational spectrum of hydrogen chloride in which the absorption induces transitions such that the rotational quantum number m changes from 7.5 to 8.5, 8.5 to 9.5, and 9.5 to 10.5. To determine the number of molecules N_i in the lower quantum state for the different lines, we may take the total concentration of the hydrogen chloride vapor and then calculate with the help of the Maxwell-Boltzmann distribution law as given by equation (186) in Chapter 8 the number in the desired level. The value of the integral $\int a_\nu d\nu$ can be determined graphically from the data presented in Czerny's curves giving the percentage transmission for a tube of given length as a function of wave length. Substituting into equation (367), we obtain for the three transitions in question the values of $B_{i \to j}$ 5.6×10^{16}, 6.8×10^{16} and 10.6×10^{16}

[13] W. Wien, *Ann. der Physik*, 73, 483 (1924).
[14] Czerny, *Zeitschrift für Physik*, **34**, 227 (1925).

everything being expressed in c.g.s. units. These values are subject to considerable uncertainty owing to the difficulties involved in Czerny's work, and to a slight degree owing to doubt as to the correct a-priori probabilities to use in calculating the values of N_i.

The above values are of special interest, since the simple dipole rotator which can be taken as the model involved has made it possible to calculate theoretically for the first time the expected absolute values $B_{i \to j}$ from the correspondence principle. This has been done[15] and the theoretical values of $B_{i \to j}$ 11.9×10^{16}, 11.8×10^{16} and 11.6×10^{16} obtained for the transitions in question. These results are also subject to some uncertainty owing to doubt as to the correct a-priori probabilities and as to the proper value to use for the electric moment of the dipole. The approximate check, however, should be noted.

It is also interesting to calculate, from the above values of $B_{i \to j}$, the values of $A_{j \to i}$ for spontaneous transitions in the opposite direction, and the values for the mean life of the molecules in the upper quantum states. We find, what at first sight must seem the surprising result, that the mean life of the atoms in the upper states involved is of the order of seconds. Thus the mean life of hydrogen chloride in the quantum state $m = 9.5$ comes out 27 seconds as calculated from the experimental value of $B_{i \to j}$ given above. This means of course that transitions between the different rotational states presumably occur in the gas under ordinary conditions much more frequently through collision than through interaction with radiation. There is no theoretical reason why such a condition of affairs should not be the case, since for example in the case of the hydrogen molecule we know from specific heat data that rotational states do exist, but from the homopolar nature of the molecule we are certain that transitions between these states by interaction with radiation will almost never take place.

221. Range of Values for Einstein's Coefficient of Absorption $B_{i \to j}$.—A knowledge of the probable range of values for the coefficient of absorption $B_{i \to j}$ will be of value for our later interpretations of the rate of chemical reactions, since it will permit us to estimate the probability that chemical activation occurs by the absorption of radiation.

By the methods illustrated above, experimental values of $B_{i \to j}$ have been calculated for a considerable number of transitions and found to range all the way from 10^{14} to 2.3×10^{19} c.g.s. units. In addition it is possible theoretically to estimate a probable upper limit for values of $B_{i \to j}$ with the help of the correspondence principle.

Consider the transition between an upper quantum state j and a lower state i. If we have N_j molecules in the upper state, then the rate of

[15] *Phys. Rev.*, **27**, 383 (1926).

spontaneous emission of energy due to the transition in question will have on the basis of the quantum theory the value

$$\frac{dR}{dt} = N_j A_{j \to i} h\nu \qquad (373)$$

On the other hand, on the basis of the classical theory the rate of spontaneous emission of N_j oscillators having the frequency ν would be

$$\frac{dR}{dt} = N_j \left(\frac{16\pi^4 e^2}{3c^3} \right) Q^2 \nu^4 \qquad (374)$$

where e is the charge of the oscillating particle and Q the amplitude.

On the basis of the correspondence principle we should expect an approximate equality between these two expressions provided we take Q as the amplitude of that particular harmonic in the internal motion of the molecule whose order numbers $\tau_1 \ldots . \tau_s$ are the same as the changes in quantum number $\Delta n_1 \ldots . \Delta n_s$ which characterize the quantum transition. Equating the two expressions and solving for $A_{j \to i}$ we obtain

$$A_{j \to i} = \frac{16\pi^4 e^2}{3hc^3} Q^2 \nu^3 \qquad (375)$$

or introducing in accordance with equation (357) the relation

$$A_{j \to i} = \frac{8\pi h \nu^3}{c^3} B_{i \to j} \qquad (376)$$

where for simplicity we have considered the system as non-degenerate so that the a-priori probabilities are unity, we can obtain

$$B_{i \to j} = \frac{2\pi^3 e^2}{3h^2} Q^2 \qquad (377)$$

The correct manner of calculating Q has not been completely specified, nevertheless we may expect that the upper limit for this quantity will be of the order of molecular dimensions. If then we take Q as 3×10^{-8} cm. and e as the charge of the electron we obtain

$$B_{i \to j} = 10^{20} \text{ c.g.s. units} \qquad (378)$$

and should not expect to encounter values of this quantity greatly in excess of this amount. We shall find a later application for this important though approximate conclusion.

Chapter 17

Collisions of the First and Second Kinds

222. **Introduction.**—The principle of microscopic reversibility, discussed in Chapter 15, makes it possible to draw some interesting and important conclusions as to the probability of collisions in which molecules are transferred from one quantum state to another. These conclusions were first obtained by Klein and Rosseland[1] by a method of reasoning similar to but not quite identical with that which we shall present.

In connection with the phenomena to be discussed it has become customary to employ the terms elastic collisions, and non-elastic collisions of the first and second kinds. By an *elastic collision* we shall mean one in which the total kinetic energy of the two colliding molecules is not changed as a result of the impact. By a *collision of the first kind* we shall understand one in which all or a portion of the relative kinetic energy of the molecules before their impact is used in raising one of the molecules from a lower to a higher quantum state. Such collisions are thus somewhat similar to the inelastic collisions of everyday mechanics. By a *collision of the second kind* we shall understand one in which one of the molecules is at the start in an activated or higher quantum state and falls to a lower state as a result of the impact, transferring all or a portion of the energy of activation into the kinetic form. Such collisions are hence "superelastic."

223. The simplest examples of these three kinds of collisions are furnished by the interaction of atoms with moving electrons.

If we have a tube containing a dilute gas, for example mercury vapor, and provided with a source of electrons such as a hot filament, it is well known that for low velocities of the electrons only *elastic collisions* will be made with the mercury atoms, the kinetic energy of the electrons being practically unchanged by the impacts, since the mercury atoms are so heavy compared with the electrons as to suffer only a negligible change in kinetic energy.[2]

[1] Klein and Rosseland, *Zeitschrift für Physik,* 4, 46 (1921). See also Fowler, *Phil. Mag.,* 47, 257 (1924).

[2] In the case of collisions between electrons and lighter atoms such as helium it sometimes becomes necessary to consider the small kinetic energy transferred to the helium atom even in an elastic collision.

By accelerating the electrons, however, through a proper potential drop they can be given sufficient kinetic energy so that on impact with the mercury atoms they can transfer the necessary energy to the mercury atom to raise the latter to a higher quantum state, and collisions of the *first kind* will then in general take place. In the case of mercury, these occur when the electrons have received the kinetic energy corresponding to a potential drop of about 4.7 volts which is sufficient to knock the atom from the normal $1S$ to the $2p_3$ state, which is the next lowest quantum state for this atom. Processes of this nature are the simplest examples of collisions of the first kind. They provide as a matter of fact the most unequivocal evidence we have as to the real existence of quantized energy levels in atoms.

The existence of these collisions of the first kind between electrons and atoms, however, also necessitates the existence of collisions of the *second kind,* since in accordance with the principle of microscopic reversibility we must expect the possibility of a reversal of the foregoing process of a nature such that excited atoms, coming in contact with slow moving electrons, drop back to a lower quantum state and give their energy of excitation up to the electrons in the form of increased velocity.

224. Collisions of these three kinds also undoubtedly occur between pairs of atoms or molecules as well as between an atom and an electron. Most of the collisions between atoms and molecules are presumably elastic, not unlike those usually considered in the older kinetic theory, but very clear cut examples of collisions of the second kind between atoms have been found; and their existence evidently guarantees, by the principle of microscopic reversibility, the occurrence of collisions of the first kind. The unequivocal demonstration of collisions of the second kind between atoms was first obtained by Cario,[8] who showed that mercury atoms raised to the $2p_2$ state by the absorption of λ 2537 could transfer a portion of their energy of excitation to thallium atoms putting the latter into a condition to emit lines of the thallium spectrum, the excess energy of the mercury atom being transferred into the kinetic form. The work of Cario also actually gave some direct evidence of the existence of collisions of the first kind between mercury and thallium atoms, since in some cases the thallium atoms were raised to quantum states of slightly higher energy than that of the excited mercury atom. This means of course that the kinetic energy of the atoms had to be called upon to supply some of the energy.

225. Probability of a Collision of the First Kind between an Atom and a Fast Electron.—We are now ready to deduce the Klein

[8] Cario, *Zeitschr. für Phys.*, **10**, 185 (1922).

and Rosseland relation between the probabilities for the occurrence of collisions of the first and second kinds between atoms and electrons. Consider a gaseous system of unit volume containing under *equilibrium* conditions N_e electrons and N_i atoms in a lower quantum state i, which can be raised to a higher quantum state j by collisions of the first kind with the electrons. Since the atoms are very heavy compared with the electrons, we can regard them as fixed and consider only the velocities of the electrons as determining the chance of collision and the amount of energy available.

In order for an electron to be able to produce a collision of the first kind, it must have at least the kinetic energy ε_{ij} which is necessary to raise the atom from the state i to the state j, and the probability of a successful collision will evidently be some function of the excess energy which it has over this amount. Let us then consider the probability of collisions of the first kind by electrons lying in the energy range from $(\varepsilon_{ij} + \varepsilon)$ to $(\varepsilon_{ij} + \varepsilon + d\varepsilon)$. The number of such electrons present in the system, from Maxwell's distribution law (see equation (94) Chapter 5), will evidently be

$$dN_e = N_e a e^{-\frac{\varepsilon_{ij} + \varepsilon}{kT}} \sqrt{\varepsilon_{ij} + \varepsilon}\, d\varepsilon \tag{379}$$

where a is a parameter depending on the temperature. The chance that one of these electrons will make *any* collision with an atom in the state i will evidently be proportional to the number of electrons and atoms, to the cross section of the atoms, and to the velocity $\sqrt{2(\varepsilon_{ij} + \varepsilon)/m}$ with which the electrons are moving. Hence for the number of collisions of the first kind occurring per second in which these electrons induce the change from the state i to j, we may evidently write

$$Z_{+i+j} = P_{+i+j} N_{+i} N_e a e^{-\frac{\varepsilon_{ij} + \varepsilon}{kT}} \sqrt{\frac{2}{m}} (\varepsilon_{ij} + \varepsilon) d\varepsilon \tag{380}$$

where the probability factor P_{+i+j} may be thought of as the product of the cross section of the atom multiplied by the chance that a collision will be effective. This quantity P_{+i+j} is evidently a function of the energy of the electron $(\varepsilon_{ij} + \varepsilon)$. (The plus signs in the subscripts have been introduced since we shall desire to distinguish between the states i and j of an atom and the reverse states $-i$ and $-j$.)

226. Probability of a Collision of the Second Kind between an Atom and a Slow Electron.—Let us now consider the reverse of the foregoing process, such that an electron with kinetic energy in the range ε to $\varepsilon + d\varepsilon$ collides with an excited atom in the reverse state, $-j$, and through a collision of the second kind receives the energy ε_{ij}, the atom falling to the reverse of the previous original state, $-i$. Evidently for the number of such processes taking place in unit time we can write in analogy to equation (380) the expression

$$Z_{-j,-i} = P_{-j-i} N_{-j} N_{\epsilon} a e^{-\frac{\epsilon}{kT}} \sqrt{\frac{2}{m}}\, \epsilon d\epsilon \tag{381}$$

where P_{-j-i} is the probability factor for the reverse change.

227. Relation between the Two Probabilities.—In accordance now with the principle of microscopic reversibility these two rates, which are for processes that are the reverse of each other, must be equal, so that by combining equations (380) and (381) and cancelling common factors, we can obtain

$$P_{+i+j} N_{+i}(\epsilon_{ij} + \epsilon) = P_{-j-i} N_{-j} \epsilon\, e^{\frac{\epsilon_{ij}}{kT}} \tag{382}$$

This equation connecting the probabilities of the two kinds of collision will be unfamiliar, since it connects the probability of transition from state i to j, not as is usual with the probability of transition from state j to i, but with the probability of transition from the reverse state $-j$ to the reverse state $-i$. The equation is, however, the most immediate conclusion that we can draw from the principle of microscopic reversibility stated in its exact form. Moreover, the method of derivation avoids the difficulty, correctly raised by Fowler (*l. c.*) as to the justification in the usual derivation of equating the number of collisions in which molecules go from state $+i$ to $+j$ with the number in which molecules go from $+j$ to $+i$, since in the case of atoms with more than two levels, say i, j, k the equilibrium might be maintained by cyclical processes of the type $i \rightarrow j \rightarrow k \rightarrow i$.

228. If now we desire to obtain an expression which makes no distinction between reverse states, we may proceed as was done in deriving Corollary *a* to the principle of microscopic reversibility, by noting that an expression similar to equation (382) must evidently hold for the pair of transitions, $-i$ to $-j$ and j to i, and indeed also for the pairs $-i$ to j, $-j$ to i and i to $-j$, j to $-i$ if these transitions occur. Hence by adding together these similar expressions we can obtain the equation

$$P_{\pm i \pm j} N_{\pm i}(\epsilon_{ij} + \epsilon) = P_{\pm j \pm i} N_{\pm j} \epsilon e^{\frac{\epsilon_{ij}}{kT}} \tag{383}$$

where the symbolism is evident, $P_{\pm i \pm j}$, for example, being the total probability of a collision of the first kind in which an atom goes from either of the states i or $-i$ to either of the states j or $-j$. Nothing, however, will now be gained by retaining the complicated symbolism so that we may rewrite equation (383) in the usual form

$$P_{ij}N_i(\varepsilon_{ij}+\varepsilon) = P_{ji}N_j\varepsilon e^{\frac{\varepsilon_{ij}}{kT}} \tag{384}$$

This expression can be further simplified by substituting for the ratio N_j over N_i, which in accordance with the Maxwell-Boltzmann distribution law has the value $(p_j/p_i)e^{-\frac{\varepsilon_{ij}}{kT}}$ giving us as our final expression

$$p_iP_{ij}(\varepsilon_{ij}+\varepsilon) = p_jP_{ji}\varepsilon \tag{385}$$

where p_i and p_j are the a-priori probabilities of the two states.

This equation connects in a simple manner the probability P_{ij} that impact between an atom in state i and an electron with kinetic energy $\varepsilon_{ij}+\varepsilon$ will result in a collision of the first kind, with the probability P_{ji} that impact between an atom in state j and an electron with kinetic energy ε will result in a collision of the second kind. The equation has sometimes been called a necessary consequence of thermodynamic reasoning. Using the terms in their English sense, however, it is evident that the equation is to be regarded as the result of statistical rather than thermodynamic theory.

229. The equation was deduced on the assumption of atoms which could be regarded as stationary compared with the electrons. It can easily be shown, however, as has been done by Fowler (*l. c.*), that the same equation will hold when we remove this restriction, providing we understand by $(\varepsilon_{ij}+\varepsilon)$ and ε the mutual kinetic energy of the atom and electron due to their motion relative to their center of gravity. This arises from the fact that the number of collisions between molecules in which the mutual kinetic energy lies in the range $d\varepsilon$, has the same form as equations (380) and (381). (See equation 122, Chapter 5.)

230. Experimental Example of a Collision of the First Kind.—At the present time the most satisfactory determination of the probability of a collision of the first kind has been carried out by Dr. George Glockler.[4] The work was done with helium, using the transition between the normal $1S$ state of parhelium and the metastable $2s$ state of orthohelium, which occurs when the impinging electrons are given the velocity corresponding to a potential drop of 19.77 volts. It was definitely shown by this work, as had been suspected before, that successful collisions of the first kind can occur only when the electrons have very close to the exact energy (19.77 volts) necessary for the transition. The electron stream used in the experiments came from a hot filament and

[4] To be published in the *Physical Review*.

hence there was the usual temperature distribution in the energies of the electrons. Treating the electrons, however, as divided into energy groups each of 0.1 volt range, it was possible to show that all the successful collisions of the first kind occurred with electrons having energies in the lowest possible range from 19.77 to 19.87 volts, electrons of greater energy causing within the limits of error no transitions to the higher state at all. It was also possible to show that even for the electrons in this range approximately 98.9 per cent of the actual collisions are elastic, only 1.1 per cent leading to transition to the higher state. This may indicate that only those electrons which have even much closer values than 0.1 volt to the right energy can produce the transition.

231. By making use of the relation which we have derived between the probabilities for collisions of the first and second kind, it is also possible to obtain from the above facts, some information as to the probability of collisions of the second kind. Referring to equation (385), it is evident, since collisions of the first kind only occur within a range of energies 0.1 volt greater than the resonance energy, that collisions of the second kind will only occur with very slow electrons, in the range between zero and 0.1 volt. This conclusion, however, does not seem entirely surprising. An electron with the energy corresponding to 0.1 volt is moving with a velocity of 1.9×10^7 centimeters per second and hence would remain in the neighborhood of the excited atom only for a very short time. If we assume that the electron in the excited orbit is making say 10^7 revolutions per second, the impinging electron would move 1.9 centimeters even in the short time taken by one revolution, and hence if very close or prolonged interaction between the two electrons should be necessary for deactivation, we see that there would be little chance for such action even when the impinging electron has as low a velocity as that due to a potential drop of 0.1 volt. This explanation is of course by no means certain or convincing, but has some interest.

232. We may also draw a conclusion as to the functional relation between the probability of a collision of the second kind and the energy of the impinging electron. Solving equation (385) for P_{ij} we obtain

$$P_{ij} = \frac{p_j}{p_i} \frac{\varepsilon}{\varepsilon_{ij} + \varepsilon} P_{ji} \qquad (386)$$

which within the range of ε of 0.1 volt can be written in the form

$$P_{ij} = \frac{p_j}{p_i} \frac{\varepsilon}{\varepsilon_{ij}} P_{ji} \qquad (387)$$

The experimental facts, however, show us that the probability of a collision of the first kind P_{ij} reaches its maximum value when ε is less than 0.1 volt, so that we must conclude that within this range the probability of a collision of the second kind P_{ji} must decrease more rapidly than inversely as the first power of the energy ε. This conclusion which seems inevitable is surprising, since it might have seemed natural to expect that the probability would decrease inversely as the velocity or one half power of the energy. Until we are able to get evidence as to the exact form of the functional relation between P_{ij} and ε, further conclusions can hardly be drawn.

233. Probability of Ionization by Electron Impact.—It is also possible to use similar methods to the foregoing to obtain information as to the relation between the probabilities of ionization and recapture by electron impact, another case of collisions of the first and second kinds first treated by Fowler (*l. c.*). Consider a system of unit volume under equilibrium conditions containing N_e electrons and N_i atoms in a quantum state i, which are capable of being ionized by electron collision. If ε_∞ is the energy necessary for ionization, we may think of the ionization process as commencing with an atom in the state i and an electron having the energy $\varepsilon_\infty + \varepsilon_1$ and ending with the atom in the ionized state, the colliding electron having dropped in energy content from $\varepsilon_\infty + \varepsilon_1$ to some new value ε_2, and the dislodged electron going off with the energy ε_3, where

$$\varepsilon_1 = \varepsilon_2 + \varepsilon_3 \tag{388}$$

(We thus rule out the conceivable case that part of the energy goes off in the form of radiation.)

For the number of electrons in the range $\varepsilon_\infty + \varepsilon_1$ to $\varepsilon_\infty + \varepsilon_1 + d\varepsilon_1$ we may write

$$dN_e = N_e a e^{-\frac{\varepsilon_\infty + \varepsilon_1}{kT}} \sqrt{\varepsilon_\infty + \varepsilon_1}\, d\varepsilon_1 \tag{389}$$

and since their velocity is $\sqrt{2(\varepsilon_\infty + \varepsilon_1)/m}$ we may write for the number of collisions per second of the nature described above, the expression

$$Z_{+i+\infty} = P_{+i+\infty} d\varepsilon_2 N_{+i} N_e a e^{-\frac{\varepsilon_\infty + \varepsilon_1}{kT}} \sqrt{\frac{2}{m}} (\varepsilon_\infty + \varepsilon_1) d\varepsilon_1 \tag{390}$$

where the probability factor $P_{+i+\infty}$ may be thought of as the product of the cross section of the atom multiplied by the chance that a collision will be successful and of such a nature that the impinging electron drops in energy from the range $d\varepsilon_1$ to the range $d\varepsilon_2$. The quantity $P_{+i+\infty}$ is evidently a function both of the energy of the impinging electron $(\varepsilon_\infty + \varepsilon_1)$ before it collides and its energy ε_2 after the completion of the collision.

234. Probability of Electron Recapture by Triple Collision.—Let us now consider the reverse of the foregoing process, which will lead to the recapture of an electron by an ionized atom. Three bodies will evidently be involved in such a recapture, an ionized atom in the state $-\infty$ which is the reverse of that produced by the previous process, an electron with energy $\varepsilon_3 = \varepsilon_1 - \varepsilon_2$ which will be recaptured, and an electron with energy ε_2 which will assist in the recapture by carrying off the excess energy. The two electrons in this reverse process will of course be moving at the start towards the atom instead of away from it, as at the end of the ionizing collision.

Since the chance that an ion and the electron to be bound are located together within any chosen small volume is proportional to the product of the numbers of each present, and the chance that such a pair will be hit by a second electron is proportional to the number of such electrons and their velocity, we may write for the number of reverse processes the expression

$$Z_{-\infty-i} = P_{-\infty-i}\, N_{-\infty} N_e a e^{-\frac{\varepsilon_3}{kT}} \sqrt{\varepsilon_3}\, d\varepsilon_3 N_e a e^{-\frac{\varepsilon_2}{kT}} \sqrt{\frac{2}{m}}\, \varepsilon_2 d\varepsilon_2 \qquad (391)$$

where the proportionality factor $P_{-\infty-i}$ may be regarded as the product of the volume of the sphere of influence within which the ion and electron to be captured must be located multiplied by the cross section of this sphere and by the chance that a collision of the impinging electron with this cross section will lead to recapture.

235. Relation between the Two Probabilities.—The numbers of the two kinds of processes occurring per second as given by equations (390) and (391) must, however, be equal in accordance with the principle of microscopic reversibility, so that by cancelling equal factors and introducing equation (388), we can obtain

$$P_{+i+\infty}(\varepsilon_\infty + \varepsilon_1) = P_{-\infty-i}\, \varepsilon_2 \sqrt{\varepsilon_3}\, \frac{N_{-\infty} N_e}{N_{+i}}\, a e^{\frac{\varepsilon_\infty}{kT}} \qquad (392)$$

connecting the probabilities of an ionizing collision and its reverse recapture.

Since a similar equation will evidently hold for the pair of transitions $-i$ to $-\infty$, ∞ to i, and indeed also for the pairs $-i$ to ∞, $-\infty$ to i, and i to $-\infty$, ∞ to $-i$, if these transitions exist, we can obtain by combination an expression

$$P_{i\infty}(\varepsilon_\infty + \varepsilon_1) = P_{\infty i}\, \varepsilon_2 \sqrt{\varepsilon_3}\, \frac{N_\infty N_e}{N_i}\, a e^{\frac{\varepsilon_\infty}{kT}} \qquad (393)$$

which neglects the distinction between reverse states.

This equation can be still further simplified. In accordance with the Maxwell distribution law, as given by equation (94) Chapter 5, we have for a the value

$$a = \frac{2}{\sqrt{\pi}\,(kT)^{3/2}} \tag{394}$$

while the equilibrium constant $N_\infty N_e/N_\iota$ for the dissociation of an atom into ion and electron has in accordance with equation (318) Chapter 14 the value

$$\frac{N_\infty N_e}{N_\iota} = \frac{(2\pi mkT)^{3/2}\,\sigma}{h^3}\,e^{-\frac{\varepsilon_\infty}{kT}} \tag{395}$$

so that by substituting into equation (393) we obtain

$$P_{\iota\infty}(\varepsilon_\infty + \varepsilon_1) = C\,P_{\infty\iota}\,\varepsilon_2\,\sqrt{\varepsilon_3} \tag{396}$$

where the constant factor C has the value

$$C = \frac{4\sqrt{2}\,\pi\, m^{3/2}\,\sigma}{h^3} \tag{397}$$

The above equation connects in a simple manner the probability factors for ionization by collision with a fast electron and recapture by the intervention of a slow electron. The factor $P_{\iota\infty}$ is evidently a function of the energy of the impinging electron $(\varepsilon_\infty + \varepsilon_1)$ and also a function of the energy of this electron ε_2 after the impact, while $P_{\infty\iota}$ is a function of ε_2 the energy of the electron which assists in the recapture and of $\varepsilon_3 = \varepsilon_1 - \varepsilon_2$ the energy of the electron which is to be bound.

236. Experimental Values of Ionization Efficiency.—As in the case of the similar equation connecting the probabilities of collisions of the first and second kinds between quantized states, we may try to use equation (396) to assist in the interpretation of the experimentally determined values for the chance of ionization by collision between an atom and a fast electron.

Solving equation (396) for $P_{\iota\infty}$ and substituting the value for ε_3 given by (388) we obtain

$$P_{\iota\infty} = \frac{C}{\varepsilon_\infty + \varepsilon_1}\,P_{\infty\iota}\,\varepsilon_2\,\sqrt{\varepsilon_1 - \varepsilon_2} \tag{398}$$

Now $P_{\iota\infty}\,d\varepsilon_2$ gives in accordance with our treatment the chance of ionization by collision with an electron of energy $(\varepsilon_\infty + \varepsilon_1)$ provided this impinging electron drops to the energy interval $d\varepsilon_2$. Hence integrating over all possible values of ε_2 from zero to ε_1, we get an expression for the total probability of ionization by an impacting electron of energy $(\varepsilon_\infty + \varepsilon_1)$, giving us

$$\int_0^{\varepsilon_1} P_{i\infty}\, d\varepsilon_2 = \frac{C}{\varepsilon_\infty + \varepsilon_1} \int_0^{\varepsilon_1} P_{\infty i}\, \varepsilon_2 \sqrt{\varepsilon_1 - \varepsilon_2}\; d\varepsilon_2 \qquad (399)$$

To try to simplify the integral on the right hand side of this equation, we recall the conclusion which could be obtained from Glockler's work that an impinging electron can lower the quantum state of an atom only when it is moving extremely slowly. Such a lowering of quantum state, however, is somewhat similar to the lowering from an unquantized orbit to a quantized state that we are now considering. Hence if we assume that $P_{\infty i}$ will have appreciable values only when the velocity and hence energy ε_2 of the impinging electron is extremely small, we see that for all except very small values of ε_1 we can rewrite equation (399) for the total probability of ionization by collision with an electron of energy $(\varepsilon_\infty + \varepsilon_1)$ in the form

$$\int_0^{\varepsilon_1} P_{i\infty}\, d\varepsilon_2 = \frac{C\sqrt{\varepsilon_1}}{\varepsilon_\infty + \varepsilon_1} \int_0^{\varepsilon_1} P_{\infty i}\, \varepsilon_2 d\varepsilon_2 \qquad (400)$$

237. Numerous measurements of the probability of ionization by electron impact have been made, and in particular the corrected values given by K. T. Compton and Van Voorhis [5] seem reliable. They show in general for a considerable number of substances that the probability of ionization rises from zero when $\varepsilon_1 = 0$ to a maximum when ε_1 becomes several times as large as ε_∞. This indicates that the value of the integral on the right-hand side of equation (400) must increase with ε_1, since the factor before the integral has its maximum at $\varepsilon_1 = \varepsilon_\infty$. The quantity behind the integral is a function of the energy of the impinging electron ε_2 and of the energy of the electron to be formed $\varepsilon_3 = \varepsilon_1 - \varepsilon_2$. Since we have reason to believe that $P_{\infty i}$ falls off very rapidly with ε_2, it may be that the increase of the integral with ε_1, indicates that electrons with high energies $\varepsilon_3 = \varepsilon_1 - \varepsilon_2$ are the most easily bound, provided they are within the assumed sphere of influence. The complicated form of the relation makes it difficult, however, to draw very satisfactory conclusions.[6]

[5] K. T. Compton and Van Voorhis, *Phys. Rev.*, **27**, 724 (1926). The attempted application of equation (400) assumes of course that no radiation is given off in the actual process of ionization by electron impact.

[6] Since the above was written, new and apparently very precise measurements of the efficiency of ionization have appeared by E. O. Lawrence, *Phys. Rev.*, **28**, 947 (1926). If the step-like rise in efficiency found by him proves to be correct, the interpretation with the help of equation 399 will have to be altered.

238. To conclude the chapter, we desire to emphasize the great importance of the qualitative conclusion that the existence of collisions of the first kind necessitates the existence of those of the second kind and vice versa. As yet the quantitative forms of the relations between the probabilities for the two kinds of collisions have not been greatly used, but this will undoubtedly come in the future, and in the meantime all our thinking in the field of chemical gas reactions has been greatly modified by the new possibilities.

Chapter 18

Introduction to the Velocity of Physical Chemical Processes. Boltzmann's H-Theorem

239. The Approach Towards the Most Probable Distribution.—In our introductory discussion of statistical mechanics, special attention was called to the fact that this science is superior to thermodynamics in leading to theoretical predictions as to the rate at which physical-chemical processes proceed, a question almost untouched by thermodynamics, and we shall now devote a considerable portion of the remainder of the book to a discussion of the velocity of such changes.

The most general prediction furnished by statistical mechanics as to the nature of the physical-chemical changes which will take place in a system which has not already come to equilibrium, is that the direction of the change will be towards a more probable distribution of the molecules. Thus for example in the case of a gas in which the molecules do not have the positions, velocities and internal conditions corresponding to the Maxwell-Boltzmann distribution law, we may expect in general a readjustment of the configuration in such a way as to conform to the Maxwell-Boltzmann distribution.

The justification for this prediction may be seen by considering the motion of the representative point for the gas in the generalized phase space (γ-space) which may be used for plotting the values of the coordinates and momenta which determine the state of the gas. Nearly the whole of the volume of this phase space corresponds to configurations in which the molecules have approximately the most probable configuration, and a very rapidly decreasing fraction of the space to the configurations that differ increasingly from this most probable or Maxwell-Boltzmann condition. Furthermore, in accordance with Liouville's theorem the representative point has no tendency to linger in any particular region of the phase space, and in accordance with the ergodic hypothesis may be regarded as travelling on a continuous path which visits substantially all regions of the phase space in succession before returning to its starting point. Hence it is evident that when the representative point spontaneously leaves the regions of the phase space which correspond to the Maxwell-Boltzmann configuration it will remain outside only for a time which is short compared with that in which the Maxwell-Boltzmann law is obeyed, and furthermore that a system arbitrarily started in a state where it is not in equilibrium when left to itself will in general move in the direction of configurations in which the Maxwell-Boltzmann distribution prevails.

The above conclusion is analogous to the thermodynamic conclusion that systems tend to readjust themselves in the direction of increasing entropy, and it is apparent that the statistical mechanical state of maximum probability corresponds to the thermodynamic state of maximum entropy. It should be noted, however, that the statistical mechanical state of maximum probability specifies the values of *microscopic* variables belonging to the individual molecules of the system, while the state of maximum entropy merely specifies the values of *macroscopic* variables belonging to the system as a whole, so that even on the basis of these partial considerations we obtain from statistical mechanics a more complete account of the result of a physical-chemical change than we do from thermodynamics.

Nevertheless, neither the thermodynamic conclusion that systems tend to change in the direction of increasing entropy, nor the statistical mechanical conclusion that they move in the direction of the Maxwell-Boltzmann configuration, provide information as to the velocity or real nature of the process by which physical-chemical changes take place, and we must now proceed to our main business of studying the actual rate and internal mechanism by which systems do approach the statistical mechanical state of maximum probability

Our first endeavor in this direction will be a presentation of Boltzmann's famous H-theorem, which gives us an expression for the rate at which a gas approaches the Maxwell-Boltzmann distribution owing to the result of the collisions between its molecules. The discussion will not, as a matter of fact, lead to numerical results, since the expression for the rate of approach contains coefficients whose experimental value is not known. Nevertheless, the investigation is one of such fundamental theoretical interest as to form a natural approach to our further discussions of rates of physical chemical change.

240. Definition of the Function *H*.—Consider a system of molecules, containing a large number of molecules of each of the kinds present. Let us consider the molecules of each kind as uniformly distributed throughout the space available, and let us specify the possible states of these molecules, as in Section **198**, Chapter 15, by assigning to each state a small finite region in the phase space, which gives the range

$$d\omega = dq_4 \dots dq_n \, dp_1 \dots . dp_n \qquad (401)$$

within which the internal coordinates $q_4 \ldots\ldots q_n$ and external and internal momenta of the molecule $p_1 \ldots\ldots p_n$ shall fall. Let N_1, N_2, $N_3 \ldots\ldots$ be the numbers of molecules in the different possible states, assumed by the different kinds, and let us define a function H depending on the way the molecules are distributed among the different states by the equation

$$H = N_1 \log N_1 + N_2 \log N_2 + N_3 \log N_3 + \ldots \tag{402}$$

241. Our problem now will be to calculate the *probable* rate with which H will change with the time, if the distribution of molecules in the different states has not already reached the steady condition determined by the Maxwell-Boltzmann law. Differentiating equation (402) with respect to the time, we obtain

$$\frac{dH}{dt} = \frac{dN_1}{dt}\log N_1 + \frac{dN_2}{dt}\log N_2 + \frac{dN_3}{dt}\log N_3 + \ldots\ldots$$
$$+ \frac{dN_1}{dt} + \frac{dN_2}{dt} + \frac{dN_3}{dt} + \ldots\ldots \tag{403}$$

or since the total number of molecules of each kind is taken as constant, we obtain

$$\frac{dH}{dt} = \frac{dN_1}{dt}\log N_1 + \frac{dN_2}{dt}\log N_2 + \frac{dN_3}{dt}\log N_3 + \ldots\ldots \tag{404}$$

242. Derivation of the *H*-Theorem.—Let us now investigate the effect of collisions in producing changes in the distribution N_1, N_2, $N_3 \ldots\ldots$ and hence changes in the value of H. Consider some particular collision, $\begin{pmatrix} b, & a \\ c, & d \end{pmatrix}$ in which a molecule in state a collides with one in state b to produce molecules in states c and d, and let us put the probability that such a collision will occur proportional to the number of molecules of the two kinds present, so that we can write for the *probable* rate of decrease of molecules in state a due to this particular collision, an expression of the form

$$-\frac{\partial N_a}{\partial t} = C^{b,\,a}_{c,\,d}\, N_a N_b \tag{405}$$

where $C^{b,\,a}_{c,\,d}$ is the probability coefficient for the collision and the sign of partial differentiation has been introduced to remind us that the number of molecules in state a can be changed by other kinds of collisions besides this particular one.

Since the above collision destroys molecules in states a and b and produces them in states c and d, we see by referring to equation (404) that the occurrence of such a collision will evidently contribute four terms to the value of $\frac{dH}{dt}$ and we may write for the part of $\frac{dH}{dt}$ due to this collision the expression

$$\left(\frac{\partial H}{\partial t}\right)_{\substack{b,\ a\\ c,\ d}} = C^{b,\ a}_{c,\ d}\, N_aN_b(\log N_c + \log N_d - \log N_a - \log N_b)$$
$$= C^{b,\ a}_{c,\ d} \log\left(\frac{N_cN_d}{N_aN_b}\right)^{N_aN_b} \qquad (406)$$

Proceeding next to the corresponding collision $\binom{d,\ c}{e,\ f}$ we shall obtain the contribution

$$\left(\frac{\partial H}{\partial t}\right)_{\substack{d,\ c\\ e,\ f}} = C^{d,\ c}_{e,\ f} \log\left(\frac{N_eN_f}{N_cN_d}\right)^{N_cN_d} \qquad (407)$$

and by adding together the contributions made by all possible collisions we shall obtain a total value for the rate of change of H produced by collisions.

In Chapter 15, however, we have shown in Section **196** that the corresponding collisions $\binom{b,\ a}{c,\ d}$, $\binom{d,\ c}{e,\ f}$, $\binom{f,\ e}{g,\ h}$, etc., form a closed cycle which will end with some term $\binom{n,\ m}{a,\ b}$ such that the molecules rearrive in their initial states a and b. Furthermore in accordance with equation (338), Section **198,** the probability coefficients $C^{b,\ a}_{c,\ d}$, $C^{d,\ c}_{e,\ f}$, etc., for such a cycle of collisions all have the same value. Hence referring again to equations (406) and (407), it is evident that we can write as a value for the total rate at which H is changing with the time, due to collisions, the expression

$$\frac{dH}{dt} = \Sigma C^{b,\ a}_{c,\ d} \log\left(\frac{N_cN_d}{N_aN_b}\right)^{N_aN_b}\left(\frac{N_eN_f}{N_cN_d}\right)^{N_cN_d}\cdots\cdots\left(\frac{N_aN_b}{N_mN_n}\right)^{N_mN_n} \qquad (408)$$

where the summation sign is introduced to indicate that we must include all possible such cycles of corresponding collisions.

For simplicity let us use the letters α, β, γ, etc., to represent the products N_aN_b, N_cN_d, N_eN_f, etc., and rewrite equation (408) in the form

$$\frac{dH}{dt} = \Sigma C \log\left(\frac{\beta}{\alpha}\right)^{\alpha}\left(\frac{\gamma}{\beta}\right)^{\beta}\left(\frac{\delta}{\gamma}\right)^{\gamma}\cdots\cdots\left(\frac{\mu}{\lambda}\right)^{\lambda}\left(\frac{\alpha}{\mu}\right)^{\mu}$$
$$= \Sigma\, C \log \alpha^{\mu-\alpha}\,\beta^{\alpha-\beta}\,\gamma^{\beta-\gamma}\,\delta^{\gamma-\delta}\ldots\ldots\mu^{\lambda-\mu} \qquad (409)$$

Now in the series of quantities α, β, γ, etc., there must be some quantity, say γ, which is not greater than either of its two neighbors. Let us then again rewrite the equation in the form

$$\frac{dH}{dt} = \Sigma C \log \left(\frac{\gamma}{\delta}\right)^{\beta-\gamma} \alpha^{\mu-\alpha} \beta^{\alpha-\beta} \delta^{\beta-\delta} \ldots . \mu^{\lambda-\mu} \tag{410}$$

Since, however, $\gamma \leqq \delta$ and $\gamma \leqq \beta$ we must conclude that

$$\left(\frac{\gamma}{\delta}\right)^{\beta-\gamma} \leqq 1 \tag{411}$$

the value unity occurring only if $\gamma = \delta$ or $\gamma = \beta$. We have thus re-expressed our original series by the product of a quantity which cannot be greater than unity and a new series of *just the same form* but one less term. Hence we can proceed taking one term at a time out of the series until we have obtained a product of quantities none of which can be greater than unity.[1] The logarithm of such a product, however, must itself be negative or zero. Hence we may conclude that

$$\frac{dH}{dt} \leqq 0 \tag{412}$$

and H is a quantity which tends to decrease with the time and reach a minimum equilibrium value only when the distribution of molecules in the different states is such that

$$\alpha = \beta = \gamma = \text{etc.}$$

or

$$N_a N_b = N_c N_d = N_e N_f = \text{etc.} \tag{413}$$

243. *H*-Theorem and Maxwell-Boltzmann Distribution Law.— It is immediately evident that the final steady distribution prescribed by equation (413) is in entire agreement with the Maxwell-Boltzmann distribution law previously derived for the steady state. In accordance with that law the numbers of molecules N_a and N_b are determined by their proportionality to the factors $e^{-\frac{\varepsilon_a}{kT}}$ and $e^{-\frac{\varepsilon_b}{kT}}$ where ε_a and ε_b are the energies of the molecules in states a and b respectively. Hence

[1] The reduction of the series will finally leave us with two terms of the form $x^{y-x}\, y^{x-y}$. Of the two quantities there will be one which is not greater than the other. Let us take x as not greater than y, and combine the two terms in the form

$$x^{y-x}\, y^{x-y} = \left(\frac{x}{y}\right)^{y-x}$$

Since $x \leqq y$, this quantity is either unity or a fraction raised to a positive power and hence less than unity.

the product N_aN_b will depend on the proportionality factor $e^{-\frac{\epsilon_a + \epsilon_b}{kT}}$ and the product N_cN_d on the factor $e^{-\frac{\epsilon_c + \epsilon_d}{kT}}$. The sum of the energies $\varepsilon_a + \varepsilon_b$ is, however, equal to the sum $\varepsilon_c + \varepsilon_d$, since the total energy of the two molecules cannot be changed by the collision. Hence the Maxwell-Boltzmann distribution law also leads to equation (413).

Moreover, not only is there such an agreement between the conclusions derived by the two methods of attack, but the H-theorem itself may be used for the derivation of the Maxwell-Boltzmann distribution law, since a minimum value for $H = N_1 \log N_1 + N_2 \log N_2 + N_3 \log N_3 + \ldots.$ will evidently give in accordance with equation (38), Chapter 4, a maximum value to the volume of the phase-space (γ-weight) in which the representative point for the system falls. And this was the principle used to derive the Maxwell-Boltzmann law.

244. Critique of the *H*-Theorem.—Before proceeding we must call attention to the important fact that the expression given by equation (408) for the change of H with the time is only an expression for the *probable* rate of change. This arises from the circumstance that the expressions, which we have used for the number of collisions of different kinds that take place per second, are evidently merely expressions for the probable number of collisions. Hence when H is appreciably displaced from its minimum value, there is a great probability that it will decrease but not a certainty. Similarly when H has reached its minimum equilibrium value, the result of collisions may sometimes lead to a temporary increase in its value.

The above consideration is of great importance in removing misconceptions as to the behavior of molecular systems. In accordance with the ergodic hypothesis the representative point for a system of molecules will in the course of time wander into all regions of the phase space (γ-space) permitted by the energy content of the system, the predominant duration of the system in states according with the Maxwell-Boltzmann distribution law being assured merely by the relatively great fraction of the total phase space which corresponds to such a distribution. To account for the occasional intervals when the system does forsake the Maxwell-Boltzmann distribution, we must evidently permit the occurrence of corresponding occasional increases in the value of H above its minimum. We now see, however, that this is permissible, since the expression we have obtained for the decrease of H with time is merely an expression for the *probable* rate of decrease, and does not rule out occasional increases.

A difficulty sometimes arises in understanding the compatibility of the occasional increases permitted to H with the great probability that

H when displaced from its minimum value will decrease. Thus for example if H_0 is the minimum value and H_1, H_2 and H_3 appreciably higher values, and we permit an occasional increase in H say from H_1 up through H_2 to H_3 and its later return back through H_2 to H_1, it may at first sight be difficult to see why the tendency of the system to pass from H_2 to H_3 on the upward course is not just as great as to pass from H_2 to H_1 on the downward. The clarification of this difficulty is particularly due to P. and T. Ehrenfest,[2] who have pointed out that we must regard the change of H with the time as a steplike process of such a nature that the attainment of any value, appreciably different from H_0, such as H_2, is far more often followed immediately by return to a lower value than by passage on upward to a higher state. Thus the change

$$\begin{array}{ccc} & H_2 & \\ H_1 & & H_1 \end{array} \qquad (414)$$

will occur far more frequently than

$$\begin{array}{ccc} & & H_3 \\ & H_2 & \\ H_1 & & \end{array} \quad \text{or} \quad \begin{array}{ccc} H_3 & & \\ & H_2 & \\ & & H_1 \end{array} \quad \text{or} \quad \begin{array}{ccc} H_2 & & H_3 \\ & H_1 & \end{array} \qquad (415)$$

Hence in the great majority of cases, when the value H_2 is attained, it forms a sharp peak of the trajectory and is immediately followed by return to a lower value. Hence taking any given system with the value H_2, there will be a great probability that H will at once move in the direction of a lower value.

These considerations also eliminate the difficulty in understanding what the behavior of a system would be if at any instant all the velocities should be reversed. If the tendency of H is to decrease in the forward motion, it has sometimes been felt that the tendency in the equally representative reverse system would be for H to increase, and the system to change into more and more improbable configurations. This misapprehension, however, is also removed by the Ehrenfestian considerations, since in the reverse motion as well as in the forward motion high values of H occur in the great majority of cases as peaks on the trajectory, so that also in the reverse motion there is a great tendency for a given system with a high value of H to move in the direction of lower values.

245. Extension of the *H*-Theorem to Include Radiation.—Boltzmann's H-theorem, as deduced above, concerns itself with only one of several processes that may be involved in changing a system of mole-

[2] P. and T. Ehrenfest, *Encyklopädie der mathematischen Wissenschaften*, IV, 32, page 42, Leipzig, Teubner, 1907-1914.

cules, which has not already come to equilibrium, into a more probable state. In the first place, since it considers a system in which the different kinds of molecules present are already uniformly distributed throughout the space available, the total process involved in reaching equilibrium must include diffusion, a matter which will be discussed in the next chapter. In addition, the deduction considers only the effect of molecular collisions in producing equilibrium, and, although the method of proof seems applicable without modification to the newly discovered kinds of collisions which can take place between quantized molecules, it was not possible in Boltzmann's time to give any satisfactory treatment to processes involving the interaction of molecules with radiation. In what follows we shall present an extension of the H-theorem in a form which includes the emission and absorption of radiation by quantized molecules, as one of the processes by which equilibrium is approached. We shall not try to consider more complicated cases, like the Compton effect in which the incident radiation is only partly absorbed by the electron, nor cases of combined collisional and radiational action. The method of treatment to be given may seem a little arbitrary and artificial, but will be of interest.

Let us consider a system of unit volume containing N_1, N_2, N_3, etc., molecules in different quantum states and n_{12}, n_{13}, n_{23}, etc., light quanta of the frequency necessary for raising the molecules by absorption from one quantum state to another as indicated by the subscripts. Furthermore, let us indicate by m_{12}, m_{13}, m_{23}, etc., the number of modes of electromagnetic vibration which on the basis of the classical theory would be the carriers for radiation of the frequency indicated. We shall now define H for the combined system of molecules and quanta by the equation

$$\begin{aligned} H = {} & N_1 \log N_1 + N_2 \log N_2 + N_3 \log N_3 + \dots \\ & + n_{12} \log n_{12} + n_{13} \log n_{13} + n_{23} \log n_{23} + \dots \\ & - (n_{12} + m_{12}) \log (n_{12} + m_{12}) - (n_{13} + m_{13}) \log (n_{13} + m_{13}) \\ & - (n_{23} + m_{23}) \log (n_{23} + m_{23}) - \dots \\ = {} & \sum_i N_i \log N_i + \sum_{ij} n_{ij} \log n_{ij} - \sum_{ij} (n_{ij} + m_{ij}) \log (n_{ij} + m_{ij}) \end{aligned} \tag{416}$$

This definition of H may seem somewhat arbitrary. The terms depending on the numbers of molecules in different states is, however, the same as that in Boltzmann's definition of H, and the choice of the new terms depending on the number of quanta of different kinds present will be justified by the results.

We must now calculate the change of H with the time. We obtain by differentiation

$$\frac{dH}{dt} = \sum_i \frac{dN_i}{dt}\log N_i + \sum_i \frac{dN_i}{dt} + \sum_{ij}\frac{dn_{ij}}{dt}\log n_{ij} + \sum_{ij}\frac{dn_{ij}}{dt}$$
$$- \sum_{ij}\frac{dn_{ij}}{dt}\log(n_{ij}+m_{ij}) - \sum_{ij}\frac{dn_{ij}}{dt}$$

or since the second term on the right hand side is evidently zero we obtain by cancellation

$$\frac{dH}{dt} = \sum_i \frac{dN_i}{dt}\log N_i + \sum_{ij}\frac{dn_{ij}}{dt}\log\frac{n_{ij}}{n_{ij}+m_{ij}} \qquad (417)$$

It is at once evident that molecular collisions will lead to a decrease of H with the time, unless the minimum value has already been reached, since the terms of the form $\frac{dN_i}{dt}\log N_i$ are the same as in the original derivation of the H-theorem and the remaining terms are not affected by collision. Hence we have only to consider the effect of the absorption and emission of radiation on the value of $\frac{dH}{dt}$.

Consider the absorption process by which a molecule goes from state i to j and the associated emission by which it goes from j to i. Using for the rates of these two processes Einstein's expressions as given in Sections **208** and **210** in Chapter 16, we evidently obtain as the net result of these two processes

$$\frac{dN_i}{dt} = -\frac{dN_j}{dt} = \frac{dn_{ij}}{dt} = N_j(B_{ji}u_{ij} + A_{ji}) - N_iB_{ij}u_{ij} \qquad (418)$$

where u_{ij} is the radiation density of the frequency involved, and the reason is evident for setting the net rate $\frac{dN_i}{dt}$ at which molecules drop from j to i equal to the rate of production $\frac{dn_{ij}}{dt}$ of the corresponding quanta.

Comparing equations (417) and (418), it is evident that the absorption and emission process considered will contribute three terms to the rate of change of H with the time, which can be combined in the form

$$\frac{\partial H}{\partial t} = [N_j(B_{ij}u_{ij} + A_{ji}) - N_iB_{ij}u_{ij}]\log\frac{N_in_{ij}}{N_j(n_{ij}+m_{ij})} \qquad (419)$$

The quantity $n_{ij}/(n_{ij}+m_{ij})$ can, however, easily be put into a better form for our purposes. n_{ij} is the number of quanta present, of the correct frequency ν_{ij} to produce the change from i to j, and m_{ij} is the number of modes of electromagnetic vibration which on the classical theory carry energy of this frequency, so that we can write

$$\frac{n_{ij}}{n_{ij}+m_{ij}} = \frac{\frac{u_{ij}}{h\nu_{ij}}\,\delta\nu}{\frac{u_{ij}}{h\nu_{ij}}\,\delta\nu + \frac{8\pi\nu_{ij}^2}{c^3}\,\delta\nu} = \frac{u_{ij}}{u_{ij} + \frac{8\pi h\nu_{ij}^3}{c^3}} \tag{420}$$

or substituting the relations between B_{ij}, B_{ji} and A_{ji} given by equation[*] (357) in Chapter 16, we obtain

$$\frac{n_{ij}}{n_{ij}+m_{ij}} = \frac{B_{ij}u_{ij}}{B_{ji}u_{ij}+A_{ji}} \tag{421}$$

Substituting in (419) and rearranging, we obtain as that part of $\frac{dH}{dt}$ which is due to the processes of absorption and emission under discussion

$$\frac{\partial H}{\partial t} = \log\left(\frac{N_iB_{ij}u_{ij}}{N_j(B_{ji}u_{ij}+A_{ji})}\right)^{[N_j(B_{ji}u_{ij}+A_{ji}) - N_iB_{ij}u_{ij}]} \tag{422}$$

This expression, however, is of the form $\log\,(x/y)^{y-x}$, and this is necessarily negative unless $x = y$, when it becomes zero. (See footnote, page 196.) Hence, since all the possible processes of absorption and emission can be combined in similar pairs, it is evident that the interaction of matter and radiation, as well as the action of collisions, will lead to a probable decrease of H with the time until the minimum value is obtained, and when this occurs we shall have an equality between the rates of emission and absorption of the quanta of any frequency capable of such interaction. We have thus obtained the desired extension to Boltzmann's H-theorem.

246. Combined Deduction of Maxwell-Boltzmann and Planck Distribution Laws.—It is also of interest to verify the fact that the attainment of the minimum value of H will lead to a condition of the system in which we have both the Maxwell-Boltzmann law for the distribution of the molecules in different states and the Planck law for the distribution of the radiation.

In accordance with our equation of definition (416), we have for H the value

$$H = \sum_i N_i \log N_i + \sum_{ij} n_{ij} \log n_{ij} - \sum_{ij}(n_{ij}+m_{ij}) \log\,(n_{ij}+m_{ij}) \tag{423}$$

[*] The a-priori probabilities p_i and p_j must be taken as equal, in agreement with the convention we have adopted in Chapter 15 for assigning equal regions of phase space to different states of a molecule when collisions are to be considered.

and if for simplicity we consider only one kind of molecules[4] we have for the total number

$$N = \sum_i N_i \tag{424}$$

while for the total energy of the system E we evidently have the value

$$E = \sum_i N_i \varepsilon_i + \sum_{ij} n_{ij} h\nu_{ij} \tag{425}$$

To obtain the minimum value of H we may set its variation equal to zero. The variation can not, however, be carried out entirely arbitrarily, since the total energy and total number of molecules must be preserved constant. Hence setting the variations for all three expressions (423), (424), and (425) equal to zero and combining by the familiar method of undetermined multipliers, we may obtain

$$\begin{aligned} 0 = {} & \sum_i \delta N_i \log N_i + \sum_{ij} \delta n_{ij} \log \frac{n_{ij}}{n_{ij} + m_{ij}} \\ & + \lambda \sum_i \delta N_i \\ & + \mu \sum_i \varepsilon_i \delta N_i + \mu \sum_{ij} h\nu_{ij} \delta n_{ij} \end{aligned} \tag{426}$$

where λ and μ are undetermined multipliers. This equation can be re-written in the form

$$\sum_i (\log N_i + \lambda + \mu \varepsilon_i) \delta N_i + \sum_{ij} \left(\log \frac{n_{ij}}{n_{ij} + m_{ij}} + \mu h\nu_{ij} \right) \delta n_{ij} = 0 \tag{427}$$

and since the introduction of the undetermined multipliers has made the variations arbitrary, we must have for each state i and each frequency ν_{ij}, the result

$$\log N_i + \lambda + \mu \varepsilon_i = 0 \quad ; \quad N_i = e^{-\lambda} e^{-\mu \varepsilon_i} \tag{428}$$

$$\log \frac{n_{ij}}{n_{ij} + m_{ij}} + \mu h\nu_{ij} = 0 \quad ; \quad \frac{n_{ij}}{n_{ij} + m_{ij}} = e^{-\mu h\nu_{ij}} \tag{429}$$

Substituting, however, the familiar value $\mu = 1/kT$, and noting that $e^{-\lambda}$ must have such a value that $\Sigma N_i = N$ we easily change equation (428) over into the familiar Maxwell-Boltzmann form

$$N_i = \frac{N e^{-\frac{\varepsilon_i}{kT}}}{\sum_i e^{-\frac{\varepsilon_i}{kT}}} \tag{430}$$

[4] The extension to more than one kind of molecules is simple and obvious.

while, substituting for $n_{ij}/(n_{ij}+m_{ij})$ the expression given by equation (420) and transforming we obtain the familiar form of the Planck distribution law

$$u_{ij}=\frac{8\pi h\nu_{ij}^3}{c^3}\frac{1}{e^{\frac{h\nu_{ij}}{kT}}-1} \tag{431}$$

We thus see that the attainment of the minimum value of H will lead both to the Maxwell-Boltzmann and Planck laws.

As indicated by the subscripts in equation (431) our proof concerns only those frequencies which can be absorbed or emitted by the molecules present, and in an actual system containing only one kind of molecules, we must often expect that the walls will play a considerable part in the attainment of radiation equilibrium.

247. Conclusion.—We cannot conclude this chapter without an expression of admiration for Boltzmann's masterful achievement in discovering and deducing the H-theorem, even though his deserved and honored fame needs from us no words. The significance and originality of inventing a function which should measure the displacement of a system of molecules from statistical equilibrium, and the ingenuity and elegance of the proof that this function will decrease with the time, alike compel attention.

The analogy between increase in the thermodynamic quantity entropy and decrease in the statistical mechanical quantity H is evident. And the attempts to relate the entropy S to the negative of H should not be overlooked. The new statistical mechanical quantity, however, certainly gives the greater insight into the inner mechanism of physical-chemical changes.

The cosmological implications of his theorem were also appreciated by Boltzmann. In whatever part of the whole universe or instant in time we may find ourselves, there is an enormous probability that the value of H will be decreasing with the time unless it has already reached its minimum value, and we shall label forward in time that direction in which H decreases and energy is dissipated, and perhaps look longingly backward to the good old days when there was a greater store of undissipated energy. But in the course of time enormous, however, compared with that which we label æons, the value of H may again increase and prepare for a new cycle of significant events.

In the following chapters we shall discuss more specific cases of the rate at which equilibrium is approached in some selected physical-chemical processes.

Chapter 19

The Rates of Some Physical Processes Transport Problems

248. Introduction.—In the present chapter we shall consider the rates of some important physical processes which are often encountered in physical or chemical experimentation. The processes to be discussed are those which concern the transport of matter, energy, momentum or electricity by molecular systems.

These processes take place of course in systems which are not in a state of equilibrium, and hence are examples of the general tendency of molecular systems to change in the direction of more probable configurations. Since the systems under consideration are not in equilibrium but in a "steady state" of change, the molecules of course do not have their coordinates and momenta distributed according to the Maxwell-Boltzmann law, and the most fundamental method of treating these problems is to investigate the actual distribution which the coordinates and momenta do assume and thus determine the behavior of the system during the change that occurs. Unfortunately this method is usually one of great mathematical complexity and often involves a great deal of information or hypothesis as to the dynamics of molecular interaction. Hence we shall employ instead the methods of the older kinetic theory which treat the system as though there were merely a superimposed disturbance which is transmitted by molecules having the Maxwell-Boltzmann distribution. The effect of the superimposed disturbance in changing the Maxwell-Boltzmann distribution is thus neglected and this sometimes introduces errors which can be shown to be of appreciable magnitude. On the other hand the uncertainty of the physical hypotheses that must be arbitrarily introduced, at our present state of knowledge, rarely justifies the more fundamental but more difficult methods of attack.[1]

Since the material in this chapter is somewhat foreign to the main theme of the book and is introduced mainly for convenience of reference,

[1] As an important example of the use of more fundamental methods see S. Chapman, *Philosophical Trans. of the Royal Society,* **216**, 279 (1916). The more fundamental methods employed in this work seem justified by the agreement that is now obtained between values of molecular diameters calculated from viscosity and from van der Waals *b*.

we shall in some cases content ourselves merely with a statement of results, the deductions for which can be found in the literature. A considerable number of deductions will be included, however, to illustrate the methods. As already noted the results are not always exact.

We shall first consider the transport of matter from one region to another in both of which the molecules may be regarded as sensibly in statistical equilibrium, the two regions being separated by a negligible layer where an abrupt change in properties takes place. (*Effusion, evaporation, solution.*)

We shall then consider the transport of matter, momentum, and energy in a gas under the simplified conditions that the density is low enough so that the collisions between molecules can be neglected compared with collisions with the walls of the apparatus. (*Phenomena of molecular flow.*)

Finally we shall consider the transport of matter, energy, momentum and electricity through regions where there is a continuous variation in conditions as we proceed in the direction of flow, and the concentration is so high that the rate of flow is determined by the collisions of the molecules among themselves. (*Diffusion, thermal conduction, viscosity, electrical conduction as dependent on the phenomenon of mean free path.*)

249. General Equation for the Rate of Transport of Matter Across a Boundary in a Gas at Equilibrium.—If we consider a gas in equilibrium, with the molecular velocities distributed in accordance with Maxwell's law, we can easily calculate the number of molecules which will cross any boundary surface in the gas by considering the average component of molecular velocity perpendicular to this surface. In the case of a gas in equilibrium, this flow will of course be compensated by an exactly equal flow in the opposite direction. But in the case the surface is a thin layer of discontinuity we can often obtain information as to the net rate of transfer of material across this boundary, with the help of such a calculation.

For convenience let us consider a surface of one square centimeter perpendicular to the x-axis. For the number of molecules per cubic centimeter, with component velocities in the range $d\dot{x}\ d\dot{y}\ d\dot{z}$, we have in accordance with Maxwell's law as given by equation (90) in Chapter 5

$$dN = N\left(\frac{m}{2\pi kT}\right)^{3/2} e^{-\frac{m}{2kT}(\dot{x}^2+\dot{y}^2+\dot{z}^2)}\, d\dot{x}d\dot{y}d\dot{z} \tag{432}$$

where N is the total number per cubic centimeter.

Components of velocity in the y and z directions will have no effect on the passage of molecules across the boundary in question, so that a parallelopiped of gas of volume $1 \times 1 \times \dot{x}$ may be regarded as supplying the molecules of velocity $\dot{x}$ which cross the boundary per second.[2]

Hence for the total number of molecules crossing the boundary per square centimeter per second, we may evidently write the integral

$$Z = \int_{-\infty}^{+\infty}\int_{-\infty}^{+\infty}\int_{0}^{+\infty} N\left(\frac{m}{2\pi kT}\right)^{3/2} e^{-\frac{m}{2kT}(\dot{x}^2+\dot{y}^2+\dot{z}^2)}\,\dot{x}\,d\dot{x}\,d\dot{y}\,d\dot{z} \qquad (433)$$

The indicated integrations can, however, easily be performed giving us

$$Z = N\sqrt{\frac{kT}{2\pi m}} = 3{,}638\,N\sqrt{\frac{T}{M}} \qquad (434)$$

where M is the ordinary molecular weight.[3]

Comparing equation (434) with equation (95) in Chapter 5 for the average velocity Ω we see that we can also write the above expression in the form

$$Z = \frac{N}{4}\Omega \qquad (435)$$

a result which was perhaps obvious, since the average component in any direction of a velocity Ω, distributed in random directions, is easily shown to be $\Omega/4$.

Substituting the gas laws in the form

$$N = \frac{p}{kT} \qquad (436)$$

we can obtain

$$Z = p\sqrt{\frac{1}{2\pi mkT}} = 3.535 \times 10^{22}\frac{p}{\sqrt{MT}}\ \text{cm.}^{-2}\ \text{sec.}^{-1} \qquad (437)$$

(p in mm. of mercury)

or for the mass of gas crossing the boundary per second

$$Zm = p\sqrt{\frac{m}{2\pi kT}} = 58.32 \times 10^{-3}\,p\sqrt{\frac{M}{T}}\ \text{gm. cm.}^{-2}\ \text{sec.}^{-1} \qquad (438)$$

(p in mm. of mercury)

These equations will be found quite useful.

[2] Molecules leaving the parallelopiped because of their y and z components of velocity are evidently substituted for by those entering the parallelopiped. In addition collisions occurring within the boundary have no net effect.

[3] A number of numerical factors given in this chapter have been taken from the very useful book of Dushman, "High Vacuum," Schenectady, 1922.

250. Rate of Effusion of Gas through a Small Opening.—If we consider now two samples of gas at different pressures separated by a thin partition with a small opening of area S, *small enough so that we can regard the gas on each side as under substantially equilibrium conditions,* we can calculate the flow through the opening as the net result of the passage of gas in the two directions as given by equation (437), and obtain for the net number of molecules passing per second [4]

$$Z = Z_2 - Z_1 = (p_2 - p_1)S\sqrt{\frac{1}{2\pi mkT}} \qquad (439)$$

If we desire the flow expressed in terms of the volume measured at unit pressure discharged per second, we obtain

$$Q = ZkT = (p_2 - p_1)S\sqrt{\frac{kT}{2\pi m}} = (p_2 - p_1)S\sqrt{\frac{RT}{2\pi M}} \qquad (440)$$

For a circular opening of diameter D, this gives us

$$Q = (p_2 - p_1)\frac{\pi D^2}{4}\sqrt{\frac{RT}{2\pi M}} = 2864\,(p_2 - p_1)D^2\sqrt{\frac{T}{M}} \qquad (441)$$

where D is in centimeters and the voluume Q per second is in cubic centimeters measured at temperature T and unit pressure on the scale used for p_2 and p_1.

These equations of course only hold when the opening is so small that we can regard the gas as substantially under equilibrium conditions on the two sides of the system. This will be true when the diameter of the hole is small compared with the mean free path of the molecules.

251. Thermal Effusion.—The method of calculation can be applied equally well when the gas on the two sides of the partition is at *different temperatures.* We then obtain from equation (437) for the number of molecules passing per second

$$Z = Z_2 - Z_1 = \frac{S}{\sqrt{2\pi mk}}\left(\frac{p_2}{\sqrt{T_2}} - \frac{p_1}{\sqrt{T_1}}\right) \qquad (442)$$

and for zero net rate of flow, we obtain the result

$$\frac{p_2}{p_1} = \frac{\sqrt{T_2}}{\sqrt{T_1}} \qquad (443)$$

If we consider an infinitesimal change in temperature across a boundary

[4] This relation was first discovered and experimentally verified by Knudsen, *Ann. der Physik*, **28**, 999 (1909).

in a tube connecting two reservoirs at different temperatures, we obtain from equation (443)

$$d \log p = \tfrac{1}{2} d \log T \tag{444}$$

and hence by integration obtain the same equation

$$\frac{p_2}{p_1} = \frac{\sqrt{T_2}}{\sqrt{T_1}} \tag{445}$$

for the steady condition of no net flow, even when the temperature drop occurs gradually in a tube connecting two reservoirs at temperatures T_2 and T_1.

This interesting relation was first discovered and tested by Knudsen,[5] and shows that *at pressures low enough so that the mean free path is large compared with the dimensions of the connecting tube,* the steady state of a gas in two connecting reservoirs is not one of uniform pressure throughout, but that the reservoir at higher temperature will also be at the higher pressure. The analogy with the Soret effect in solutions should be noted.

252. Rate of Evaporation.—Our simple equation (434) for the number of molecules reaching a given surface can also be used to investigate the rate of evaporation from a solid or liquid. If we consider a surface of one square centimeter of some material in equilibrium with its vapor, the number of molecules reaching that surface per second from the gas phase will be

$$Z = N\sqrt{\frac{kT}{2\pi m}} = p_0\sqrt{\frac{1}{2\pi mkT}} \tag{446}$$

where p_0, being the pressure of the gas at equilibrium, is the vapor pressure of the material.

If r is the fraction of these molecules which are reflected on striking this surface, and $(1-r)$ the fraction that condense, then it is evident that to maintain equilibrium the rate of evaporation per square centimeter must be

$$Z_{\text{evap.}} = (1-r)p_0\sqrt{\frac{1}{2\pi mkT}} \tag{447}$$

and since we can in general expect the rate of evaporation to be independent of the pressure of vapor, equation (447) can be used directly for the rate of evaporation in to a vacuum.

Experimentally the fraction r of molecules reflected is often found to be very small. Thus in the case of a *clean* surface of mercury

[5] Knudsen, *Ann. der Physik,* **31**, 205 (1910). See also for a more satisfactory demonstration Smoluchowski, *Ann. der Physik,* **33**, 1559 (1910).

Knudsen[6] found that less than one per cent of the impinging molecules were reflected. This makes it possible as a first approximation to neglect the value of r in equation (447) in comparison with unity and use the rate of evaporation into a vacuum to determine the vapor pressure p_0 of highly involatile substances. This has been done by Langmuir[7] and Langmuir and McKay[8] for tungsten, platinum, and molybdenum. The method is a very beautiful and simple one for the solution of problems that could not otherwise be attacked.

Since the current from a hot filament measures the rate of evaporation of electrons from it, equation (447) can also be used to obtain a relation between the value of the saturation current from a filament and the vapor pressure of the electron gas which would be in equilibrium with it at the temperature under consideration. This has made it possible to apply thermodynamic reasoning to the problem of thermal emission[9] and to show that the entropy of electron gas is given by the same formula as for other monatomic gases.[10]

For the *net rate* of evaporation into a vessel containing vapor at pressure p, we should evidently expect from equation (446) to obtain the value

$$Z_{\text{net}} = (1 - r)\sqrt{\frac{1}{2\pi mkT}}\,(p_0 - p) \tag{448}$$

253. Rate of Solution.—The process of solution is analogous to that of evaporation. We have, however, imperfect information as to the velocities of the solute molecules and the nature of the boundary between solid and solution. Nevertheless, we are not surprised to find that the net rate of solution is proportional to the difference between the concentration at saturation C_0 and the instantaneous concentration C in accordance with the equation

$$\frac{dC}{dt} = K(C_0 - C) \tag{449}$$

which is entirely analogous to equation (448) for evaporation. The equation was first tested experimentally by Noyes and Whitney.[11]

254. Transfer of Momentum at Very Low Pressures.—Let us now turn our attention to some problems concerning the behavior of gases at very low pressures, which can be treated quite simply if the density

[6] Knudsen, *Ann. der Physik*, **47**, 697 (1915).
[7] Langmuir, *Phys. Rev.*, **2**, 329 (1913).
[8] Langmuir and McKay, *Phys. Rev.*, **4**, 377 (1914).
[9] See Richardson, "The Emission of Electricity from Hot Bodies," Longmans Green, London, 1921.
[10] See Tolman, *Jour. Amer. Chem. Soc.*, **43**, 1592 (1921).
[11] Noyes and Whitney, *Zeits. f. Physik. Chemie*, **23**, 689 (1897).

is low enough so that the collisions between molecules can be neglected compared with collisions with the walls of the apparatus.

Consider two parallel plates A and B separated by a distance d and immersed in a gas at low enough pressure so that the mean free path is long compared with the separation d. Then if the molecular velocities are assumed to be distributed in accordance with Maxwell's law with an average value Ω, it is evident, since $\frac{\Omega}{2}$ will be the average velocity towards one plate or the other, that a molecule in the space between the plates will make on the average $\Omega/4d$ "round trips" per second between the plates. And if the concentration of molecules per cubic centimeter is N, the total number of round trips per second per square centimeter of plate area will be

$$Z = Nd\frac{\Omega}{4d} = \frac{N\Omega}{4} \tag{450}$$

Suppose now that one of the plates A is in motion in its own plane with the velocity u, then if we assume that each molecule striking A itself acquires a component of velocity u in the direction of motion of the plane, but that otherwise the Maxwell distribution is not disturbed, we shall have for the momentum transferred per square centimeter per second

$$M = \frac{N\Omega}{4}mu \tag{451}$$

and this of course is the force per square centimeter of surface necessary to keep the plane A in its state of steady motion. Hence substituting for Ω its value as given by equation (95), Chapter 5,

$$\Omega = \sqrt{\frac{8kT}{\pi m}} \tag{452}$$

we have for the force per square centimeter necessary to keep the plane A in motion parallel to B, with the velocity u

$$F = \frac{Nmu}{4}\sqrt{\frac{8kT}{\pi m}} = p\sqrt{\frac{m}{2\pi kT}}\,u \tag{453}$$

Unlike the condition of affairs in ordinary viscosity experiments with gases at higher pressures, as will be seen in Section **267**, the force is ***independent*** of the distance between the plates, and at a given temperature is ***proportional*** to the pressure of the gas instead of independent of the pressure.

Of course if the molecules on striking plate A do not receive on the average the component of velocity u the same as the velocity of the plate, the amount of momentum transferred will be smaller and the force itself smaller. As a result of experiments, however, Knudsen believes that the molecules do acquire on the average substantially the velocity of the plate.

The principle exemplified by equation (453) has been used by Dushman[12] in constructing a gauge for measuring low gas pressures, in which momentum is transferred from a rotating disk to parallel disk hung above it by a quartz fibre. If the molecules do acquire on the average the velocity of the moving disk the instrument could be made an absolute one, but in any case we should expect the torque to be proportional to $p\sqrt{m/kT}\,u$. The instrument has been used for measuring pressures as low as 10^{-8} mm. of mercury.

255. Flow of Gas Through a Tube at Very Low Pressures.— The foregoing method can be applied to investigate the flow of gases through tubes. If the pressure is low enough so that we can neglect the collisions between molecules as compared with collisions with the walls, the only momentum lost by the gas flowing through the tube will be that transferred to the walls, and we can equate this loss of momentum per second to the force exerted on the moving gas by the pressure drop along the tube. If we assume then that the molecules moving through the tube have the Maxwell distribution of velocities, with an average superimposed velocity u, we can write with the help of equation (434), for a tube of diameter D, the expression

$$\frac{\pi D^2}{4}\,dp = muN\sqrt{\frac{kT}{2\pi m}}\,\pi D\,dl \tag{454}$$

where dp is the drop in pressure in the length dl, mu is the momentum transferred per collision, and $N\sqrt{kT/2\pi m}\,\pi D\,dl$ is the number of collisions per second in the length dl. In a state of steady flow, however, the product Nu, which is the number of molecules passing any boundary per square centimeter per second, must be a constant along the length of the tube, so that we can easily integrate over a total length L corresponding to the pressure drop $(p_1 - p_2)$ and obtain

$$\frac{\pi D^2}{4}\,(p_1 - p_2) = Nu\sqrt{\frac{mkT}{2\pi}}\,\pi DL \tag{455}$$

For the total number of molecules passing any boundary in the tube per second we evidently have $Nu\pi D^2/4$ or for the volume Q measured at unit pressure kT times this quantity, so that by substituting in (455) and solving for Q we can obtain

[12] Dushman, *Phys. Rev.*, **5**, 212 (1915).

$$Q = (p_1 - p_2)\frac{\pi D^3}{16L}\sqrt{\frac{2\pi kT}{m}} \tag{456}$$

The derivation is evidently not exact since we have arbitrarily treated the molecules as having the Maxwell distribution of velocities with a uniformly superimposed velocity u along the tube. Knudsen,[13] by assuming that the molecules of high absolute velocity c have a proportionally high superimposed velocity u, obtains the slightly different result

$$Q = \frac{8}{3\pi}(p_1 - p_2)\frac{\pi D^3}{16L}\sqrt{\frac{2\pi kT}{m}} = 3809\,(p_1 - p_2)\frac{D^3}{L}\sqrt{\frac{T}{M}} \tag{457}$$

and this same result is also obtained by Smoluchowski[14] using the more fundamental methods of investigating the actual form of the law for the distribution of the velocities. Smoluchowski also considers the possibility that only a fraction f of the molecules are diffusely reflected from the walls, losing on the average the component of velocity u, the remainder $(1-f)$ being specularly reflected. In such a case the rate of flow would of course be greater than given by (457), and this is found to be in the ratio $(2-f)/f$. Equation (457) is found, however, to express the facts at low pressures extremely well. At higher pressures the flow through tubes is governed by Poiseuille's law. (See Section **268.**)

Equations (441) and (457) are sometimes combined in the form

$$Q = \frac{9.118 \times 10^3}{\dfrac{2.394\,L}{D^3} + \dfrac{3.184}{D^2}}(p_2 - p_1)\sqrt{\frac{T}{M}} \tag{458}$$

to calculate the flow through a tube connecting two reservoirs at pressures p_2 and p_1, allowing both for the resistance at the entrance to the tube and the resistance of the tube itself. All these equations hold of course only at low enough pressures so that the mean free path is large compared with the diameter of the tube. Q is the volume discharged per second measured at temperature T and at unit pressure on the same scale as p_2 and p_1.

256. Transfer of Momentum at Very Low Pressures between Plates at Different Temperatures.—Similar methods can also be used to investigate the behavior at low pressures of molecules between parallel plates at different temperatures T_1 and T_2. We may consider the molecules between the plates as divided into two groups N_1 per cubic centimeter which have rebounded from the hotter plate at temperature T_1 and are moving towards the colder plate at T_2, and N_2 moving from the

[13] Knudsen, *Ann. der Phys.*, **28**, 75 (1909).
[14] Smoluchowski, *Ann. der Phys.*, **33**, 1559 (1910).

colder plate towards the hotter plate. If we assume as a first approximation that these molecules have velocities distributed in accordance with Maxwell's law, with average velocities respectively Ω_1 and Ω_2, we can evidently write

$$N_1\Omega_1 = N_2\Omega_2 = \frac{N}{2}\Omega \tag{459}$$

where N and Ω are the number of molecules per cubic centimeter and average velocity for the gas surrounding the plates, equation (459) merely expressing the requirement that the number of molecules moving from one plate to the other must be balanced by the number returning, and that the number escaping from between the plates must also be balanced by the number returning from the neighboring gas if we are to maintain a steady state.

For the pressure between the plates we can evidently write in accordance with equation (105) Chapter 5, the expression

$$p_{\text{Int.}} = \frac{N_1}{3} mG_1^2 + \frac{N_2}{3} mG_2^2 \tag{460}$$

and for the pressure outside the plates

$$p = \frac{N}{3} mG^2 \tag{461}$$

where G_1, G_2 and G are the mean square velocities corresponding to the average velocities Ω_1, Ω_2 and Ω. Hence we shall obtain per square centimeter of plate surface an unbalanced force of amount

$$F = \frac{N_1}{3} mG_1^2 + \frac{N_2}{3} mG_2^2 - \frac{N}{3} mG^2 \tag{462}$$

tending to force the plates apart, or substituting equations (459) and (461), this can easily be rewritten in the form

$$F = p\left(\frac{1}{2}\,\frac{\dfrac{G_1^2}{\Omega_1} + \dfrac{G_2^2}{\Omega_2}}{\dfrac{G^2}{\Omega}} - 1\right) \tag{463}$$

In accordance, however, with our assumption of the Maxwellian distribution, we should have

$$\frac{G_1}{\Omega_1} = \frac{G_2}{\Omega_2} = \frac{G}{\Omega} \tag{464}$$

and may rewrite (463) in the form

$$F = p\left(\frac{1}{2}\frac{G_1 + G_2}{G} - 1\right) \tag{465}$$

The interpretation of equation (465) would now be very simple, if the molecules rebounding from the two plates at T_1 and T_2 could be regarded as assuming the mean square velocities, which can be denoted by the primed letters G_1' and G_2', which correspond to the temperatures T_1 and T_2 in accordance with the equation

$$\frac{G_1'}{G_2'} = \frac{\sqrt{T_1}}{\sqrt{T_2}} \tag{466}$$

Experiments on the conduction of heat, however, have shown this not to be the case but that the molecules rebounding from the cold plate have not dropped to as low a velocity as would be required and vice versa those rebounding from the hot plate have not completely attained the required high velocity. Under the circumstances Knudsen has assumed that the molecules of mean square velocity G_2 on arriving at the hot plate acquire only a fraction a of the required velocity increase $(G_1' - G_2)$ and those reaching the cold plate lose only the same fraction a of the required decrease $(G_1 - G_2')$, giving us the equations

$$G_1 = G_2 + a(G_1' - G_2) \tag{467}$$

$$G_2 = G_1 - a(G_1 - G_2') \tag{468}$$

where a has been named by Knudsen the accommodation coefficient.

Combining equations (467) and (468) we obtain

$$G_1 + G_2 = G_1' + G_2' \tag{469}$$

and substituting this relation together with (466) into (465) we can obtain the desired expression for the force of repulsion per square centimeter

$$F = p\left(\frac{1}{2}\frac{\sqrt{T_1} + \sqrt{T_2}}{\sqrt{T}} - 1\right) \tag{470}$$

where p and T are the pressure and temperature of the surrounding gas, and T_1 and T_2 the temperatures of the two plates.

This equation was made by Knudsen [15] the basis of his "absolute gage" for measuring the pressure of highly rarefied gases.

257. Transfer of Energy at Very Low Pressures between Plates at Different Temperatures.—By similar methods, we can also investigate the transfer of energy by a highly rarefied gas between two plates at different temperatures T_1 and T_2. We may again treat the gas be-

[15] Knudsen, *Ann. der Phys.*, 32, 809 (1910).

tween the two plates as consisting of N_1 molecules per cubic centimeter moving from the hotter to the cooler plate and N_2 per cubic centimeter moving in the opposite direction and write the equation

$$N_1\Omega_1 = N_2\Omega_2 = \frac{N}{2}\Omega \tag{471}$$

where the symbols have the same significance as in the last section.

For the average time taken by a molecule to make a round trip between the plates we shall evidently have $2d/\Omega_1 + 2d/\Omega_2$, and hence for the number of round trips per second per square centimeter, can write

$$Z = \frac{(N_1 + N_2)d}{2d\left(\frac{1}{\Omega_1} + \frac{1}{\Omega_2}\right)} = \frac{(N_1 + N_2)\Omega_1\Omega_2}{2(\Omega_1 + \Omega_2)} \tag{472}$$

and substituting equation (471) can obtain

$$Z = \frac{N}{4}\Omega \tag{473}$$

just as in equation (450) for the case where the two plates are at the same temperature.

Let us consider now first the *internal* energy which the molecules transport from the hotter plate to the cooler plate. On the average this will evidently stand in the ratio $(C_v - 3/2R)/(3/2R)$ to the kinetic energy of the molecules, where C_v is the molal specific heat at constant volume. Hence combining with equation (473) we may write for the rate of transfer of internal energy per square centimeter per second

$$Q_{\text{int.}} = \frac{N}{4}\Omega\frac{C_v - \frac{3}{2}R}{\frac{3}{2}R}\frac{1}{2}m(G_1^2 - G_2^2) \tag{474}$$

Turning now to the *kinetic energy of translation* of the molecules, we cannot calculate this directly by multiplying the number of transfers by the mean kinetic energy, since the molecules which travel across in the shortest time are also the ones which carry the most kinetic energy. A simple computation shows, however, that this rate of transfer has under our assumptions the value

$$Q_{\text{kin.}} = \frac{4}{3}\frac{N}{4}\Omega\frac{1}{2}m(G_1^2 - G_2^2) \tag{475}$$

which combined with equation (474) gives us for the total rate of transfer of energy per square centimeter per second

$$Q_{\text{total}} = \frac{N}{24} m\Omega \left(\frac{2C_v + R}{R}\right)(G_1^2 - G_2^2) \tag{476}$$

The quantities G_1^2 and G_2^2 occurring in this formula are the mean square velocities of the two streams of gas which we have postulated as moving in the two directions between the plates at T_1 and T_2. Experiment shows, however, that we cannot assume these quantities proportional to T_1 and T_2, since complete temperature equalization does not take place on reflection. If we again make the assumption of Knudsen as given by the former equations (467) and (468) we can obtain

$$G_1^2 - G_2^2 = \frac{a}{2-a}(G_1'^2 - G_2'^2) \tag{477}$$

where a is Knudsen's accommodation coefficient and $G_1'^2$ and $G_2'^2$ are the mean square velocities that actually correspond to the temperatures T_1 and T_2. Making the substitution in (476) and inserting the known relations between p, Ω, G and T given by the gas law and equations (95) and (96), Chapter 5, we obtain

$$Q_{\text{total}} = \frac{a}{2-a}\frac{p}{4}\sqrt{\frac{2k}{\pi m T}}\left(\frac{2C_v + R}{R}\right)(T_1 - T_2) \tag{478}$$

The derivation of this formula is due to Knudsen.[16] It suffers, as did the work in the last section, from uncertainty as to the best method of allowing for the lack of equalization of temperature when molecules are reflected. The justification for the particular form of the equations (467) and (468), where the accommodation coefficient was first introduced, does not seem certain. The equations are certainly valuable, however, in giving an idea of the nature of the phenomena to be expected. Experiment shows that the accommodation coefficient for a given surface and gas is relatively independent of the temperature. It has a higher value, as would be expected, for a platinized than for a smooth platinum surface. When the two surfaces are of different materials or are not parallel, the theory must be modified along evident lines. For further details Knudsen's paper (*l. c.*) must be consulted.

258. Transport as Determined by Mean Free Path.—In the case of gases at higher pressures than those that we have just been considering, the motion of the molecules engaged in transferring matter, energy or momentum will no longer be terminated in general by collision with the walls of the apparatus, but rather by collision with other molecules of the gas, and to handle transport problems at higher pressures we must introduce the idea of the mean free path which molecules travel between

[16] Knudsen, *Ann. der Physik*, **34**, 593 (1911).

their collisions. Having a value for this free path, we can then calculate the rate at which any superimposed disturbance is carried through the gas by the motion of the molecules, under the assumption that the velocities and free paths are the same as they would be under equilibrium conditions. We must again emphasize, however, that such calculations are only approximate in nature, since they do not allow for the effect of the disturbance itself in changing the Maxwell-Boltzmann distribution.

The expressions to be obtained can be expected to have validity only in a range of pressure high enough so that collisions with the walls may be neglected in comparison with collisions with other molecules, but not so high that the effect of triple collisions between molecules will have to be taken into consideration.

259. Probability for a Path of a Given Length.—Before making any calculations as to the dependence of free path on size or velocities of the molecules, we may first consider some general characteristics of the mean free path.

Let us denote by $f(l)$ the probability, under equilibrium conditions that a molecule of a given class (i.e. having a given shape, size, and velocity) will traverse the distance l without making a collision. For the chance that it will traverse the longer distance $l + dl$, we shall then have $f(l + dl)$. We may also, however, set the chance that it will make a collision in the distance dl equal to adl where a is a constant of proportionality, and hence the chance that it will not make collision in the distance $l + dl$ equal to the product of the two probabilities $f(l)$, that it will travel the distance l multiplied by the probability $(1 - adl)$, that it will travel the further distance dl. Equating, we obtain

$$f(l + dl) = f(l)(l - adl)$$

or

$$f(l) + \frac{df(l)}{dl}\,dl = f(l) - af(l)dl$$

which gives us on integration

$$f(l) = e^{-al} \tag{479}$$

the constant of integration being determined by the fact that $f(l)$ must evidently be unity when l itself is zero.

To interpret the constant a we may write for the chance of *collision* within the distance l

$$1 - f(l) = 1 - e^{-al} \tag{480}$$

or by differentiation for the chance of collision in the range l to $l + dl$

$$-df(l) = ae^{-al}\,dl \tag{481}$$

The mean free path λ for the molecules under consideration, however, will evidently be given by integrating over all possible distances l, the product of l times the chance of termination in the particular range l to $l + dl$, giving us

$$\lambda = \int_0^\infty a e^{-al}\, l dl \tag{482}$$

Performing the integration and substituting the limits we obtain

$$\lambda = \frac{1}{a} \tag{483}$$

Hence for the probability dw that the free path of a molecule will terminate in any particular interval dl we have

$$dw = \frac{dl}{\lambda} \tag{484}$$

and referring back to equations (479) and (481), we may write for the chance of traversing a path l without collision

$$f(l) = e^{-\frac{l}{\lambda}} \tag{485}$$

and for the chance of a free path of length between l and $l + dl$

$$-df(l) = e^{-\frac{l}{\lambda}} \frac{dl}{\lambda} \tag{486}$$

We note at once that the probability of a free path much longer than the mean is very small. Thus from equation (485) the chance of travelling a path twice λ is only 0.135, and the chance of travelling a path five times λ only 0.0067.

260. Mean Free Path and Path to Go.—We must now consider a matter of interpretation which is sometimes puzzling. The quantity λ, which was introduced above and called the mean free path for molecules of the class in question, is from its definition by equation (482) evidently the free path that any chosen molecule still has to travel on the average after it has been picked out. Owing, however, to the assumption that the chance of collision adl is the same for any range dl, the length of its path up to the time of choice has no effect on the probable length of the succeeding path. Hence, since we can if desired pick the molecule immediately after a collision, the quantity λ will also be the mean free path travelled by molecules of the class in question between collisions.

Furthermore, it must be noted if we pick molecules of the class in question arbitrarily, that λ will also be on the average the free path which they have travelled *since* their last collision, owing to the fact that the chance of collision in any previous element dl also had the value $a dl$.

A simple example due to Boltzmann will perhaps make the condition of affairs clearer. If we consider the throws of a cubical die, the average number of throws between two throws of a six spot will evidently be five. In addition, however, at any time in a series of throws the average number of throws until we get a six spot will be five and the average number since we got a six spot will be five.

261. Mean Free Path of Clausius.—The quantity λ introduced above was the mean free path for molecules of a particular class. If we desire now an average for molecules of all the classes present, we shall obtain different expressions for the value, depending on the way the average is taken. The particular expression for the free path to be chosen for subsequent discussions will of course be determined by the kind of average that is demanded by the problem under discussion.

The first calculation of mean free path is due to Clausius, and was made on the assumption that all the molecules had the *same velocity*, distributed in random directions. Postulating rigid spherical molecules the value thus obtained was

$$\lambda_{\text{Clausius}} = \frac{3}{4\pi N\sigma^2} = \frac{0.75}{\pi N\sigma^2} \tag{487}$$

where N is the number of molecules per cubic centimeter and σ is the diameter of a molecule. The quantity given is the average distance travelled between collisions by any molecule of the assumed system.

262. Mean Free Path for Molecules of a Given Velocity.—In the case of an actual gas with molecular velocities distributed according to Maxwell's law, the free path for molecules of higher velocities will be somewhat greater than for those of lower velocity, and the problem is much more complicated. It can be shown,[17] however, that for molecules having velocities in the range c to $c + dc$ the mean free path will be given by the expression

$$\lambda_c = \frac{\dfrac{mc^2}{2kT}}{\sqrt{\pi}\, N\sigma^2\, \psi\left(c\sqrt{\dfrac{m}{2kT}}\right)} \tag{488}$$

[17] Derivations for the above and succeeding expressions for mean free path are best found in Jeans "Dynamical Theory of Gases," Cambridge University Press.

where $\psi(x)$ has the form [18]

$$\psi(x) = xe^{-x^2} + (2x^2+1)\int_0^x e^{-y^2}dy \tag{489}$$

263. Maxwell's Mean Free Path.—For a system having the equilibrium distribution of velocities, the mean free path for all the molecules of the system was defined by Maxwell as the total distance travelled in one second by all the molecules present divided by the total number of free paths described in that time.

For the total distance L travelled by the molecules of the system we may take the product of the number of molecules N and their average velocity Ω giving us

$$L = N\Omega = N\sqrt{\frac{8kT}{\pi m}} \tag{490}$$

To calculate the total number of free paths described per second, we note from equation (484) that the chance of collision for molecules having velocities between c and $c + dc$ is l/λ_c per centimeter length of path, or c/λ_c per second. Furthermore, each collision terminates two mean free paths. Hence for the total number of mean free paths described per second we may take

$$Z = 2\int_0^\infty \frac{c}{\lambda_c}\, dN_c \tag{491}$$

where dN_c is the number of molecules having velocities in the range c to $c + dc$.

Substituting for λ_c the value given by equations (488) and (489) and for dN_c the value given by equation (92) in Chapter 5, it is found possible to integrate equation (491) and obtain the result [19]

$$Z = 4N^2\sigma^2\sqrt{\frac{\pi kT}{m}} = \sqrt{2}\,\pi N\sigma^2\Omega \tag{492}$$

Dividing equation (490) by (492) we obtain for Maxwell's value of the mean free path

[18] Tables for the value of this function were calculated by Tait, Edin. Trans., **33**, 74 (1886) and may be found in Jeans (*l. c.*).
If c is one half the average velocity $\lambda_c/\lambda_{\text{Maxwell}}$ drops to 0.6411, while if c is infinity $\lambda_c/\lambda_{\text{Maxwell}}$ rises only to 1.4142. Hence it is evident that the mean free paths are closely concentrated for molecules of all speeds.

[19] This is of course twice the expression we have obtained in Chapter 5, equation (121), for the total number of collisions per second. The calculation is given in the present form to show the analogy with the following calculations of other types of mean free paths.

$$\lambda_{\text{Maxwell}} = \frac{1}{\sqrt{2}\pi N\sigma^2} = \frac{0.707}{\pi N\sigma^2} \tag{493}$$

In the case of spherical molecules of more than one kind, Maxwell's value for the mean free path for the molecules of a particular kind of mass m_1 assumes the form

$$\lambda_1 = \frac{1}{\pi \sum_s N_s \sigma_{1s}^2 \sqrt{1 + \frac{m_1}{m_s}}} \tag{494}$$

where σ_{1s} is the average diameter of molecules of kind 1 and kind s, and the summation is to be taken for all kinds of molecules present.

264. Tait's Mean Free Path.—Again considering a system of molecules with the Maxwell distribution of velocities, we have a definition due to Tait of rather clearer physical significance than the foregoing. Tait takes the mean free path as the average value of λ_c for the molecules of all the different velocities present. Hence if we consider a given instant of time, Tait's mean free path will be the average distance travelled by all the molecules before their next collision.

In accordance with the definition we may write

$$\lambda_{\text{Tait}} = \frac{1}{N}\int_0^\infty \lambda_c dN_c \tag{495}$$

On substituting the value of λ_c given by equations (488) and (489) and the value of dN_c given by equation (92) in Chapter 5, an expression is obtained which must be integrated by quadrature. This has been done both by Tait and Boltzmann and leads to the value

$$\lambda_{\text{Tait}} = \frac{0.677}{\pi N\sigma^2} \tag{496}$$

265. Mean Free Path for Transport Problems.—Finally we may define a special value of the mean free path, which will be useful in interpreting the transport of disturbances, as the average value of $\lambda_c c$ divided by the average velocity Ω in accordance with the equation

$$\lambda_{\text{Transport}} = \frac{1}{\Omega}\frac{1}{N}\int_0^\infty \lambda_c c\, dN_c \tag{497}$$

Substituting the values of λ_c and dN_c, this also leads to an expression which has been evaluated by Boltzmann and Tait by quadrature, leading to the value

$$\lambda_{\text{Transport}} = \frac{0.743}{\pi N \sigma^2} \tag{498}$$

This is the last of the methods of defining the mean free path in a system of molecules that we shall consider. Other definitions would of course give other values which would sometimes be of interest. Owing to the fact that λ_c does not vary rapidly with the velocity c, different methods of defining the mean free path lead to results which do not differ greatly as will be seen by comparing equations (487), (493), (496), and (498).

266. General Equation for the Transport of a Disturbance as Dependent on Mean Free Path.—We may now use the foregoing information concerning mean free paths to investigate the transport of matter, energy and momentum, through gases, which are held at pressures such that the free motions of the molecules are in general terminated by simple collisions between pairs of molecules.

Consider a gas sensibly in a state of equilibrium but with a superimposed disturbance which is being transported in the direction of the z axis owing to a variation in the intensity of the disturbance as we proceed along that axis. This variation makes the molecules coming from one direction carry a greater amount of the disturbance than those coming from the opposite direction and thus leads to the transfer of the disturbance through the gas.

To calculate the rate of transport of the disturbance let us consider a plane, perpendicular to the direction of flow located at the point $z = z_0$, and investigate the motion of the molecules from the two sides through this plane. We shall first fix our attention on molecules having velocities in the range c to $c + dc$ which pass through this plane at an angle to the normal between ϑ and $\vartheta + d\vartheta$. The number so crossing per second, from one side to the other of the plane, will be proportional to the solid angle $2\pi \sin\vartheta d\vartheta$ in which the direction of motion must lie, proportional to their component of velocity $c \cos\vartheta$ parallel to the z axis, and proportional to the number dN_c per cubic centimeter which have the velocity under consideration, so that we may write for the number of such crossings per second per square centimeter of the surface the quantity

$$dZ = \frac{2\pi \sin \vartheta d\vartheta}{4\pi} c \cos \vartheta \, dN_c = \frac{c \sin \vartheta \cos \vartheta}{2} dN_c d\vartheta \tag{499}$$

If now λ_c is the mean free path for molecules of velocity c, the molecules under consideration will on the average have made their last collision at a distance [20] $\lambda_c \cos \vartheta$ from the plane $z = z_0$. Hence, if we denote

[20] See Section 260.

by $G(z)$ the average value of the disturbance per molecule for gas in the layer $z = z$, the molecules denoted above may be assumed to bring with them the quantity of disturbance $G(z_0) + \lambda_c \cos\vartheta \frac{\partial G}{\partial z}$ if they come from the direction of larger values of z, and the quantity $G(z_0) - \lambda_c \cos\vartheta \frac{\partial G}{\partial z}$, if they come from the opposite direction. Combining with equation (499) for the number of crossings from each side, we obtain for the net rate of transport of the disturbance, in the direction of decreasing values of G, due to the molecules under consideration

$$d\Gamma_{\Theta,c} = \lambda_c c \sin\vartheta \cos^2\vartheta \frac{\partial G}{\partial z} dN_c d\vartheta \tag{500}$$

and integrating for all possible values of ϑ from 0 to $\pi/2$, we easily obtain

$$d\Gamma_c = \frac{\lambda_c c}{3} dN_c \frac{\partial G}{\partial z} \tag{501}$$

for the transport produced by molecules having velocities in the range c to $c + dc$.

Finally to obtain the total rate of transport for molecules of all velocities $c = 0$ to ∞, we must substitute the value of λ_c given by equation (488) in Section **262**, and for dN_c the value given by equation (92) in Chapter 5 and integrate from zero to infinity. From the definition of $\lambda_{\text{Transport}}$ given in the preceding section, however, it is evident that this will give us for the total rate of transport per square centimeter per second

$$\Gamma = \frac{1}{3}\frac{\partial G}{\partial z}\int_0^\infty \lambda_c c\, dN_c = \frac{N\Omega}{3}\lambda_{\text{Transport}}\left(\frac{\partial G}{\partial z}\right) \tag{502}$$

and for rigid spherical molecules of diameter σ this gives us in accordance with equation (498)

$$\Gamma = \frac{0.743}{3\pi\sigma^2}\Omega\left(\frac{\partial G}{\partial z}\right) = \frac{115}{\sigma^2}\sqrt{\frac{T}{M}}\left(\frac{\partial G}{\partial z}\right) \tag{503}$$

It should be specially noticed that the rate of transport at these higher pressures is independent of the density of the gas at constant temperature. This seemingly surprising result, which agrees with experiment, was regarded as a great triumph of the older kinetic theory. The result is due of course to the fact that although an increase in density furnishes more molecules to transport the disturbance, it decreases in the same ratio the free path which the molecules travel without collision. At very low pressumes, as we have already seen, where the motions of the molecules are terminated by the walls of the apparatus, rates of transport become proportional to the pressure.

267. The Transport of Momentum (Viscosity).—Equation (502) may now be used to investigate the transport of momentum through a gas and thus to obtain an expression for its viscosity.

Consider a gas in which the mass velocity u in the x direction varies from point to point as we go along the z axis. Then for the average amount of directed momentum per molecule in any layer perpendicular to the z axis we may write

$$G = mu \tag{504}$$

and substituting in equation (502), obtain for the rate of transport of momentum per square centimeter across a given surface perpendicular to the z axis

$$\Gamma = \frac{Nm\Omega}{3} \lambda_{\text{Transport}} \left(\frac{\partial u}{\partial z}\right) \tag{505}$$

The rate of transport of momentum per square centimeter per unit velocity gradient is, however, by definition the viscosity η of the gas, so that we obtain for the viscosity the simple expression

$$\eta = \frac{Nm\Omega}{3} \lambda_{\text{Transport}} = \frac{1}{3} \varrho \, \Omega \, \lambda_{\text{Transport}} \tag{506}$$

In case we assume rigid spherical molecules of diameter σ, by substituting equation (498) for $\lambda_{\text{Transport}}$ and equation (95), Chapter 5, for Ω we can obtain for the viscosity in c.g.s. units

$$\eta = 1.89 \times 10^{-21} \frac{\sqrt{TM}}{\sigma^2} \tag{507}$$

so that a measurement of viscosity leads to an experimental method of determining the diameters of molecules.

268. Flow of Gases at Moderate Pressures through a Tube.— The great difference in the mechanism by which momentum is transferred through gases at high and low pressures produces a corresponding difference in the formulæ for flow through tubes. Thus at moderate pressures we find the flow in cubic centimeters measured at unit pressure governed by Poisseuille's law

$$Q = \frac{\pi D^4 (p_1^2 - p_2^2)}{256 \, \eta L} \tag{508}$$

instead of by equation (457) in Section **255**, valid at very low pressures, which makes the flow proportional to the cube rather than the fourth power of the diameter. For intermediate pressures, empirical laws such as given by Knudsen (*l. c.*) must be employed.

269. The Transport of Energy (Thermal Conductivity).—We may next use equation (502) to investigate the transport of energy through a gas and thus to obtain an expression for the thermal conductivity of the gas.

Consider a gas in which the temperature T varies as we proceed along the z axis. For the average energy per molecule in any layer perpendicular to the z axis, we may then write

$$G = \frac{C_v}{A} T \qquad (509)$$

where C_v is the heat capacity per mol at constant volume, and A is Avogadro's number.

Substituting in equation (502), we obtain for the rate of transport of energy per square centimeter across a given surface perpendicular to the z axis

$$\Gamma = \frac{N\Omega}{3} \lambda_{\text{Transport}} \frac{C_v}{A} \left(\frac{\partial T}{\partial z} \right) \qquad (510)$$

The coefficient of the temperature gradient $(\partial T/\partial z)$ is, however, by definition the thermal conductivity of the gas $\varkappa$, giving us

$$\varkappa = \frac{N\Omega}{3} \lambda_{\text{Transport}} \frac{C_v}{A} \qquad (511)$$

or combining with (506) and noting that C_v/Am is evidently the specific heat c_v we obtain

$$\varkappa = \eta c_v \qquad (512)$$

a relation between viscosity and thermal conductivity which formed one of the early triumphs of the kinetic theory.

The derivation just given suffers of course from the general error introduced, as emphasized in Sections **248** and **258**, by neglecting the effect of the superimposed disturbance in altering the distribution from the equilibrium Maxwell-Boltzmann law. In addition, moreover, the present derivation makes no allowance for the fact that the fastest moving molecules are the ones that bring the most translational energy, which will tend to make the value of $\varkappa$ larger than that given by equation (512). In view of other uncertainties, however, it does not seem important to correct the calculation so as to allow for this. Experimentally, the right-hand side has to be modified by introducing a factor which varies from 1.5 to 2.5 for different gases. The more complex the molecule under consideration, the smaller the value of this experimental factor is found to be, agreeing with the natural hypothesis that the internal energy is transported more nearly as required by equation (510). Using the more refined methods previously mentioned, Chapman (*l. c.*) finds for elastic spheres the multiplier 2.522 and this is not far from the value experimentally found for monatomic gases.

270. The Transport of Matter (Diffusion).—A satisfactory investigation of the diffusion of gases into each other by elementary methods using the concept of mean free paths is difficult. The following relatively simple treatment which leads to the reasonably good formula of the Stefan-Maxwell theory is perhaps as elaborate as the inherent weaknesses of the method would justify.

Consider a gaseous system containing

$$N = N_1 + N_2 \tag{513}$$

molecules per cubic centimeter, N_1 of the first kind and N_2 of the second kind. We shall assume that the proportions N_1 and N_2 vary as we proceed along the z axis, thus leading to the process of diffusion. By Avogadro's principle, however, since the pressure throughout the system is constant, the total number N per cubic centimeter will be constant, which leads by differentiation of equation (513) to the result

$$\frac{\partial N_1}{\partial z} = -\frac{\partial N_2}{\partial z} \tag{514}$$

For simplicity we shall consider all the molecules of each kind as having identical velocities Ω_1 and Ω_2 distributed in random directions, and in addition a component w_0 along the z axis which corresponds to any mass flow in that direction that may result from the diffusion process. The quantities Ω_1 and Ω_2 will be taken as the average thermal velocities corresponding to the temperatures and the quantity w_0 will be determined in the course of the demonstration. The free paths of the two kinds of molecules will be denoted by the symbols λ_1 and λ_2.

Consider now a plane perpendicular to the direction of flow at the point $z = z_0$ moving itself with the velocity w_0 and let us fix our attention on molecules of the *first kind* which pass through this plane at an angle to the normal between ϑ and $\vartheta + d\vartheta$. The number so crossing per second, from one side of the plane to the other, will be proportional to the solid angle $2\pi \sin\vartheta d\vartheta$ in which the direction of motion must lie, proportional to their component velocity $\Omega_1 \cos \vartheta$ parallel to the z axis and to the number of molecules per cubic centimeter, $N_1 \pm \lambda_1 \cos\vartheta \frac{\partial N_1}{\partial z}$, at the average distance $\pm \lambda_1 \cos\vartheta$, from which they have come since their last collision. So that we may write for the *net* passage of molecules per second per square centimeter in the direction of increasing values of z

$$dZ_1 = -\frac{2\pi \sin\vartheta d\vartheta}{4\pi} \Omega_1 \cos\vartheta\, 2\lambda_1 \cos\vartheta \frac{\partial N_1}{\partial z} \tag{515}$$

which on integration over all possible values of ϑ from 0 to $\pi/2$ easily gives us

$$Z_1 = -\frac{1}{3} \Omega_1 \lambda_1 \frac{\partial N_1}{\partial z} \tag{516}$$

This is the number of molecules crossing a plane which is itself moving with the velocity w_0 in the z direction, i.e. which has the mass velocity of the gas. Hence the total number crossing a stationary plane in the positive direction per square centimeter per second will be

$$Z_1' = N_1 w_0 - \frac{1}{3} \Omega_1 \lambda_1 \frac{\partial N_1}{\partial z} \tag{517}$$

Similarly the number of molecules of the second kind crossing the plane in the same direction will be

$$Z_2' = N_2 w_0 - \frac{1}{3} \Omega_2 \lambda_2 \frac{\partial N_2}{\partial z} \tag{518}$$

Since in accordance with equation (513), however, the total number of molecules on each side of the plane must be constant, these two quantities must be equal and opposite in sign, permitting us to obtain for w_0 the value [21]

$$w_0 = \frac{\Omega_1 \lambda_1 \frac{\partial N_1}{\partial z} + \Omega_2 \lambda_2 \frac{\partial N_2}{\partial z}}{3(N_1 + N_2)} \tag{519}$$

and substituting this value in equations (517) and (518), noting also the relation (514), we obtain

$$Z_1' = -\frac{N_1 \Omega_2 \lambda_2 + N_2 \Omega_1 \lambda_1}{3(N_1 + N_2)} \frac{\partial N_1}{\partial z} \tag{520}$$

and

$$Z_2' = -\frac{N_1 \Omega_2 \lambda_2 + N_2 \Omega_1 \lambda_1}{3(N_1 + N_2)} \frac{\partial N_2}{\partial z} \tag{521}$$

[21] The quantity w_0 may be regarded as the mass velocity necessary to preserve the constancy of $N = N_1 + N_2$.

By definition, however, the coefficient for the concentration gradient in the expression for the rate of diffusion is the diffusion coefficient D for the substance in question, giving us

$$D=\frac{N_1\Omega_2\lambda_2+N_2\Omega_1\lambda_1}{3(N_1+N_2)} \tag{522}$$

the same for both gases in the mixture.

To interpret equation (522) we must next inquire what values to use for the mean free paths λ_1 and λ_2. For a mixture of two gases, Maxwell's value for the mean free path takes, in accordance with equation (494), the form

$$\lambda_1=\frac{1}{\pi N_1\sigma_{11}{}^2\sqrt{2}+\pi N_2\sigma_{12}{}^2\sqrt{1+\dfrac{m_1}{m_2}}} \tag{523}$$

On examination this expression shows that the free path is limited both by the concentration N_1 of the molecules of the first kind present in the mixture and by the concentration N_2 of molecules of the second kind, corresponding to the two terms in the denominator. It seems hardly reasonable, however, to suppose that the rate of diffusion of the molecules of a given kind will be hindered by the collision of molecules of this kind among themselves, since the total momentum of the molecules in the z direction will not be changed by such collisions. Hence following the suggestion of Stefan and Maxwell we may use for λ_1 the value it would have if N_1 were zero, and similarly for λ_2 the value obtained with N_2 equal to zero. Substituting these values in equation (522), we easily obtain as our final expression for the coefficient of diffusion

$$D=\frac{\sqrt{m_1}\Omega_2+\sqrt{m_2}\Omega_1}{3\pi(N_1+N_2)\sigma_{12}{}^2\sqrt{m_1+m_2}} \tag{524}$$

or introducing the proportionality $\Omega_1{}^2/\Omega_2{}^2=m_2/m_1$, we can write

$$D=\frac{\sqrt{\Omega_1{}^2+\Omega_2{}^2}}{3\pi(N_1+N_2)\sigma_{12}{}^2} \tag{525}$$

The deduction of this result has been plausible rather than certain in character. Nevertheless the formula agrees reasonably well with experimental data, and considerably better than it would if we substituted the full expression for λ_1 and λ_2 into equation (522).

271. The Transport of Electricity (Electrical Conductivity).— The phenomenon of electrical conduction in material systems is due to the presence of electrically charged particles, electrons or ions, which move under the influence of the applied electromotive force, thus leading to the transport of electricity in the direction of decreasing potential. The magnitude of the electrical conductivity of a given substance can be quantitatively accounted for, in a somewhat formal manner, in terms of the concentrations and charges of the various ions present and their mobilities.

Consider a system containing per cubic centimeter N_1, N_2, N_3, etc., ions carrying the charges $\nu_1 e$, $\nu_2 e$, $\nu_3 e$, etc., which are multiples of the electronic charge e. For the case of a neutral system we must evidently have the total charge equal to zero, giving us

$$(N_1\nu_1 + N_2\nu_2 + N_3\nu_3 + ..) = 0 \tag{526}$$

but our considerations do not need to be limited to this case.

Let the mobilities of the various ions present, that is the average velocities which they attain in the direction of the field under unit potential gradient, be represented by U_1, U_2, U_3, etc. It is then evident that the current flowing per square centimeter under unit potential gradient will be

$$I = N_1\nu_1 eU_1 + N_2\nu_2 eU_2 + N_3\nu_3 eU_3 + \ldots \tag{527}$$

where it should be noted that positive values of ν will evidently be associated with positive values of U and vice versa. Since the current flowing per square centimeter under unit potential gradient is by definition the specific conductivity of the material in question, we have thus obtained an expression for that quantity.

The condition for Ohm's law to hold will be a constancy in the number of ions present per cubic centimeter, and a velocity proportional to the potential gradient, that is the same mobility for different gradients. In the case of gaseous conductors where the number of ions can be increased at high potential gradients by ionization through collision, we have of course marked deviations from Ohm's law. In the case of electrolytic solutions, however, where the ionic velocities are small and the retarding force may be regarded as of a quasifrictional character, so far no deviations from Ohm's law have been observed. In the case of metals, with very high potential gradients corresponding to current densities of 5×10^6 amperes per square centimeter, increases in resistance of the order of one per cent have been found in gold and silver by Bridgman.[22]

272. Relation between Mobility and Coefficient of Diffusion.— A simple relation between the mobility of an ion and its diffusion coefficient can be derived by elementary methods which have sufficient

[22] Bridgman, *Phys. Rev.*, **19**, 387 (1922).

plausibility to justify their application in view of the complexity and uncertainty of more rigorous treatment.

Let Z be the number of molecules crossing a given surface per square centimeter per second under the influence of a concentration gradient $\partial N/\partial z$, and connected with this gradient and the diffusion coefficient D by the equation

$$Z = -D\frac{\partial N}{\partial z} \tag{528}$$

Let us now regard this process of diffusion as due to the force arising from the gradient in the partial pressure of the component in question. Substituting the expression for pressure

$$p = NkT$$

we can write equation (528) in the form

$$Z = -\frac{D}{kT}\frac{\partial p}{\partial z} \tag{529}$$

The quantity $(\partial p/\partial z)dz$ can, however, be regarded as the *directed* force acting on the molecules in a layer of area one square centimeter and thickness dz, so that we can write for the directed force *per* molecule

$$F = \frac{1}{N}\frac{\partial p}{\partial z} \tag{530}$$

and substituting in equation (529) obtain

$$Z = -\frac{ND}{kT}F \tag{531}$$

which connects the number of molecules crossing the surface with the force acting on a molecule.

On the other hand in the case of an applied electromotive force the number of ions crossing a surface of one square centimeter per second will be given by the expression

$$Z = -NU\frac{\partial E}{\partial z} \tag{532}$$

where N is the number per cubic centimeter, U their mobility and $(\partial E/\partial z)$ the potential gradient. The directed force acting on each ion, however, is

$$F = \nu c\frac{\partial E}{\partial z} \tag{533}$$

so that equation (532) can be rewritten in the form

$$Z = -\frac{NU}{\nu e} F \tag{534}$$

Identifying now the coefficients of the force F in equations (531) and (534) we obtain the desired relation between mobility and diffusion coefficient in the form

$$D = \frac{U}{\nu e} kT \tag{535}$$

273. Diffusion of Electrolytes.—If we assume that the osmotic pressure (diffusion pressure) in solutions is also given by the relation $p = NkT$, equation (535) should also hold in electrolytic solutions where it can easily be tested. The possibility of such a treatment was first appreciated by Nernst.

As an example let us consider a binary electrolyte completely dissociated into ions with charges $+e$ and $-e$ and mobilities U and $-V$. Since the independent diffusion of the two ions at different rates would lead to strong electrostatic forces, the ions will diffuse at the same rate, a potential gradient $(\partial E/\partial z)$ being set up by the attempt of the faster ion to get ahead which makes them actually proceed at the same velocity, and keeps the concentration N of the two ions the same in any layer of the diffusing column.

Under the combined action of diffusion and the potential gradient, we may now evidently write for the number of positive ions crossing unit area per second

$$Z_1 = -D_1 \frac{\partial N}{\partial z} - NU \frac{\partial E}{\partial z}$$

and for the number of negative ions

$$Z_2 = -D_2 \frac{\partial N}{\partial z} + NV \frac{\partial E}{\partial z}$$

where D_1 and D_2 are the diffusion coefficients. These two rates, as we saw above, however, must be equal, so that by substituting equation (535), we can write

$$Z_1 = Z_2 = -\frac{U}{e} kT \frac{\partial N}{\partial z} - NU \frac{\partial E}{\partial z} = -\frac{V}{e} kT \frac{\partial N}{\partial z} + NV \frac{\partial E}{\partial z}$$

or eliminating $(\partial E/\partial z)$

$$Z_1 = Z_2 = -\frac{2UV}{e(U+V)} kT \frac{\partial N}{\partial z} \tag{536}$$

Hence the diffusion coefficient for the salt as a whole has the value

$$D = \frac{2UV}{e(U+V)} kT \tag{537}$$

Since ion mobilities can be determined from conductivity measurements, this relation is susceptible of easy investigation, and has been satisfactorly tested over a wide range by Öholm.[23] A similar equation has been derived by Noyes[24] for the general case of an electrolyte giving more than two ions.

274. Concluding Remarks on the Method of Mean Free Paths.— As already emphasized in several places, the method of mean free paths suffers from a logical deficiency in treating the molecules of the system under consideration as though they had the equilibrium Maxwell-Boltzmann distribution of coordinates and momenta, in spite of the fact that the system is not in equilibrium but is in a disturbed condition. The method has the great advantage, however, of relative simplicity, and of sticking close to the qualitatively correct physical picture of the mechanism of transport by molecular motions. In addition the results obtained are usually approximately correct and almost always give the right functional relationship between the physical quantities involved even though experiment shows that the value of the numerical constants occurring in the equations must be altered.

In addition to the purely logical fallacy mentioned above, it will of course be appreciated that many physical hypotheses have to be made in order to treat actual problems by this or any other method. A few remarks with respect to such hypotheses will not be out of place.

In considering the problems of viscosity and thermal conductivity, it was assumed that molecules crossing a given plane, having made their last collision on the average at a distance $\lambda \cos\vartheta$ from this plane, will bring with them an amount of the disturbance in question corresponding to the mean value of the disturbance at the distance $\lambda \cos\vartheta$. It is evident, however, since bodies tend to continue in the same line of motion after collision, that molecules coming from collision in the layer at $\lambda \cos\vartheta$ will on the average have come previous to this collision from a still greater distance from the plane in question. This tends to increase the rate of transport over that calculated by the simple theory.

The question of persistence of velocity has been elaborately treated by Jeans. For rigid spherical molecules of the same mass he finds on the average that molecules retain about forty per cent of their velocity in the original direction. For rigid molecules of different masses the

[23] Öholm, *Zeits. f. physik. Chem.*, **50**, 309 (1904); **70**, 378 (1910).
[24] See Haskell, *Phys. Rev.*, **27**, 145 (1908).

persistence can vary from zero to unity, according as the molecule under consideration is very light or heavy compared with the one with which it collides. For non-rigid molecules the definition of a collision becomes arbitrary and the problem of persistence loses its significance.

With the help of his calculations of persistence of velocity, considerable progress has been made by Jeans in bettering the older kinetic theory. It should be remarked, however, that too great an elaboration of the older theory does not prove profitable when the complications assume a difficulty comparable with those encountered in more rigorous methods of treatment. It may also be noted in passing that in the Stefan-Maxwell treatment of gas diffusion given in the present text, the effect of persistence of velocities was in some measure allowed for by neglecting collisions between molecules of the same kind.

Another physical hypothesis that should be considered is the assumption of rigid molecules in the calculation of mean free paths. This of course is only justifiable as a first approximation, and the experimental fact that the viscosity of gases increases more rapidly than the square root of the temperature T as given in equation (507), indicates that we must regard the molecules as compressible, with lower values of the effective diameter σ when the temperature and hence average relative velocity of collision is greater. Appreciable success in this direction has been achieved by Sutherland[25] by taking the effective diameter as given by a formula of the form

$$\sigma^2 = \sigma_\infty^2\left(1 + \frac{C}{T}\right) \tag{538}$$

Viscosity measurements at moderate temperatures agree well with this formula but it fails at very low temperatures.[26]

In conclusion it should be pointed out that cases may arise where the extent of our physical knowledge does not permit a satisfactory application of the method of mean free paths. Thus the Drude theory of metallic structure which assumes the electrons in a metal to behave like the molecules of a gas must be abandoned in its simple form. With the help of this hypothesis Drude was able to calculate the ratio of the thermal to the electrical conductivity of metals by taking the same mean free path for electrons engaged in either kind of conduction. In this way he obtained a value of the ratio at ordinary temperatures which agreed approximately with that found experimentally by Wiedemann and Franz. The phenomenon of super conductance at low temperatures, however, indicates that a considerable change in theory, perhaps modified by quantum considerations, must be made.[27]

[25] Sutherland, *Phil. Mag.*, **36**, 50 (1893).

[26] Kammerlingh Onnes and Sophus Weber, *Communications Leiden Phys. Lab.*, **134b**, p. 18.

[27] See Bridgman, *Phys. Rev.*, **17**, 161 (1921).

Chapter 20.

The Rate of Chemical Reaction.

275. Introduction.—The rate at which chemical reactions proceed is one of profound interest both for theory and practice. With the help of thermodynamics it is now possible in large measure for chemists to predict in advance just which chemical reactions can take place, namely those accompanied by an increase in entropy, and to calculate how far they will go before equilibrium is attained. The laws controlling the specific rates at which these reactions proceed are, however, beyond the scope of thermodynamic reasoning, and hence the thermodynamist is not in a position to foresee what will actually happen when chemical materials are mixed together, since his knowledge of the thermodynamically possible reactions does not suffice for the prediction of the actual reactions which do occur. Indeed by a strange perversity, it often seems as if the reactions which are accompanied by the largest increases of entropy, as for example the combination of hydrogen and oxygen to form water, take place at the slowest rates,[1] while reactions which have almost no thermodynamic tendency to proceed, as for example the dissociation of nitrogen tetroxide into the dioxide, take place at speeds so great that they still defy experimental measurement.

The ultimate solution for such problems of reaction velocity cannot be looked for in the field of thermodynamics which entirely neglects the atomic and molecular structure of matter. The mechanism of chemical reaction seems certainly determined by the nature of the interacting atoms themselves. In the case of the individual processes which occur, the question is one of atomic and molecular mechanics, while for the overall rate of change in collections of molecules, the problem is one of sta-

[1] Reactions having a large thermodynamic tendency to proceed are in general exothermic, and in some cases the slow rate of reaction may be due to the difficulty of providing an escape for the excess energy from the molecule formed. Thus the combination of monatomic hydrogen to form molecular hydrogen in its normal state

$$H + H = H_2$$

involves the emission of an enormous amount of energy, and if we assume that the non-polar character of the hydrogen molecule prevents the emission of this energy in the form of radiation, a three-body collision would be necessary for the reaction. This hypothesis is in agreement with the slow rate of recombination of monatomic hydrogen and its catalysis by solid substances which can conduct away the heat.

tistical mechanics. In the three following chapters we shall present the progress in this field, in some ways meagre but in many ways significant, which has already been made with the help of statistical mechanics.

276. Classification of Chemical Reactions.—Chemical reactions may for convenience be classified first of all as homogeneous, heterogeneous, and explosive, depending on the spatial location of the actual region in the system where the chemical changes are taking place. Secondly as catalytic or non-catalytic, depending on the necessity for the presence of an intermediate substance to promote the change. And thirdly as photochemical or thermal, depending on the presence or absence of radiation from an external source.

277. In a *homogeneous* reaction, the reacting substances are all present in a single phase and the chemical change proceeds at a uniform rate throughout the whole mixture. In particular no catalytic effect of the walls of the apparatus must occur, or, if reaction does proceed more rapidly at the walls, correction for this must be made in order to obtain the true homogeneous rate.

There are of course many examples of homogeneous chemical change. The reactions most familiar of all to the chemist are the changes taking place in aqueous solution, and these, except when they actually involve steps such as precipitation or solution, are for the most part homogeneous, since rates of diffusion to the walls are so slow that contact catalysis can usually play no appreciable part even if there is a tendency in that direction. Of more interest for statistical mechanics, however, are the very considerable number of homogeneous gas reactions which have been investigated. These include the first-order decompositions of nitrogen pentoxide and sulfuryl chloride, the transformation of trimethylene to propylene, the second-order decompositions of ozone, nitrous oxide, and chlorine monoxide, the reaction between hydrogen and bromine, and a considerable proportion of the reaction that takes place between hydrogen and iodine as well as the reverse change of hydrogen iodide back into its elements.

278. In a *heterogeneous* reaction, the actual chemical changes take place at the surface of discontinuity between two phases. This may either result from the circumstance that the substances which are to interact are present in separate phases and can only come in contact at the boundary between these phases, or may be due to the necessity for the catalytic action of the surface. In the former class are of course many of the most important reactions of industry, including the combustion of fuels and the calcination and smelting of ores. In the latter class are the increasingly important number of gas reactions which are accelerated by contact catalysis, including the contact process for the manufacture of sulfuric acid, the synthesis of ammonia from its elements, the oxidation of ammonia for the manufacture of nitric acid, the hydrogenation of hydrocarbons, etc.

279. In an *explosive* reaction, the chemical changes take place in a reaction zone which propagates itself through an originally homogeneous system. The simplest examples are furnished by the spread of combustion in an explosive mixture of gases where the luminosity of the reaction zone makes it possible to follow its motion through the mixture using photographic methods. It is of course apparent that the distinction between homogeneous and explosive reactions may sometimes be profitably disregarded by considering the moving reaction zone as consisting of a set of infinitesimal layers of mixture of varying concentration and temperature.

280. The distinction between *catalyzed* and *non-catalyzed* reactions is one of great practical importance but not of very deep theoretical significance. In a non-catalyzed reaction the chemical substances enter directly into the changes which lead to the final products. In a catalyzed reaction, however, one of the reacting substances must first interact with the *catalyst* either to form a definite "chemical" compound or a looser more "physical" aggregate such as an adsorbed layer, the catalyst later being regenerated in the course of the further changes. The separate steps thus occurring in a catalyzed reaction are themselves chemical or physical processes capable of individual study. As pointed out by Perrin, the statement sometimes made that all reactions have to be catalyzed is a meaningless one, since the reaction between catalyst and reactant would then have to be catalyzed by a second catalyst, and the action of this catalyst promoted by a third catalyst and so on *ad infinitum*.

281. *Photochemical* reactions are to be distinguished from *thermal* reactions by the circumstance that the former take place under the influence of radiation from an external source which is not in temperature equilibrium with the reacting mixture. In carrying out thermal reactions on the other hand, the enclosure containing the reacting substances is kept in surroundings at constant temperature which can supply or remove energy in the form of radiation as well as by conduction. The attempt is thus made to maintain both the radiation and matter inside the reaction vessel in a condition which corresponds as nearly as possible to temperature equilibrium. This of course does not mean that in thermal reactions no use is made of the radiation present, nor, on the other hand, that it is actually possible to maintain the radiation of all frequencies at the densities required by the Planck radiation law. These are matters which must be specifically investigated.

282. In the classification of chemical reactions it should finally also be noted that the word reaction is sometimes used to signify an individual chemical step in a chain or collection of chemical processes and sometimes to signify the total result of all the steps. It is on the whole most logical, when any confusion from this source might arise, to use the word reaction as applying to each separate step, and to speak of *isolated reactions* when there is only one step and a set of *simultaneous reactions* when several steps are involved.

283. In the present chapter we shall discuss thermal reactions. The material presented in the first part of the chapter will apply to homogeneous reactions and mainly to gaseous reactions, since they are the ones most amenable to statistical mechanical treatment. A brief treatment of heterogeneous reactions will then be given and a few remarks concerning explosive reactions will also be included. In the next chapter we shall discuss the temperature coefficient of reaction rate, and in the following chapter shall treat photochemical reactions. These latter are, as a matter of fact, really simpler theoretically than thermal reactions, since the mechanism producing the changes is more completely specified and controllable.

284. The Order of Chemical Reactions.—In the study of reaction velocities a careful distinction must be made between the terms *order of reaction* and *mechanism of reaction,* the first of these being a matter of direct experimental measurement, and the latter usually a matter of more or less uncertain theoretical interpretation.

The order of a reaction with respect to one of the reacting substances is the power to which the concentration of that substance enters in the expression connecting rate of reaction with concentration. Thus if a reaction between substances A, B, C, etc., is found experimentally at a given temperature to proceed at a rate which can be expressed by the equation

$$-\frac{dC_A}{dt} = kC_A{}^x C_B{}^y C_C{}^z \ldots . \tag{539}$$

the reaction is said to be of the order x, y, z, etc., with respect to the substances A, B, C, etc., where C_A, C_B, C_C, etc., are the instantaneous concentrations [2] of the reacting substances, and k is a constant.

[2] It is often convenient and customary to express the concentration of a substance by its chemical formula enclosed in brackets, and this practice will be employed in what follows.

In writing this equation of order it is often convenient to use ***equivalent*** concentrations in expressing the quantities C_A, C_B, C_C, etc., since then the rate of change of concentration will have the same numerical value for the different substances involved. For purposes of statistical mechanical treatment, however, it is more satisfactory to use ***molal*** concentrations which are proportional to the number of molecules present, or in many cases to express the concentrations directly, as ***numbers of molecules*** per *cubic centimeter,* since the formulæ of the kinetic theory can then be immediately applied to calculate the number of collisions between molecules that are taking place.

The quantity k occurring in equation (539) is called the *specific reaction rate,* and has the dimensions of $[t^{-1}C^{x+y+z+\cdots-1}]$. It is most conveniently expressed using time in seconds and concentrations in mols per cubic centimeter, and it is most unfortunate that the literature contains so much work in which inconvenient or even unspecified units are employed.

285. Many reactions proceed throughout the major part of their course with a simple and constant order, the exponents being constant integers, and the specific reaction rate k being also a constant. The exponents are, however, not necessarily integers and expressions of more complicated form than (539) are sometimes more convenient.[3] In the case of a reaction of constant order, the specific rate k will be a constant, but many reactions occur in which the order varies as the reaction proceeds. In such a case either the value of k or the values both of k and of the exponents x, y, z, etc., may be regarded as functions of the concentrations if we desire to express the actual facts by an equation of the form of (539). The specific rate k is in general a function of the temperature, usually increasing as the latter increases, and indeed often doubling or trebling for a 10° rise in temperature.

286. Relation between the Mechanism and Order of Chemical Reactions.—By the mechanism assigned to a chemical reaction we try to express the actual nature of the individual chemical processes which are occurring in the reaction mixture. Thus we call a reaction ***unimolecular*** if we believe that only a single molecule of the reactant is involved in the individual step, as for example in radioactive decomposition, *bimolecular* if two molecules enter into the individual step, as in

[3] Thus the rate of the reaction $H_2 + Br_2 = 2HBr$ can best be expressed by an equation of the form

$$-\frac{dC}{dt} = k\,\frac{(H_2)\,(Br_2)^{\frac{1}{2}}}{1 + k'\dfrac{(HBr)}{(Br_2)}}$$

the combination of hydrogen and iodine to form hydrogen iodide, and *trimolecular, quadrimolecular,* etc., for larger numbers of molecules.

The determination of the actual mechanisms by which reactions do take place is of course often a very uncertain and speculative business. The mechanism postulated must of course be one which will explain the order of the reaction as experimentally found, and any other facts that we can command, but more than one hypothetical mechanism may often explain all the facts at our disposal and the present status of our experimental knowledge in the field of reaction rates is all too meagre.

287. In the case of reactions of simple and constant order, the obvious hypothesis is to assume a direct correspondence between order and mechanism. Thus a reaction of the *first order* is presumably *unimolecular* in character, since such a mechanism would evidently lead to a rate of decomposition proportional to the number of molecules present. Similarly a reaction of the *second order* with respect to any component presumably results from a *bimolecular* process, since the number of collisions between the molecules of this component will evidently be proportional to the square of the concentration. And in similar fashion we can proceed to reactions of higher orders.

288. This assumed correspondence between order and mechanism, which finds its rational explanation on the basis of the kinetic theory of molecular motions, was of course the original hypothesis of Guldberg and Waage, in the formulation of the law of mass action. Since their time several circumstances have tended to obscure the clarity and fundamentality of the idea. In the first place, it is often found that the order and mechanism of a reaction do not correspond to the usual stoichiometric equation which is most naturally written to express the total chemical change. In the second place, further investigation has often shown that the rate of a reaction originally supposed to be of a simple definite order really depends on the concentration of substances which had not previously been varied. And finally the discovery of many reactions which are catalyzed by traces of impurity, such for example as water vapor, has unfortunately led some investigators to become fearful of the whole basis of chemical kinetics and to suspect that catalytic complications lurk behind every molecule like prowling wolves. The rational procedure, however, seems to be to regard the order of any simple reaction as *prima-facie* evidence of its mechanism, and not introduce complications until these are made necessary by the facts. In addition, it is to be appreciated that even in the case of a complicated set of simultaneous reactions, the individual steps of the total process are themselves simple reactions of definite order and mechanism, so that a study of the theory of isolated reactions is the natural preliminary for the study of all branches of chemical kinetics.

289. A couple of illustrations of the possibility of drawing premature conclusions as to the order and mechanism of chemical reactions will be of interest.

Thus the rate of transformation of a given sample of aceto-acetic ester from the enol to the keto form is found to be of the first order with respect to the ester present. This might lead to the conclusion that the transformation was a simple unimolecular change. It has been found, however, that the change is catalyzed by minute traces of hydroxide present and will not proceed at all if kept in quartz vessels and hence free from alkali. Hence we must conclude that some intermediate interaction with hydroxide is one of the steps in the total process.

As another interesting example, the decomposition of ozone is found to be of the second order with respect to the concentration of ozone. It would be incorrect, however, to conclude that the reaction is the bimolecular combination of two molecules of ozone to form three of oxygen, since further investigation shows that the reaction is inhibited by the presence of oxygen, and that the more complete formulation of the rate can be expressed by the equation

$$-\frac{d(O_3)}{dt}=k\frac{(O_3)^2}{(O_2)}$$

This result would be in accord with the mechanism proposed by Jahn

$$O_3 \rightleftarrows O_2+O$$

$$O_3+O \longrightarrow 2O_2$$

where the first reaction is assumed fast enough to maintain the concentration of the monatomic oxygen at its equilibrium value

$$(O)=K\frac{(O_3)}{(O_2)}$$

while the overall rate of the decomposition is determined by the second reaction in accordance with the expression

$$-\frac{d(O_3)}{dt}=k'(O_3)(O)=k'K\frac{(O_3)^2}{(O_2)}=k\frac{(O_3)^2}{(O_2)}$$

It is of course to be recognized that this result does not prove Jahn's mechanism, but shows it to be a possible one. It may be again emphasized, however, that complications such as the above do not destroy the fundamental importance of the relation between order and mechanism postulated by Guldberg and Waage, since our attempt will always be to analyze the total set of reactions into separate steps of simple order and mechanism. Both of the examples illustrate the desirability of determining the effect of the concentration of all materials present on the rate of a reaction.

290. Unimolecular Reactions and the Fraction Decomposed.—In the case of unimolecular reactions, of which the decomposition of gaseous nitrogen pentoxide is one of the best examples, it is found at any given temperature that a constant definite fraction of the material present decomposes in each second.

For unimolecular chemical processes it is observed in general that the fraction undergoing change per second increases rapidly with the temperature. From this it seems necessary to conclude that the normal state of the molecule is a stable one, but that certain higher quantum states of the molecule are unstable. Hence with rising temperature we have an increased rate of reaction, since the tendency to go over into higher states increases with the temperature in a manner which can be determined from statistical mechanical considerations. This matter of the temperature coefficient of reaction rates will form the special subject of the next chapter.

In the case of radioactive decompositions, the fraction decomposing per second is found over the wide temperature ranges that have been tested to be constant. The most natural inference would be to conclude that the internal motions of electrons or protons in the atom occasionally result in an unstable configuration which thus leads to a rate of decomposition in accordance with the laws of chance. There is also the possibility, which has been advanced by Perrin, that the decomposition is due to the occasional absorption of very penetrating radiation of extremely short wave length of extraterrestrial origin. If this should be the case, however, the radiation would have to be enormously more penetrating even than the penetrating radiation recently studied by Kohlhörster and by Millikan in order to account for the constancy in the rate of decomposition which has been found.

291. Bimolecular Reactions and Number of Collisions.—In the case of bimolecular reactions, we naturally inquire into the relation between the number of molecules which collide and the number of times that reaction results. The exact specification of what we shall mean by a collision in this connection is of course difficult. Nevertheless, experiment and theory alike agree in confirming the supposition that the forces around neutral atoms and molecules fall off with a very high power of the distance. Hence as a first approximation it will not be unreasonable, just as in the kinetic theory of gases, to consider molecules as nearly rigid spheres, and compare the number of times reaction occurs with the calculated number of collisions using the kinetic theory values of diameters.

For the number of collisions between molecules of the same kind occurring per cubic centimeter per second we have

$$Z = 2N^2\sigma^2\sqrt{\frac{\pi kT}{m}} \tag{540}$$

where N is the number per cubic centimeter, σ their diameter, and m their mass. And for the number between molecules of different kinds, we have

$$Z = 2N_1N_2\sigma_{12}^2\sqrt{2\pi kT\,\frac{m_1+m_2}{m_1m_2}} \tag{541}$$

where N_1 and N_2 are respectively the numbers of the two kinds present per cubic centimeter, σ_{12} the average diameter, and m_1 and m_2 their masses. (See Chapter 5.)

For purposes of calculation in connection with reaction rates, it is often convenient to express the above results in terms of the number of mols per cubic centimeter which suffer collision per second; we then obtain for the two cases

$$\frac{Z}{A} = 10^{28.2922}\,C^2\sigma^2\sqrt{\frac{T}{M}} \tag{542}$$

and

$$\frac{Z}{A} = 10^{28.4427}\,C_1C_2\sigma_{12}^2\sqrt{T\,\frac{M_1+M_2}{M_1M_2}} \tag{543}$$

where A is Avogadro's number, the concentrations are in mols per cubic centimeter, the diameters in centimeters and the masses expressed as ordinary molecular weights.

These equations apply of course strictly only to gases. It seems, nevertheless, probable that we can safely regard them as giving upper limits to the numbers of collisions between solute molecules in a solution.

292. Using these equations for the number of collisions and substituting values for the molecular diameters of the order of 10^{-8} to 10^{-7} centimeters, we always find for bimolecular reactions in the *usual case where the rate has a high temperature coefficient* that the number of collisions enormously exceeds the number of times a reaction occurs. The factor is so great, for example of the order 10^{11} in the case of the bimolecular decomposition of nitrous oxide at 1000° absolute studied by Hinshelwood and Burk,[4] that we cannot reasonably conclude that the failure to

[4] Hinshelwood and Burk, *Proc. Roy. Soc.*, 106A, 284 (1924).

react is predominantly due to lack of proper orientation of the molecules at the time of collision. It seems more reasonable to assume either that the molecules must be in a higher quantum state to react on collision, or that only collisions of high kinetic energy can lead to reaction. These matters are of course susceptible to statistical mechanical treatment and will be considered in the next chapter.

293. In the case of bimolecular reactions, in which the rate does not have a large temperature coefficient, we can conclude that activation to a higher quantum state is not necessary and hence shall expect a closer correspondence between number of collisions and number of times a reaction occurs. Experimental information concerning such reactions is at present necessarily very scarce, owing to the probability of a high rate comparable to the number of collisions and owing to the fact that the lack of appreciable temperature coefficient makes it impossible to vary the rate in the customary manner and thus find a convenient range to work in. By indirect methods, however, it has been concluded by Bodenstein and Lücktmeyer [6] that the rate of recombination of monatomic bromine has a value which is at least nearly independent of the temperature and of the concentration of other gases present, and that about one collision out of 1250 leads to recombination. Such a factor might easily result from the necessity for proper orientations of the atoms at the time of collision.

294. It would of course be very surprising if we found a case of bimolecular reaction in which the rate of reaction was greater than the calculated number of collisions. The interpretation of such a circumstance could only be that the range of action of the "chemical" forces surrounding molecules is much greater than we had suspected. It seems doubtful if such a state of affairs will be found; it is very important, however, that calculations similar to the above should be made, whenever the data present themselves.

295. Reactions of Higher Order and Number of Multiple Collisions.—It is a significant experimental fact that reactions in which the rate is of higher order than the second have so far been but rarely encountered. Even in the case where the stoichiometric equation for the change would lead us to expect a reaction of higher order, this is seldom found. A reasonable explanation of this fact lies in the probable circumstance that triple or higher order encounters with a proper orientation to permit a reaction between the molecules will be very rare,

[6] Bodenstein and Lücktmeyer, *Zeitschr. für Physik. Chem.*, **114**, 233 (1924).

and hence the chemical change that does take place will usually occur as the result of a series of bimolecular steps.

Reactions of higher order than the second are, however, sometimes found. The polymerization of cyanic acid to cyamelide is a third order heterogeneous gas reaction, the absolute rate depending on the surface of the solid cyamelide present and being proportional to the cube of the gas pressure. The reduction in solution of ferric iron to ferrous iron by stannous chloride is a third order reaction which apparently takes place in accordance with the mechanism

$$2Fe^{+++} + Sn^{++} \rightarrow 2Fe^{++} + Sn^{++++}$$

and the reaction in solution between ferrous iron and chlorate ion which is catalyzed by acid is also of the third order, being proportional to all three concentrations of ferrous ion, chlorate ion, and hydrogen ion.

As a reaction of the fourth order, we have the interaction of bromide and bromate which appears as though dependent on the step

$$2H^{+} + Br^{-} + BrO_3^{-} \rightarrow HBrO + HBrO_2$$

and as a reaction of the eighth order the interaction of hydrochloric and chloric acids which appears as though dependent on the step

$$4H^{+} + 2ClO_3^{-} + 2Cl^{-} \rightarrow 2H_2ClO_3 + Cl_2$$

Four very interesting gas reactions have also been studied which are of the third order and apparently homogeneous. These all involve quite interestingly the same molecule NO. The stoichiometric equations for these reactions are [6]

$$2NO + O_2 = 2NO_2$$

$$2NO + Cl_2 = 2NOCl$$

$$2NO + Br_2 = 2NOBr$$

$$2NO + 2H_2 = N_2 + 2H_2O$$

The first three of these gas reactions are third order in agreement with the mechanism which would be assumed from the stoichiometric equation.

The fourth reaction is found to be second order with respect to NO, and first order with respect to H_2. This experimental fact would be in agreement with the mechanism

$$2NO + H_2 = N_2 + H_2O_2 \quad \text{(rate determining step)}$$

$$H_2O_2 + H_2 = 2H_2O \quad \text{(fast reaction)}$$

[6] Bodenstein, *Zeitschr. f. physik. Chem.*, **100**, 68 (1922). Trautz, *Zeitschr. f. anorg. Chem.*, **88**, 285 (1914). Trautz and Dalal, *Zeitschr. f. anorg. Chem.*, **102**, 149 (1918). Hinshelwood and Green, *Jour. Chem. Soc.*, **129**, 730 (1926).

or also with the mechanism

$$2NO + H_2 = N_2O + H_2O \quad \text{(rate determining step)}$$
$$N_2O + H_2 = N_2 + H_2O \quad \text{(fast reaction)}$$

296. Reactions such as the above in which the rate is certainly found experimentally to be of higher order than the second do occur. The interpretation of these higher orders as indicating the simultaneous collision of three or more molecules is, however, not certain, since the existence of a small equilibrium concentration of a compound formed from the reacting substances is not ruled out, and this compound could then enter into the reaction leaving the order unchanged. Thus the reaction between three substances A, B and C to form their final products might take place in the following two steps

$$A + B \rightleftarrows AB$$
$$AB + C \rightarrow \text{Products}$$

Since the compound AB is assumed to be in equilibrium with A and B, its concentration would be maintained proportional to the product of the concentrations A and B, and the overall rate of the reaction would be that of the second step. The observed rate of the reaction would thus be of the third order in spite of the fact that it is brought about by two bimolecular steps.

297. The question of intermediate compound formation versus collision between three or more molecules is to some extent amenable to theoretical treatment. If the molecules of substances were *rigid* elastic spheres, as we assume for a first approximation, it is immediately evident that the simultaneous collision of three or more molecules could never occur, since the time during which two molecules remain in contact would be zero. It is evident, however, that for purposes of estimating the chance of chemical reaction, we are not interested in a hypothetical instantaneous contact between molecules, but rather in calculating the number of times per second that the molecules which are to react come within a small range of each other.

298. To investigate this problem, we shall first need to calculate the mean length of time that the surfaces of *two* colliding molecules remain within a short distance of each other δ. This distance δ may be regarded as the separation that the molecules can have and chemical action still be possible, and the length of time during which they remain within this distance may be called the duration of collision. We shall treat the molecules as rigid spheres and assume that δ is small compared with the diameters of these spheres.

If now we consider a collision between two of the molecules in which the *component of relative velocity parallel to the line of centers* has the value R, it is evident that before the collision the two molecules will be within the range δ for a time equal to δ/R. Furthermore it is evident from the principle of microscopic reversibility that on the average the time before and after collision during which the molecules are within this range will have the same values. Hence to determine the *mean duration* of these collisions, we merely have to determine the mean value of $2\delta/R$ for all collisions.

We have, however, in accordance with equation (123b) in Chapter 5 for the fraction of all collisions in which the component of relative velocity parallel to the line of centers lies in the range R to $R + dR$, the expression

$$\frac{dZ}{Z} = \frac{m_1 m_2}{m_1 + m_2} \frac{1}{kT} e^{-\frac{m_1 m_2}{m_1 + m_2} \frac{R^2}{2kT}} R dR$$

and hence we can obtain the mean duration τ for all collisions by an integration over all values of R in accordance with the expression

$$\tau = \int_0^\infty \frac{2\delta}{R} \frac{m_1 m_2}{m_1 + m_2} \frac{1}{kT} e^{-\frac{m_1 m_2}{m_1 + m_2} \frac{R^2}{2kT}} R dR \tag{544}$$

Simplifying and performing the integration we easily obtain

$$\tau = \delta \sqrt{\frac{2\pi m_1 m_2}{kT(m_1 + m_2)}} \tag{545}$$

or

$$\tau = \delta \sqrt{\frac{\pi m}{kT}} \tag{546}$$

in the case the two molecules have the same mass.

The reciprocals of these quantities

$$\frac{1}{\tau} = \frac{1}{\delta} \sqrt{\frac{kT(m_1 + m_2)}{2\pi m_1 m_2}} \tag{547}$$

and

$$\frac{1}{\tau} = \frac{1}{\delta} \sqrt{\frac{kT}{\pi m}} \tag{548}$$

will evidently be the chance per unit time that a pair of two molecules within a distance δ apart will break up. The result contains a small approximation, since it neglects the behavior of molecules which come within the distance δ without actually colliding.

299. With the help of these expressions for the chance per unit time that a pair of molecules within the reaction distance δ will break up, we can now determine the number of collisions suitable for a given reaction which will occur per second between any desired number of molecules. To do this we first calculate by the laws of chance, the number of molecular constellations of the kind that could lead to reaction which are present in the system at any time. With the help of equations (547) and (548), we then determine the chance per second that some one of the necessary pairs of molecules in the constellation will break up, and by multiplication with the number of constellations present obtain the rate per second at which the constellations themselves would be breaking up if reaction did not occur, and this can of course be taken as the rate at which they are forming.

300. The above method can best be illustrated by examples since the necessity of considering the particular configuration which could be expected to lead to reaction in any given case makes the development of general formulæ not very practicable.

As a hypothetical example, let us consider a system containing N_1, N_2 and N_3 molecules per cubic centimeter of three different kinds having the masses m_1, m_2 and m_3, and let us assume that reaction could be expected to have a chance of taking place when a collision occurs in which simultaneously a molecule of the first kind is within the distance δ of one of the second kind, and this in turn within the distance δ of one of the third kind.

If now we single out a given molecule of the first kind, it is evident that the chance that a molecule of the second kind will lie, within the distance δ from it, is equal to the ratio of the volume $4\pi\sigma_{12}^2\delta$ within which the center of the second molecule must lie to the total volume, substantially unity, in which it may lie, multiplied by the number of molecules of the second kind N_2. Hence for the number of pairs present per cubic centimeter at any instant we should expect

$$N_{12} = N_1 N_2 4\pi\sigma_{12}^2\delta \qquad (549)^*$$

* The above derivation of this equation contains approximations which are only valid at high enough dilutions. We may proceed more rigorously as follows. If we single out a *specified* pair of molecules of the first and second kinds, the probability that they will lie within the range δ, all other molecules of the first and second kinds lying *outside* the range δ of the specified molecules of the second and first kinds respectively, will be

$$P_1 = \frac{4\pi\sigma^2_{12}\delta}{1}\left(\frac{1-4\pi\sigma^2_{12}\delta}{1}\right)^{(N_1-1)}\left(\frac{1-4\pi\sigma^2_{12}\delta}{1}\right)^{(N_2-1)}$$

The molecule of the second kind, however, could now be chosen in N_2 *independent* ways to give us N_2 *mutually exclusive* configurations, each having the probability

The chance, however, that a molecule of the third kind will be within the distance of any given molecule of the second kind will be $N_3 4\pi\sigma_{23}^2\delta$. Hence neglecting the steric hindrance of the molecules, we should expect for the number of favorable triplets present at any time per cubic centimeter

$$N_{123} = N_1 N_2 N_3 (4\pi\sigma_{12}^2)(4\pi\sigma_{23}^2)\delta^2 \tag{550}$$

Such a triplet, however, will be broken up and reaction prevented if either of the pairs 12 or 23 breaks up. Hence referring to equation (547) for the chance per unit time that a pair will break up, we can write for the chance per unit time that a triplet will break up

$$\frac{1}{\tau} = \frac{1}{\delta}\sqrt{\frac{kT(m_1+m_2)}{2\pi m_1 m_2}} + \frac{1}{\delta}\sqrt{\frac{kT(m_2+m_3)}{2\pi m_2 m_3}} \tag{551}$$

The product, however, of the number of triplets present by their chance of breaking up evidently gives us the number of triplets disintegrating and reforming per second, so that we can write

$$Z = N_1 N_2 N_3 (4\pi\sigma_{12}^2)(4\pi\sigma_{23}^2)\delta\left\{\sqrt{\frac{kT(m_1+m_2)}{2\pi m_1 m_2}} + \sqrt{\frac{kT(m_2+m_3)}{2\pi m_2 m_3}}\right\} \tag{552}$$

for the number of triple collisions per second per cubic centimeter, of the particular kind which might lead to reaction.[7]

To use the above expression in interpreting a given reaction of the third order, we shall have to compare the quantity Z with the experimental rate

$$-\frac{dN_1}{dt} = kN_1 N_2 N_3 \tag{553}$$

given above, so that the probability that a specified molecule of the first kind will be thus associated with any one of the second kind is

$$P_1 = N_2 \frac{4\pi\sigma^2_{12}\delta}{1}\left(\frac{1-4\pi\sigma^2_{12}\delta}{1}\right)^{(N_1-1)}\left(\frac{1-4\pi\sigma^2_{12}\delta}{1}\right)^{(N_2-1)}$$

This, however, is the fraction of the time that a given molecule of the first kind is thus associated and hence the number of such pairs we may expect present at any time is

$$N_{12} = N_1 N_2 \frac{4\pi\sigma^2_{12}\delta}{1}\left(\frac{1-4\pi\sigma^2_{12}\delta}{1}\right)^{(N_1-1)}\left(\frac{1-4\pi\sigma^2_{12}\delta}{1}\right)^{(N_2-1)}$$

For dilute systems, however, the above expression reduces to equation 549. Furthermore, in such dilute systems the number of pairs in which more than one molecule of one kind is associated with a molecule of the other kind can be neglected.

Similar considerations apply to the derivation of equation 550.

[7] This treatment may be compared with the calculations of Smoluchowski, Boltzmann Festschrift, p. 626, and Syrkim, *Physik. Zeitschr.*, **24**, 236 (1923), where for the most part, however, the assumption of point particles has been made.

It is immediately evident that the reaction can be accounted on the basis of triple collisions only in case we can satisfy the relation

$$Z > kN_1N_2N_3 \qquad (554)$$

If on substituting reasonable values of the order 10^{-8} centimeters for σ_{12} and σ_{23} we do not find it possible to satisfy this inequality, taking δ as a very small fraction of the molecular diameter, we shall conclude that the mechanism of the reaction cannot be one depending on triple collisions, but must necessarily involve preliminary association of the molecules.

301. The application of these methods to actual experimental cases is at present limited by the nature of the data available. Most of the higher order reactions, that we know, occur in aqueous solution between ionized substances, and thus introduce difficulties. The hydration of the ions which presumably occurs may be expected to make the chances of break up less than we calculate from equations (547) and (548), and to introduce unknown steric hindrances. In addition the chance that two ions will be within a distance δ is increased if the ions are of the opposite sign and decreased if they are of the same sign, and this would have to be allowed for by the methods used by Milner and Debye in developing the interionic attraction theory.

As an experimental case to use, merely by way of illustration, we shall choose the gas reaction between nitric oxide and oxygen

$$2NO + O_2 \longrightarrow 2NO_2$$

and shall try to test the possibility that this reaction could occur as the result of triple collisions, in which two NO molecules are simultaneously within some small distance δ from an oxygen molecule.

In accordance with our previous method of treatment, the number of such triplets present per cubic centimeter at any time would be

$$N = \frac{N_1^2 N_2}{2} (4\pi\sigma_{12}^2\delta)^2 \qquad (555)$$

where N_1 and N_2 are respectively the numbers of NO and NO_2 molecules per cubic centimeter, and σ_{12} is the mean diameter of the two ions, and the reason for the factor 1/2 is evident.

Such a triplet can be broken up if either NO molecule leaves the O_2 molecule and hence in accordance with equation (547) we shall have for the chance of destruction per unit time

$$\frac{1}{\tau} = \frac{2}{\delta}\sqrt{\frac{kT(m_1 + m_2)}{2\pi m_1 m_2}} = \frac{2}{\delta}\sqrt{\frac{RT(M_1 + M_2)}{2\pi M_1 M_2}} \qquad (556)$$

Combining with equation (555), we get for the number of triple collisions per second per cubic centimeter which could lead to reaction

$$Z = \frac{N}{\tau} = N_1^2 N_2 (4\pi\sigma_{12}^2)^2 \delta \sqrt{\frac{RT(M_1 + M_2)}{2\pi M_1 M_2}} \tag{557}$$

Taking the mean diameter σ_{12} as 3×10^{-8} centimeters, the temperature as 273° absolute and substituting for M_1 and M_2 the molecular weights of NO and O_2, we obtain

$$Z = 1.95 \times 10^{-24} \, \delta N_1^2 N_2 \quad \text{collisions per second} \tag{558}$$

This is to be compared with the value obtained by Bodenstein for the rate of the reaction at 0°,

$$-\frac{dN_2}{dt} = 2.37 \times 10^{-38} \, N_1^2 N_2 \tag{559}$$

Comparing (558) and (559) we see that the value of δ could be as small as 10^{-18} centimeters and still have the number of collisions greater than the rate of reaction. Hence our calculation does not rule out the possibility of triple collisions as a possible mechanism for this third-order reaction, as was also concluded by Bodenstein (*l. c.*) using a different method of estimating the number of triple collisions.

If we assumed such a mechanism, then the small negative temperature coefficient actually found for the rate of this reaction would have to be accounted for by the hypothesis that triple collisions of high velocity are less likely to lead to reaction than those of low velocity. Such an assumption, however, would not seem out of the question. (Compare Chapter 17, Section 231.)

Although, as we have seen, our calculation has not eliminated the possibility of triple collision as the mechanism for the reaction between NO and O_2, nevertheless, as is so often the case in chemical kinetics, we can not conclude that this is the actual mechanism of the reaction. The reaction may be due to intermediate chemical steps of which we now have no knowledge.

302. Heterogeneous Reactions.—The foregoing treatment of order and mechanism applies to homogeneous reactions and specially to those taking place in gaseous systems. Some discussion must now be given of the rates of heterogeneous reactions. This is a field in which experimental data and theoretical interpretation are rapidly accumulating, and at the present stage of development it will be wise to treat only a few simple cases. We shall consider primarily reactions which occur at a solid-gas boundary.

In the case of heterogeneous reactions, two factors are operative in determining the rate of change, in the first place the rate at which the reacting substances come to the boundary where the reaction occurs, and in the second place the nature of this surface and the mechanism of the changes that it induces.

In the last chapter we discussed the first of these factors and obtained a simple expression (434) for the rate at which the molecules of a gas at a given temperature strike a surface of unit area. And in that chapter we applied this expression to investigate the rates of simple heterogeneous processes like evaporation and solution which are usually considered as physical rather than chemical. The equation, however, is of course also of value in interpreting the rates of chemical changes that take place at a surface.

As to the second factor mentioned above, it has been emphasized by Langmuir, in agreement with our knowledge of crystal structure as obtained from X-ray analysis, that we must regard the molecules at the surface of a solid as definitely oriented, forming a checkerboard-like arrangement or two-dimensional lattice. The exposed atoms or molecules on this surface are found by Langmuir, as perhaps might be expected, to have unsaturated fields of force on the side away from the solid, and hence to have a great tendency to adsorb gaseous molecules, presumably in definite positions on the different elementary spaces provided by the two dimensional lattice, to form a layer one molecule deep. Layers of greater depth apparently do not form until the gas is brought to pressures and temperatures near the condensation point. Under these circumstances it is evident that reactions at the surface will be largely determined by the behavior of these adsorbed films of unimolecular depth. And we shall confine our further considerations principally to cases in which such concepts apply.

303. The Rates of Adsorption and Reevaporation.—We shall first consider the rates of adsorption and reevaporation for two of several possible cases discussed by Langmuir. By equating the rates of adsorption and reevaporation in these cases, we shall then obtain equations connecting the amount of gas adsorbed at equilibrium with the pressure of the gas.

Case I.—As the first case let us consider adsorption on a plane surface containing N_0 elementary spaces per square centimeter, each capable of holding one molecule of adsorbed gas.

For the number of molecules striking this surface per second, we shall have in accordance with equation (437) of the preceding chapter

$$Z = \frac{p}{\sqrt{2\pi mkT}} \tag{560}$$

And if a is the chance (usually nearly unity) that a molecule will condense when it strikes an uncovered portion of the surface, and ϑ is the fraction of the surface which is bare, we shall have for the rate of adsorption

$$Z_{\text{ads.}} = a\vartheta Z \tag{561}$$

On the other hand, the rate of reevaporation will evidently be proportional to the number of molecules on the surface, giving us

$$Z_{\text{evap.}} = \nu N_0 \vartheta_1 \tag{562}$$

where ν is the chance per second at the given temperature that a molecule will reevaporate, and ϑ_1 is the fraction of all the N_0 spaces which contain an adsorbed molecule.

We evidently have, however, for the total surface

$$\vartheta + \vartheta_1 = 1 \tag{563}$$

and for the number of molecules N adsorbed on the surface

$$N = N_0 \vartheta_1 \tag{564}$$

Equating (561) and (562) to obtain the condition for equilibrium, and substituting (563) and (564), we easily obtain for the number of molecules adsorbed at equilibrium

$$N = N_0 \frac{\left(\frac{a}{\nu N_0}\right) Z}{1 + \left(\frac{a}{\nu N_0}\right) Z} \tag{565}$$

Or substituting σ for the ratio $a/\nu N_0$ we can write

$$N = N_0 \frac{\sigma Z}{1 + \sigma Z} \tag{566}$$

Since Z is related to the pressure of the gas by expression (560), this equation gives the relation between the number of molecules adsorbed at equilibrium and the pressure of the gas at a given temperature.

For the adsorption of a number of gases on mica and glass this equation was found by Langmuir[8] to express the actual experimental facts. The experimental value of N_0 in that equation was for all gases found to be somewhat less than the total number of molecules of solid calculated to be on the surface of the glass or mica. This shows that the assumption of layers not greater than one molecule thick is justified and indicates that some of the surface spaces did not have the proper fields of force to produce adsorption.

[8] Langmuir, *Jour. Amer. Chem. Soc.*, 40, 1361 (1918).

The quantity

$$\sigma = \frac{a}{\nu N_0} \tag{567}$$

occurring in equation (566) is capable of physical interpretation. a is the chance of condensation when a molecule hits a vacant space and νN_0 is the rate of evaporation from one square centimeter of occupied spaces. Hence σ is a measure of the binding force of adsorption and has been called by Langmuir the "relative life" of the molecules on the surface. The relative life of the molecules on glass and mica was found to be of an order several thousand to a million times greater than the relative life of the same molecules on the surface of condensed liquid at the same temperature. This illustrates the great strength of the stray fields of force from the solid surface and explains the lack of tendency to form a second layer on the solid surfaces except when near the condensation point.

Case II.—As a second case let us consider the adsorption of a diatomic gas on a surface such that each atom is held on a separate one of the elementary spaces, and assume a large value for the heat of dissociation so that the rate of evaporation of individual atoms will be small compared with the rate of evaporation of two atoms from adjacent spaces in the form of a molecule.

In the state of equilibrium Langmuir assumes a haphazard distribution of the atoms over the surface. If ϑ is the fraction of the surface that is bare, then the chance that any given space is vacant is ϑ and the chance that two given adjacent spaces are vacant ϑ^2, so that the rate of adsorption of molecules, which can only occur with two adjacent spaces vacant will be proportional to ϑ^2. On the other hand the chance of re-evaporation will be determined by the chance that two adjacent spaces are filled and hence will be proportional to $\vartheta_1{}^2$, where ϑ_1 is the fraction filled. Hence equating the two rates to determine the conditions at equilibrium we obtain

$$a\vartheta^2 Z = \nu_1 \vartheta_1{}^2 \tag{568}$$

where a and ν_1 are constants and Z is the rate, given by equation (560), at which molecules strike the surface.

For the number of atoms adsorbed, however, we can substitute $N_0\vartheta_1$ and for the number of molecules N one half this value, giving us

$$N = \frac{N_0\vartheta_1}{2} \quad ; \quad \vartheta_1 = \frac{2N}{N_0} \quad ; \quad \vartheta = 1 - \frac{2N}{N_0} \tag{569}$$

and substituting in (568) we obtain

$$a\left(1 - \frac{2N}{N_0}\right)^2 Z = \nu_1 \left(\frac{2N}{N_0}\right)^2 \tag{570}$$

For a simple case of practical interest where the fraction covered is small compared with unity, we obtain

$$N = \frac{N_0}{2}\sqrt{\frac{aZ}{\nu_1}} \tag{571}$$

There is evidence to show that this type of atomic adsorption occurs for certain surfaces and gases, and we shall use equation (571) in interpreting heterogeneous reaction rates.

For n-atomic molecules condensing in n separate elementary spaces we obtain

$$N = \frac{N_0}{n}\left(\frac{aZ}{\nu_1}\right)^{\frac{1}{n}} \tag{572}$$

304. Reaction between a Gas and Solid.—The simplest imaginable case of a heterogeneous chemical change would perhaps be the reaction between a solid and a gas, in which a certain fraction of the molecules coming up to the surface combine with the solid, the product formed being immediately removed by evaporation. The case is quite nearly met by the reaction studied by Langmuir[9] between a hot tungsten filament and oxygen to form the yellow oxide WO_3, which distills off as soon as formed. The rate of the reaction was found at any temperature to be proportional to the pressure of oxygen.

Experiments on the effect of oxygen on the electron emission of the filament as well as other data, indicated that the filament in these measurements was almost completely covered with a unimolecular oxide film. The number of molecules of oxygen hitting this film per second is evidently proportional to the pressure in accordance with equation (560), and hence if we assume at any given filament temperature that a definite fraction of the impinging molecules react together with oxygen present in the film to form WO_3, we should expect the rate to be proportional to the pressure of the oxygen as was found experimentally by Langmuir. The fraction of the impinging molecules reacting was found to increase with the temperature of the filament from 0.00033 at 1070° K to 0.15 at 2770° K. The fraction reacting was, however, unaffected by the temperature of the gas (i.e. bulb temperature), the number reacting at a given filament temperature being always proportional to the number colliding with the film as given by equation (560).

305. Reaction of a Gas at a Solid-Solid Interface.—In case the product formed by reaction between a solid and gas is itself a solid not miscible with the first, we must regard the system as containing three phases, two solid and one gaseous. For such a system it can easily

[9] Langmuir, *Jour. Amer. Chem. Soc.*, 37, 1139 (1915).

be shown theoretically that the reaction rate cannot in general be determined by the area of the exposed surfaces of the two solids, but that the chemical change must take place at the solid-solid interface between them.

As a typical example let us consider the reaction between copper oxide and hydrogen in accordance with the equation

$$H_2 + CuO = H_2O + Cu$$

Suppose now we should assume that the two opposing reactions which are both taking place at equilibrium had rates proportional to the exposed surfaces S_1 and S_2 of copper oxide and copper, and to the impacts made by hydrogen and water vapor per square centimeter of surface, we should then have in accordance with equation (560) for the number of impacts, the condition for equilibrium

$$p_1 S_1 = K p_2 S_2 \tag{573}$$

where K is a constant and p_1 and p_2 are the partial pressures of hydrogen and water vapor. We should thus find that the ratio of pressures depended on the ratio of the exposed surfaces. We know, however, from the principles of chemical equilibrium that the ratio of the pressures is independent of the amounts of the two phases whenever they are present in appreciable quantities. Hence we conclude that at equilibrium the rates of the two reactions can only depend on the extent of the common interface between the two solids, and that the reactions take place at this boundary. We are thus led to the belief that, even when equilibrium has not been reached, reactions in such cases will take place at an interface. Similar conclusions in the case of other types of reactions involving two separate solid phases can of course also be drawn.

These conclusions have been satisfactorily confirmed by the measurements of Pease and Taylor [10] on the rate of the above reaction, by Jones and Taylor [11] on the rate of reduction of copper oxide by carbon monoxide and by many other examples of the autocatalytic effect produced by the increase in interface area as reactions of such kind proceed.

306. Catalytic Reactions of Gases at a Solid Surface.—The theory of adsorption to form unimolecular layers has been used by Langmuir in interpreting rates of gas reactions which are catalyzed by a solid surface. In discussing such reactions it is convenient to treat the two limiting cases in which the surface is only slightly covered by adsorbed gas, and in which the surface is nearly completely covered.

[10] Pease and Taylor, *Journ. Amer. Chem. Soc.*, **43**, 2179 (1921).
[11] Jones and Taylor, *Journ. Phys. Chem.*, **27**, 623 (1923).

Case I.—If we have a case in which the surface of the catalyst is only slightly covered, we shall have at any temperature, in accordance with equations (572) and (560), for the number of molecules of adsorbed gas of any kind per square centimeter

$$N = \frac{N_0}{n}\left(\frac{aZ}{\nu_1}\right)^{\frac{1}{n}} = Kp^{\frac{1}{n}} \tag{574}$$

where n is the number of elementary spaces occupied by the molecule in question, K is a constant and p is the partial pressure of the gas in question.

This equation can immediately be used to interpret a number of catalytic reactions. Thus the dissociation of hydrogen by a hot filament may be regarded as due to the evaporation of atoms which are held on separate elementary spaces on the filament surface. This will give us $n = 2$, and the rate is actually found experimentally to be proportional to the square root of the pressure. As another example the experiments of Bodenstein and Ohlmer [12] on the rate of reaction between carbon monoxide and oxygen on *crystalline* quartz surfaces can be interpreted if we assume that the carbon monoxide condenses on a single elementary space and the oxygen atomically on two elementary spaces, since the rate was experimentally found to be proportional to the partial pressure of the monoxide and the square root of the partial pressure of the oxygen. This is just what we should expect if the reaction occurs between an adsorbed carbon monoxide molecule and a neighboring oxygen atom.

The mode of application of such considerations to other cases where the fraction of the surface covered is small will be evident.

Case II.—If we have a case in which the surface is largely covered, the treatment of heterogeneous catalysis is again easily handled.

The simplest example will be the effect of an adsorbed layer acting as a catalytic poison. Thus the strong adsorption of oxygen on a tungsten filament, previously mentioned, makes a trace of oxygen a catalytic poison for the decompositions of hydrogen, methane and ammonia by actually preventing these gases from reaching the metal surface. Only after the oxygen has been used up to form WO_3 as previously discussed does the decomposition of the other gas commence.

As a more complicated case, we may take the reaction between carbon monoxide and oxygen on a platinum surface.[13] At low temperatures, 500° to 700° K, and pressures of the order of 0.1 mm. or less, the reaction rate is proportional to the pressure of oxygen and *inversely* proportional to the pressure of carbon monoxide and has a high temperature coefficient.

[12] Bodenstein and Ohlmer, *Zeitschr. für physik. Chem.*, **53**, 175 (1805).
[13] Langmuir, *Jour. Amer. Chem. Soc.*, **37**, 1139 (1915); Trans. Faraday Soc., **17**, 621 (1921-22).

At higher temperatures, 750° to 1050° K, the rate is practically independent of the temperature. With an excess of oxygen the rate at these higher temperatures is proportional to the pressure of the carbon monoxide, and with an excess of carbon monoxide proportional to the pressure of the oxygen.

These facts can be explained if we assume a tendency for the formation of unimolecular layers, in the case of oxygen each *atom* being held by one elementary space and in the case of carbon monoxide each molecule being held by two elementary spaces corresponding to the two unsaturated valences of the carbon atom. We must further assume that the reaction occurs only when an adsorbed oxygen atom is hit by a carbon monoxide molecule from the gas, and that at low temperatures the surface is largely covered by carbon monoxide and at high temperatures covered by oxygen in the presence of an excess of that gas but otherwise nearly bare.

If ϑ is the fraction of the surface which is bare and ϑ_1 the fraction covered by carbon monoxide, we may equate under equilibrium conditions the rates of adsorption and evaporation of carbon monoxide in accordance with the equation

$$a_1\vartheta^2 Z_1 = \nu_1\vartheta_1 \tag{575}$$

On the other hand in the case of oxygen we may equate the rate of adsorption $a_2\vartheta^2 Z_2$ to the rate of reaction ω, since we shall assume that adsorbed oxygen atoms remain until removed by reaction with carbon monoxide, giving us

$$\omega = a_2\vartheta^2 Z_2 \tag{576}$$

Furthermore, we can obviously also express the rate of formation of carbon dioxide by the equation

$$2\omega = \varepsilon Z_1 \vartheta_2 \tag{577}$$

where ϑ_2 is the fraction of the surface covered by oxygen and ε the chance of reaction by an impinging carbon monoxide molecule. And finally we can write for the total surface

$$\vartheta + \vartheta_1 + \vartheta_2 = 1 \tag{578}$$

These four equations permit a general solution for the four variables ϑ, ϑ_1, ϑ_2 and ω.

For the low temperature case, however, we can assume the surface nearly covered by the monoxide, giving us $\vartheta_1 = 1$, and can combine equations (575) and (576) to give at once

$$\omega = \frac{\nu_1 a_2 Z_2}{a_1 Z_1} \tag{579}$$

Since Z_1 and Z_2 by equation (560) are proportional to the pressures of carbon monoxide and oxygen, we thus have an immediate agreement with the experimental facts. The large temperature coefficient of the reaction rate at the lower temperatures is to be explained by the increasing fraction of the surface left free of carbon monoxide as the temperature rises.

At higher temperatures, in the presence of excess of oxygen, the platinum is nearly covered with oxygen so that the rate is determined by the number of collisions of carbon monoxide with this adsorbed film and is hence proportional to the pressure of carbon monoxide as experimentally found. At higher temperatures in the presence of excess of carbon monoxide, the surface is largely bare and the rate is determined by the rate at which oxygen strikes the surface, since any oxygen that condenses remains until it is removed by reaction with carbon monoxide. The rate is then proportional to the pressure of oxygen as experimentally found.

Almost the same considerations were found to hold for the reaction between hydrogen and oxygen on platinum surfaces, and similar treatment can be successfully given to a number of other catalytic reactions.

307. Explosive Reactions.—In concluding this chapter a brief mention may be made of the beautiful results obtained by F. W. Stevens[14] on the velocity of explosion waves in mixtures of carbon monoxide and oxygen. The experiments were performed by blowing soap bubbles containing the two gases in different proportions, touching off the mixture electrically at the center, and following the propagation of the luminous reaction zone photographically. The reaction thus takes place under conditions of constant pressure.

The remarkable result was found that the speed of propagation of this explosive wave was proportional to the concentration of the oxygen and the square of the concentration of the carbon monoxide in accordance with the equation

$$s = k(O_2)(CO)^2 \tag{580}$$

This result presumably bears some relation as yet unknown to the order which would be assumed from the stoichiometric equation

$$O_2 + 2CO = 2CO_2$$

The whole field is a very interesting one for further investigation. It would be specially interesting to investigate the temperature coefficient of such speeds of propagation.

[14] F. W. Stevens, *Zeitschr. des Ver. deutschen Ingenieure*, No. 20, p. 659 (1926); *Jour. Amer. Chem. Soc.*, **48**, 1896 (1926).

Chapter 21.

The Temperature Coefficient of Reaction Rate.

308. Introduction. The Arrhenius Postulate.—In the preceding chapter the temperature coefficient of reaction rates was only incidently mentioned and the subject will be treated more fully in the present chapter. The discussion will apply mainly to homogeneous gas reactions.

As already mentioned reaction velocities are usually found to have a very considerable temperature coefficient, the rate often doubling or trebling for a ten degree rise in temperature. The fundamentality of this fact, however, must not be unduly exaggerated, since it applies of course to the data already made available by observation, and reactions in which the rate cannot be controlled by temperature would be difficult to investigate if the rate happened to be very large or very small. Thus as noted in the preceding chapter, if the rate of bimolecular reactions without appreciable temperature coefficient should be of the order of the number of collisions occurring, the rates would be too fast to measure except by indirect methods.

To explain the high temperature coefficient of known reaction rates, it was postulated by Arrhenius that the molecules involved in the reaction are unreactive when in their normal states, and that reaction can only occur between molecules in activated states. Since the concentration of the molecules in an activated state would be proportional to their total concentration, this postulate would leave the order of the reaction the same as though molecules in their normal state could react. Furthermore, since the concentration of activated molecules would increase rapidly with the temperature, we should thus be able to account for the high temperature coefficient of reaction rate.

This hypothesis of Arrhenius with its later modifications has proved extraordinarily suggestive. The activated states, thus postulated by him long before the modern development of the quantum theory, may now be identified when occasion requires with higher quantum states of the molecule. The process of activation in other cases, however, can consist in a dissociation or combination of a kind more familiar in the older chemistry. In addition, as we shall presently show, it is possible to calculate on the basis of the hypothesis the average energy of the activated molecules from a knowledge of the temperature coefficient of the reaction rate itself.

309. The chief obstacle at present encountered by the Arrhenius picture of the mechanism of chemical reaction lies in the difficulty of discovering processes by which sufficient energy can be transmitted to molecules in their normal states rapidly enough so that the quota of activated molecules is replenished as fast as is necessary to account for observed rates of reaction. We shall give special attention to this feature of the problem in a later part of the chapter. Whether this difficulty means that the Arrhenius picture must be abandoned or subjected to further serious modification is not now evident. It may be mentioned in advance, however, that other quantum processes are certainly known, in which energy is supplied at a rate far too fast to be easily explained.

310. We shall first investigate the relation between temperature coefficient and energy of activation which would be expected on the basis of the Arrhenius hypothesis. To do this it will be advantageous to give separate treatments to four possible different cases: (1) unimolecular reactions in which the rate of activation is fast enough to maintain the equilibrium concentration of activated molecules; (2) unimolecular reactions in which the rate of activation is not fast enough to maintain the equilibrium concentration of activated molecules; (3) bimolecular reactions between previously activated molecules; and (4) bimolecular reactions in which the energy of activation is the kinetic energy of the colliding molecules themselves.

311. Unimolecular Reactions in Which the Activated Molecules Are Maintained at the Equilibrium Concentration.—Let us consider a first-order unimolecular reaction in which the rate at which molecules can be activated is so high compared with the rate at which the activated molecules decompose or otherwise react that the equilibrium concentration of molecules in the activated states is always maintained.

For the number of molecules in any activated state a, we can then write in accordance with the Maxwell-Boltzmann distribution law

$$N_a = \frac{Np_a e^{-\frac{\epsilon_a}{kT}}}{\sum_i p_i e^{-\frac{\epsilon_i}{kT}}} \tag{581}$$

where N is the total number of molecules, p_a the a-priori probability of the state a, ϵ_a the energy of a molecule in that state, and the summation in the denominator is to be taken over all states.

If now we assign to the molecules in each activated state a a chance of spontaneous reaction of magnitude k_a per second, we should obtain for the total rate of reaction

$$-\frac{dN}{dt} = \frac{N\sum\limits_a k_a p_a e^{-\frac{\varepsilon_a}{kT}}}{\sum\limits_i p_i e^{-\frac{\varepsilon_i}{kT}}} \tag{582}$$

where the summation in the numerator is to be taken over all the states a which contribute appreciably to the total rate of change.

It is evident that the postulated mechanism leads to a first-order reaction as assumed, the specific rate of reaction being

$$k_1 = \frac{\sum\limits_a k_a p_a e^{-\frac{\varepsilon_a}{kT}}}{\sum\limits_i p_i e^{-\frac{\varepsilon_i}{kT}}} \tag{583}$$

An actual knowledge of the values of k_a for the different activated states would now permit a theoretical prediction of the actual specific rate. This is of course for the present beyond our powers. We can, however, obtain information as to the temperature coefficient of the specific reaction rate by differentiating with respect to T. It will be most advantageous to take a logarithmic differentiation, giving us

$$\frac{d\log k_1}{dT} = \frac{1}{k_1}\frac{dk_1}{dT} = \frac{1}{k_1}\frac{\sum\limits_a k_a p_a e^{-\frac{\varepsilon_a}{kT}}\frac{\varepsilon_a}{kT^2}}{\sum\limits_i p_i e^{-\frac{\varepsilon_i}{kT}}} - \frac{\sum\limits_i p_i e^{-\frac{\varepsilon_i}{kT}}\frac{\varepsilon_i}{kT^2}}{\sum\limits_i p_i e^{-\frac{\varepsilon_i}{kT}}} \tag{584}$$

The first term on the right-hand side of equation (584) is evidently the sum of the values of ε_a/kT^2 for the molecules that react divided by the number so reacting and hence gives the average energy of the molecules that enter into reaction. The second term on the other hand gives the average energy of all the molecules, so that we may rewrite equation (584) in the simple form

$$\frac{d\log k_1}{dT} = \frac{\overline{\overline{\varepsilon}} - \overline{\varepsilon}}{kT^2} \tag{585}$$

where $\overline{\overline{\varepsilon}}$ is the average energy of the molecules that react and $\overline{\varepsilon}$ is the average energy of all the molecules.

312. In the cases of usual interest, the average energy of *all* the molecules and the average energy of the *unactivated* molecules will not be appreciably different, since very few molecules will be in the activated states because of their high energy content. Hence the difference $(\bar{\bar{\varepsilon}} - \bar{\varepsilon})$ will be the average energy that has to be supplied to the molecules that react and may be called the *energy of activation*. We thus see that for this case the temperature coefficient of reaction rate immediately gives the energy of activation.

313. Unimolecular Reactions in Which the Rate of Activation Is Not Fast Enough to Maintain the Equilibrium Concentration of Activated Molecules.—We must next consider first-order unimolecular reactions in which the rate of reaction is so fast compared with the rate at which molecules can be activated that the equilibrium concentration of activated molecules is not maintained. As a limiting case this includes the merging of the processes of activation and reaction, every molecule that is activated entering into the reaction.

314. In order to carry out our treatment we shall first need to obtain an important *relation between the rates of activation and deactivation of molecules*, which can be derived with the help of the principle of microscopic reversibility.

Consider a system in equilibrium, and let N_i and N_a be the number of molecules per cubic centimeter in two states i and a. For convenience we may call i the initial state and a an activated state. Now we may obviously put the rate at which molecules are passing from state i to state a proportional to the number in state i in accordance with the equation

$$Z_{ia} = \beta_{ia} N_i \tag{586}$$

where we may call β_{ia} the *coefficient of activation*.

Consider now the *reverse* processes by which molecules pass from state $-a$ to $-i$. Evidently the rate of such reverse processes will be proportional to the number of molecules N_{-a} in the state $-a$, so that we can write for the number of molecules passing per second from state $-a$ to $-i$

$$Z_{-a-i} = \alpha_{-a-i} N_{-a} \tag{587}$$

where α_{-a-i} may be called a *coefficient of deactivation*.

In accordance with the principle of microscopic reversibility, however, these two rates must be equal if the system is at equilibrium so that we can write

$$\beta_{ia} N_i = \alpha_{-a-i} N_{-a} \tag{588}$$

The two concentrations, however, are connected in accordance with the Maxwell-Boltzmann distribution law by the equation

$$\frac{N_{-a}}{N_i} = \frac{p_{-a}}{p_i} e^{-\frac{\epsilon_{i-a}}{kT}} \tag{589}$$

where p_{-a} and p_i are the a-priori probabilities of the two states and ε_{i-a} is the difference in energy of the two states.

Substituting in equation (588) we obtain

$$\beta_{ia} = \alpha_{-a-i} \frac{p_{-a}}{p_i} e^{-\frac{\varepsilon_{i-a}}{kT}} \tag{590}$$

as our relation between the coefficients of activation from states i to a, and deactivation from states $-a$ to $-i$.

The a-priori probability p_{-a} is, however, evidently equal to the a-priori probability p_a of its reverse state and the energy difference ε_{i-a} between states i and $-a$ is evidently equal to the energy difference ε_{ia} between states i and a. This permits us to rewrite equation (590) in the more convenient form

$$\beta_{ia} = \alpha_{-a-i} \frac{p_a}{p_i} e^{-\frac{\varepsilon_{ia}}{kT}} \tag{591}$$

This important equation makes no assumption as to the mechanism of the activational and deactivational processes, and the coefficients β_{ia} and α_{-a-i} may be dependent on the total concentration of molecules present if the mechanism is collisional, or on the concentration of radiant energy of different frequencies if this is involved, or on whatever factors may be operative. The equation may be regarded as a generalization of the relation between probabilities of absorption and emission discussed in Chapter 16, and the relation between probabilities of collisions of the first and second kind discussed in Chapter 17.

It must be noted that equation (591) has been rigorously derived only for a system in statistical equilibrium. We should usually expect, however, that the specific rates at which molecules pass from one state to another will not be greatly affected by a deviation of the system from equilibrium. In what follows we shall use equation (591) as valid at any stage of the reaction. This is the less questionable since it only necessitates the constancy of the ratio β_{ia} to α_{-a-i}, rather than the individual constancy of the two terms.

315. With the help of this equation we are now ready to treat the more general case of first-order unimolecular reactions where the rate of activation is not necessarily fast enough to maintain the equilibrium concentration of activated molecules.

For the rate at which molecules are being activated we may write

$$\frac{dN_{act.}}{dt} = \sum_i \sum_a \beta_{ia} N_i \tag{592}$$

where the summation is to be taken over all inactive states i and activated states a.

A certain fraction of the molecules which arrive in any activated state a may react and the remainder return into unactivated states. For the fraction ϑ_a that react we may write

$$\vartheta_a = \frac{\sum\limits_f a_{af}}{\sum\limits_f a_{af} + \sum\limits_i a_{ai}} \tag{593}$$

where $\sum\limits_f a_{af}$ is the total *specific* rate at which molecules pass from the active state *forward* to a final unactivated state of the product, and $\sum\limits_i a_{ai}$ is the total specific rate of deactivation *back* to any initial state i. In case the unimolecular reaction is an isomeric change the final states f will be those of a single molecule, but in the case of dissociation f stands for a simultaneous condition of more than one molecule. The formulation neglects the passage of a molecule from one activated state to another activated state, but this is not serious.

Combining equation (592) for the rate of activation with the fraction of activated molecules which react, we may now write for the rate of the unimolecular reaction

$$-\frac{dN}{dt} = \sum_i\sum_a \vartheta_a \, \beta_{ia} \, N_i \tag{594}$$

316. This equation contains no hypothesis except our general picture of the mechanism of unimolecular reaction, and if we knew the value of the factors we could use it to calculate the actual rate. This is of course not now possible but we can use the equation to obtain information as to the temperature coefficient of the reaction rate.

For β_{ia} we can substitute the expression derived above, given by equation (591), and, assuming statistical equilibrium for the *unactivated* molecules as seems justifiable, we can substitute for N_i the usual Maxwell-Boltzmann expression. We obtain

$$-\frac{dN}{dt} = \frac{\sum\limits_i\sum\limits_a \vartheta_a \, a_{a \cdot i} \frac{p_a}{p_i} e^{-\frac{\varepsilon_{ia}}{kT}} N p_i e^{-\frac{\varepsilon_i}{kT}}}{\sum\limits_i p_i e^{-\frac{\varepsilon_i}{kT}}} \tag{595}$$

where N is the total number of molecules in unactivated states, which is sensibly equal to the total number of molecules that have not reacted. Noting that we have the obvious relation $\varepsilon_{ia} = \varepsilon_a - \varepsilon_i$ we can rewrite this in the form

$$-\frac{dN}{dt} = \frac{\sum\limits_i \sum\limits_a \vartheta_a \, a_{\cdot a \cdot i} p_a \, e^{-\frac{\epsilon_a}{kT}}}{\sum\limits_i p_i e^{-\frac{\epsilon_i}{kT}}} N \qquad (596)$$

and obtain for the specific reaction rate

$$k_1 = \frac{\sum\limits_i \sum\limits_a \vartheta_a \, a_{\cdot a \cdot i} p_a \, e^{-\frac{\epsilon_a}{kT}}}{\sum\limits_i p_i e^{-\frac{\epsilon_i}{kT}}} \qquad (597)$$

Taking as in the previous treatment a logarithmic differentiation of the specific reaction rate with temperature, we can now write

$$\frac{d \log k_1}{dT} = \frac{1}{k_1}\frac{dk_1}{dT} = \frac{1}{k_1} \frac{\sum\limits_i \sum\limits_a \vartheta_a \, a_{\cdot a \cdot i} p_a \, e^{-\frac{\epsilon_a}{kT}} \left(\frac{d \log \vartheta_a}{dT} + \frac{d \log a_{\cdot a \cdot i}}{dT} + \frac{\varepsilon_a}{kT^2} \right)}{\sum\limits_i p_i e^{-\frac{\epsilon_i}{kT}}}$$

$$- \frac{\sum\limits_i p_i e^{-\frac{\epsilon_i}{kT}} \frac{\varepsilon_i}{kT^2}}{\sum\limits_i p_i e^{-\frac{\epsilon_i}{kT}}} \qquad (598)$$

It is evident from the principle of averaging, however, applied in the same manner as in Section **311**, that this can be rewritten in the form

$$\frac{d \log k_1}{dT} = \overline{\overline{\frac{d \log \vartheta_a}{dT}}} + \overline{\overline{\frac{d \log a_{\cdot a \cdot i}}{dT}}} + \frac{\overline{\overline{\varepsilon_a}}}{kT^2} - \frac{\overline{\varepsilon_i}}{kT^2} \qquad (599)$$

where the double line indicates an average for the molecules that actually enter into reaction, and the single line indicates the average for the unactivated molecules.

317. In using this expression to interpret actual temperature coefficients, it should be noted that the first two terms on the right-hand side of this equation are presumably small. Referring to equation (593) for the value of ϑ_a we see that both these two first terms depend merely on the temperature coefficients of the rates at which molecules fall from higher to lower quantum states. Such deactivations can occur, however, only by collisions of the second kind with other molecules or by the emission of radiation.

The first of these processes cannot have a large temperature coefficient, since the number of collisions is proportional to $T^{1/2}$ and hence at the temperatures of interest does not change rapidly. This conclusion might have to be modified, however, if the quantum state of the deactivating molecule should be of major importance in determining the probability of a deactivating collision.

The second process can only take place by spontaneous and induced emission and can also have only a small temperature coefficient. To show this we note in accordance with Chapter 16 that we have for the combined rate of emission by the two processes

$$\alpha_{at} = B_{at}\frac{8\pi h\nu^3}{c^3}\left(1 + \frac{1}{e^{\frac{h\nu}{kT}} - 1}\right) \qquad (600)$$

and by simplification and differentiation obtain

$$\frac{d \log \alpha_{at}}{dT} = \frac{h\nu}{kT^2}\frac{1}{e^{\frac{h\nu}{kT}} - 1} \qquad (601)$$

Developing $e^{\frac{h\nu}{kT}}$ into a series, however, we see that the value of this can never be greater than $1/T$.

Hence in the cases of special interest where the temperature coefficient of the reaction rate is large, we may neglect the first two terms in (599) and again obtain the simple equation for temperature coefficient found in Section **311**

$$\frac{d \log k_1}{dT} = \frac{\bar{\bar{\varepsilon}} - \bar{\varepsilon}}{kT^2} \qquad (602)$$

where $\bar{\bar{\varepsilon}} - \bar{\varepsilon}$ is the *energy of activation* or average energy that has to be supplied to the molecules that react.

318. Bimolecular Reactions between Previously Activated Molecules.—It is even simpler to obtain the temperature coefficient for reaction rate in the case of second order bimolecular reactions.

If the reaction results from the collision of molecules which are already activated in higher quantum states, it is evident that the reaction can remain second order over a range of concentrations, only if the fractions of the two kinds of molecules in their activated states remain constant. Towards the end of a second order reaction, however, the fraction of either component removed per second by reaction becomes very small so that we are forced to the conclusion, in the case of such a mechanism, that the process of activation is fast enough at all stages of reaction to maintain the equilibrium concentration of activated molecules. Hence if N and N' are the numbers of molecules of the two kinds per cubic centimeter we can write for the concentrations of activated molecules in any activated states a and a', the Maxwell-Boltzmann expressions

$$N_a = \frac{Np_a e^{-\frac{\epsilon_a}{kT}}}{\sum\limits_i p_i e^{-\frac{\epsilon_i}{kT}}} \quad \text{and} \quad N_a' = \frac{N'p_a' e^{-\frac{\epsilon_a'}{kT}}}{\sum\limits_i p_i' e^{-\frac{\epsilon_i'}{kT}}} \tag{603}$$

Furthermore, since the number of collisions between molecules is proportional to the square root of the absolute temperature T, we may evidently write for the rate of the bimolecular reaction

$$-\frac{dN}{dt} = T^{1/2} \sum_a \sum_{a'} k_{aa'} N_a N_a' \tag{604}$$

where $k_{aa'}$ is a constant for each pair of activated states a and a', depending on the factors determining collision frequency given in equation (541) of the last chapter, and on the chance of reaction when molecules in these particular activated states collide. Substituting expressions (603) and solving for the specific reaction rate k_2, we obtain

$$k_2 = -\frac{1}{NN'}\frac{dN}{dt} = \frac{T^{1/2}\sum\limits_a \sum\limits_{a'} k_{aa'} p_a p_a' e^{-\frac{\epsilon_a + \epsilon_a'}{kT}}}{\left(\sum\limits_i p_i e^{-\frac{\epsilon_i}{kT}}\right)\left(\sum\limits_i p_i' e^{-\frac{\epsilon_i'}{kT}}\right)} \tag{605}$$

319. Here again we do not have the necessary information to use this equation for the direct prediction of specific reaction velocities, but by differentiation can obtain information as to the temperature coefficient. Carrying out a logarithmic differentiation with respect to the temperature T, we obtain

$$\frac{d \log k_2}{dT} = \frac{d \log T^{1/2}}{dT} + \frac{1}{k_2} \frac{T^{1/2}\sum\limits_a \sum\limits_{a'} k_{aa'} p_a p_a' e^{-\frac{\epsilon_a + \epsilon_a'}{kT}} \frac{\varepsilon_a + \varepsilon_a'}{kT^2}}{\left(\sum\limits_i p_i e^{-\frac{\epsilon_i}{kT}}\right)\left(\sum\limits_i p_i' e^{-\frac{\epsilon_i'}{kT}}\right)}$$

$$- \frac{\sum\limits_i p_i e^{-\frac{\epsilon_i}{kT}} \frac{\varepsilon_i}{kT^2}}{\sum\limits_i p_i e^{-\frac{\epsilon_i}{kT}}} - \frac{\sum\limits_i p_i' e^{-\frac{\epsilon_i'}{kT}} \frac{\varepsilon_i'}{kT^2}}{\sum\limits_i p_i' e^{-\frac{\epsilon_i'}{kT^2}}} \tag{606}$$

This, however, can evidently be rewritten in the form

$$\frac{d \log k_2}{dT} = \frac{1}{2T} + \frac{\overline{\overline{\varepsilon_a + \varepsilon_a'}} - \overline{\varepsilon_i + \varepsilon_i'}}{kT^2} \tag{607}$$

where the double line indicates an average taken for the molecules that actually react and the single line an average taken for all molecules. Here again, since the average energy of all the molecules and the average energy of the molecules in unactivated states is approximately the same, the quantity $\overline{\overline{\varepsilon_a + \varepsilon_a'}} - \overline{\varepsilon_i + \varepsilon_i'}$ will be the average energy that has to be supplied to the molecules that react, so that we again have a simple relation between temperature coefficient of reaction rate and energy of activation.

320. Bimolecular Reaction in Which the Molecules are Activated by Collision.—Another possibility for bimolecular reaction would be that normal molecules will react provided they collide with sufficient relative velocity. Such a mechanism could be regarded as merging the processes of activation and reaction.

Assuming this mechanism we cannot of course specify the exact nature of a collision that would lead to reaction; it would seem reasonable, however, to make it dependent on the relative velocity of the molecules parallel to their line of centers at the time of impact. We have, however, in accordance with equation (123b) in Chapter 5 for the number of collisions in which the component of relative velocity parallel to the line of centers lies in the range R to $R + dR$

$$dZ = Z \frac{m_1 m_2}{m_1 + m_2} \frac{1}{kT} e^{-\frac{m_1 m_2}{m_1 + m_2} \frac{R^2}{2kT}} R dR \tag{608}$$

where Z is the total number of collisions, and hence on the basis of the assumed mechanism could write for the rate of reaction

$$-\frac{dN}{dt} = Z \frac{m_1 m_2}{kT(m_1 + m_2)} \int_0^\infty k_R e^{-\frac{m_1 m_2}{m_1 + m_2} \frac{R^2}{2kT}} R dR \tag{609}$$

where k_R is the chance of reaction if the component of relative velocity has the value R.

321. Without an actual knowledge of k_R this equation cannot of course be used to predict actual reaction rates. We can, however, obtain information from it, as to the temperature coefficient which would be expected on the basis of the assumed mechanism. Since the total number of collisions Z is proportional to the product $N_1 N_2 T^{\frac{1}{2}}$ we can obtain from (609) for the specific reaction rate

$$k_2 = AT^{-\frac{1}{2}} \int_0^\infty k_R e^{-\frac{m_1 m_2}{m_1 + m_2} \frac{R^2}{2kT}} R dR \tag{610}$$

where A is a constant.

Taking a logarithmic differentiation with respect to temperature we obtain

$$\frac{d \log k_2}{dT} = -\frac{1}{2T} + \frac{AT - \frac{1}{2}\int_0^\infty k_R e^{-\frac{m_1 m_2}{m_1 + m_2}\frac{R^2}{2kT}} \left(\frac{m_1 m_2}{m_1 + m_2}\frac{R^2}{2kT^2}\right) R dR}{k_2} \tag{611}$$

The second term, however, is evidently the average of the quantity in parentheses for the molecules that react, so that we can rewrite (611) in the form

$$\frac{d \log k_2}{dT} = +\frac{1}{2T} + \frac{\overline{\overline{\frac{m_1 m_2}{m_1 + m_2}\frac{R^2}{2}}} - kT}{kT^2} \tag{612}$$

It can easily be shown, however, from the laws of the conservation of energy and momentum that $\frac{m_1 m_2}{m_1 + m_2}\frac{R^2}{2}$ is the maximum kinetic energy available for transformation into other forms by a head-on collision of two particles with the relative velocity R. Furthermore, it can also easily be shown that kT is the average value of the above quantity for all collisions. So that the difference can be regarded as the average energy of the collisions that lead to reaction minus the average energy of all collisions and this can again be called the *energy of activation*, so that we have obtained a relation between temperature coefficient and energy of activation entirely similar to (607) of the preceding section.

322. Summary as to the Relation between Temperature Coefficient and Energy of Activation.—Examining the results of the preceding sections, we see that for all the mechanisms considered we can write for the temperature coefficient of the specific reaction rate of first-order unimolecular reactions

$$\frac{d \log k_1}{dT} = \frac{\overline{\overline{\varepsilon}} - \overline{\varepsilon}}{kT^2} \tag{613}$$

and for the specific reaction rate of second-order bimolecular reactions

$$\frac{d \log k_2}{dT} = \frac{1}{2T} + \frac{\overline{\overline{\varepsilon}} - \overline{\varepsilon}}{kT^2} \tag{614}$$

where the *energy of activation* $\overline{\overline{\varepsilon}} - \overline{\varepsilon}$ is at least *approximately* the difference between the average energy of the molecules that react and the average energy of all the molecules.

If we neglect the small change in the energy of activation with the temperature, the above equations can be integrated to give the following expressions for the rates of the two kinds of reaction

$$-\frac{dN}{dt} = k_0\, e^{-\frac{\bar{\bar{\epsilon}} - \bar{\epsilon}}{kT}} N \tag{615}$$

or

$$-\frac{dC}{dt} = k_0\, e^{-\frac{E_A}{RT}} C \tag{616}$$

where k_0 is the constant of integration and E_A the energy of activation per mol, and

$$-\frac{dN}{dt} = k_0 T^{1/2}\, e^{-\frac{\bar{\bar{\epsilon}} - \bar{\epsilon}}{RT}} NN' \tag{617}$$

or

$$-\frac{dC}{dt} = k_0' T^{1/2}\, e^{-\frac{E_A + E'_A}{RT}} CC' \tag{618}$$

where k_0 and k_0' are constants, and $E_A + E_A'$ is the sum of the energies of activation per mol brought into the reaction by the two substances. The method of using equations (617) and (618) for the case that the two molecules are alike will be evident.

323. From the known values of the temperature coefficients of reaction rate, it is now possible with the help of the above equations to calculate the actual values of the energies of activation if the Arrhenius hypothesis is correct. For typical reactions in which the rate doubles or trebles per ten degree rise at ordinary temperatures, this leads to energies of activation of the order which are known from spectral data to correspond to the oscillations of atoms inside a molecule. This seems entirely reasonable, since the weakening of a chemical bond between atoms might very probably be due to such oscillations. The result thus gives some small evidence in support of the Arrhenius picture.

324. The Mechanism of Activation.—If we are willing to continue on the basis of a tentative acceptance of the Arrhenius hypothesis, we must next inquire into the mechanism of the processes by which the molecules can be activated and thus prepared to enter into the reaction. Since the activated molecules of high energy content are continually being removed by the reaction, we have as a minimum requirement the necessity of activating the molecules as fast as they react.

In addition the order of the reaction may impose further requirements on the rate of activation. Thus if the reaction is first-order and unimolecular, the rate of activation must itself either be a first-order process, or if not first order, fast enough to maintain the equilibrium concentration of molecules in the activated state. If the reaction is second order and bimolecular, taking place between previously activated molecules, as we saw in Section **318**, the rate of activation must be fast enough to maintain the equilibrium concentration of the activated molecules.

We shall now consider the possibilities of activation by ordinary collision,—by collisions of the second kind,—by the absorption of radiation of a single frequency,—and by the absorption of radiation of more than one frequency.

325. Activation by Ordinary Collision.—The activation of a molecule by collision with another molecule of high enough velocity is one of the most obvious hypotheses to propose.

For a head-on collision between two particles, it can easily be shown from the principles of conservation of energy and momentum that the maximum kinetic energy which can be transformed into other forms is

$$\frac{m_1 m_2}{m_1 + m_2} \frac{R^2}{2}$$

where m_1 and m_2 are the masses of the two particles and R is their relative velocity. Hence for a collision between molecules it would seem fairly reasonable to take this same expression for the energy available, provided we take R as the component of the relative velocity of the two molecules parallel to their line of centers at the instant of contact. Hence if ε is the energy we must give a molecule in order to activate it, we shall take activation by collision as being possible only when

$$\frac{m_1 m_2}{m_1 + m_2} \frac{R^2}{2} > \varepsilon \tag{619}$$

In accordance with equation (123a) of Chapter 5, however, the fraction of all collisions Z in which the relative component of velocity has a value greater than a given value R_0 is $e^{-\frac{m_1 m_2}{m_1 + m_2} \frac{R_0^2}{2kT}}$ so that we may set as a reasonable upper limit for the rate of activation by collision

$$\frac{dN_{\text{act.}}}{dt} = Ze^{-\frac{\varepsilon_{\text{act.}}}{kT}} \tag{620}$$

where Z is the total number of collisions of the molecules which are to be activated and $\varepsilon_{\text{act.}}$ is the energy on the average that has to be supplied to them.

We must now inquire by an appeal to experimental facts whether such a rate of activation would be fast enough to account for known rates of reaction.

326. In the case of *unimolecular first-order reactions* it seems certain that the rate would not be fast enough. The clearest cut example of a first-order homogeneous gas reaction is the decomposition of nitrogen pentoxide. Comparing the rate of decomposition of this gas as determined by Daniels and Johnston at 25° with an initial concentration of 1.5×10^{-5} mols per cubic centimeter, with the rate of activation calculated from equation (620) substituting 10^{-7} as the diameter of the molecules, the decomposition is found as a matter of fact to take place approximately 10,000 times as rapidly as the calculated activation.[1] Similar results would be found for other first-order reactions with large temperature coefficients, and hence we cannot assume activation by ordinary collision as the mechanism for such reactions.

327. In the case of *bimolecular second-order reactions* the condition of affairs is quite different and it seems possible that activation by ordinary collision might account for known reaction rates. A very good example of a homogeneous gas reaction of this kind is given by the decomposition of nitrous oxide studied by Hinshelwood and Burk.

If we assume, in accordance with Section **318**, reaction between previously activated molecules, and *assume that each molecule brings half the necessary energy of activation,* then using 3.32×10^{-8} cm. as the diameter of a molecule it is found by calculation from equation (620) that the number of molecules activated per second would be 20,000 times as great as the number actually decomposing at the most unfavorable temperature.[2] On the other hand, if we should assume that *one* of the two reacting molecules had to bring *all* the energy of activation, the rate of activation by collision could not possibly be great enough.

Turning now to the possibility, discussed in Section **320**, that the reaction occurs between nitrous oxide molecules which collide with sufficient relative velocity we can immediately apply the methods of that section. The actual rate of the reaction was found to be given by an expression which can be put in the form [3]

$$-\frac{dC}{dt} = 7.60 \times 10^{13}\, T^{\frac{1}{2}}\, e^{-\frac{58450}{RT}}\, C^2$$

On the other hand, by taking the diameter of the nitrous oxide molecule as having the rather large value 1.1×10^{-7} centimeters, we obtain from equation (123a) in Chapter 5 *half* this figure for the number of mols of nitrous oxide colliding per cubic centimeter per second with an available energy greater than 58,450 calories per mol. Since the reaction would decompose *two* molecules of oxide per collision, it also seems possible to account for the reaction on this basis.

[1] See Tolman, *Jour. Amer. Chem. Soc.*, **47**, 1533 (1925).
[2] *Ibid.*, 1535.
[3] *Ibid.*, 1536.

328. We thus find that the bimolecular decomposition of nitrous oxide can be accounted for on the basis of activation by ordinary collisions. This also seems to be true for other *bimolecular* gas reactions. Indeed there seems to be some theoretical justification for expecting this always to be true.

To show this let us consider a simple case of bimolecular reaction between molecules of the same kind, and assume reaction for every collision where each of the molecules is in the activated state a. The rate of the reaction will then be

$$-\frac{dN}{dt} = 2Z_0 \left(\frac{Np_a e^{-\frac{\epsilon_a}{kT}}}{\sum\limits_i p_i e^{-\frac{\epsilon_i}{kT}}} \right)^2 \tag{621}$$

where Z_0 is the number of collisions at unit concentration. On the other hand, for the rate of activation we have in accordance with equation (620)

$$\frac{dN_{\text{act.}}}{dt} = Z_0 e^{-\frac{\epsilon_a}{kT}} N^2 \tag{622}$$

Since the ratio of (622) over (621), however, contains the large quantity $e^{\frac{\epsilon_a}{kT}}$ we may expect in general that the rate of activation is large compared with the rate of reaction, if each molecule brings half the total energy of activation.

329. Activation by Collisions of the Second Kind.—Since ordinary collisions will not account for the mechanism of activation for unimolecular reactions, we must investigate other possibilities.

It has been suggested by Christiansen and Kramers[4] that the mechanism of activation consists in collisions of the second kind, between activated molecules of the product formed in the reaction and unactivated molecules of the original reactant, thus raising these latter to an activated state and preparing them to enter the reaction.

This theory can best be understood with the help of the following schematic representation of the course of a unimolecular chemical change in which a molecule of reactant R breaks down into products P and Q.

$$R \rightleftarrows R' \rightarrow \begin{cases} P' \rightarrow P \\ Q' \rightarrow Q \end{cases} \text{(Decomposition)} \tag{A}$$

$$P' + R \rightarrow P + R' \quad \text{(Collision of second kind)} \tag{B}$$

In accordance with expression (A) the reactant can exist in two forms, the normal form R and an activated form R'. The activated form breaks

[4] Christiansen and Kramers, *Zeitschr. für physik. Chem.*, **104**, 451 (1923).

down spontaneously into activated forms of the products P' and Q' and these then change into the normal forms P and Q. This latter change, however, can take place, at least in the case of one of the products, through a collision of the second kind with a molecule of the reactant R, thus leading to the activation of a new molecule through the mechanism indicated by expression (B).

In case the reaction deviates from a steady first-order course the number of such activations need not be the same as the number of molecules that decompose, but in case the reaction actually takes a first-order course, we must introduce the special assumption that each decomposition in accordance with (A) is *always immediately* followed by a successful activation in accordance with (B). When this latter assumption is true, it is evident that the full quota of activated molecules R' allowed by the Maxwell-Boltzmann law will be maintained, since the loss of a molecule by decomposition is immediately made good by the specially assumed mechanism, and, hence, the ordinary agencies responsible for the transitions $R \rightleftarrows R'$ are able to maintain the same statistical equilibrium that would exist in the absence of reaction.

Since this hypothesis would appear to provide us with a mechanism of activation which would always maintain the full Maxwell-Boltzmann quota of activated molecules, we could immediately apply the methods of Section **311**, and would have no difficulty in accounting for the temperature coefficient of unimolecular reactions. The method of application to bimolecular reactions is also obvious.

330. Although this simple and ingenious assumption of activation by collisions of the second kind has much to recommend it, there are undoubted difficulties which would have to be overcome before it could be regarded as the probable mechanism of activation in typical thermal reactions.[5]

In the first place in order to account on this basis for a unimolecular reaction that retains its first-order character throughout the reaction, we should have to assume a very high efficiency in the process of reactivation by collisions of the second kind as represented by expression (B). This means of course that activated molecules R' could only give up their energy to molecules of reactant R. Now as a matter of fact collisions of the second kind have been found to be pretty specific in their nature, as shown by the work of Stuart[6] on the quenching of the fluorescence radiation from mercury in the $2p_2$ state by the addition of a variety of gases. Nevertheless, it seems unreasonable to assume that molecules

[5] In cases of photochemical catalysis, collisions of the second kind are of prime importance.

[6] Stuart, *Zt. für Phys.*, 32, 262 (1925).

of the activated product P' could make a large number of collisions with a variety of molecules of different kinds without losing their energy, and thus being incapacitated for the activation of the original reactant R.

Now as a matter of fact, the decomposition of pure nitrogen pentoxide,[7] nitrogen pentoxide mixed with its own decomposition products, nitrogen pentoxide mixed with 1000 times as many molecules of oxygen,[8] and nitrogen pentoxide mixed with 76,000 times as many molecules of nitrogen [9] all take place at the same rate. In addition the decomposition of nitrogen pentoxide dissolved in carbon tetrachloride and in chloroform [10] take place at nearly the same rate. These facts make it very difficult to accept the proposal that collisions of the second kind account for the mechanism of activation in the decomposition of nitrogen pentoxide.

A second difficulty that would have to be met by the hypothesis of activation by collisions of the second kind arises in the case of endothermic reactions, since in such reactions a molecule of product in dropping to its normal state would not give up enough energy to raise a molecule of the reactant from its normal to its activated state. Hence, in the case of endothermic reactions, collisions of the second kind would presumably have to be supplemented by an additional activational step. If this supplementary step should occur before the collision of the second kind, then the concentration of molecules of reactant in a state suitable to make successful collisions with the activated product would be much lower than the total concentration of reactant, and the number of collisions during which the activated product would have to withhold its energy would be greatly increased. On the other hand, if the collisions of the second kind should be followed by a supplementary activation, it is evident that this step would have to be one of very high specific velocity, since the concentration of molecules in the intermediate state would be much lower than the total concentration of reactant. In either case it is evident that activation by collisions of the second kind gives no easy explanation of the mechanism of *endothermic* reactions.

In this connection it is important to note that the decomposition of nitrogen pentoxide is *endothermic* to the extent of 12,290 calories when it goes as it does at the beginning of reaction to nitrogen dioxide, and apparently still endothermic to the extent of 2,690 calories even when it goes to the tetroxide.[11] The thermochemistry of the actual individual steps of the reaction is of course unknown, since the steps themselves are

[7] Daniels and Johnston, *Jour. Amer. Chem. Soc.*, **43**, 53 (1921).
[8] White and Tolman, *Ibid*, **47**, 1240 (1925).
[9] Hunt and Daniels, *Ibid*, **47**, 1602 (1925).
[10] Lueck, *Ibid*, **44**, 757 (1922).
[11] Calculated from thermochemical values of Thomsen, "Thermochemistry," Longmans, Green and Co., 1920, p. 218, and Berthelot, *Ann. Chim. phys.*, **6**, 170, 171 (1875).

unknown. The matter does not look hopeful, however, for the proposed mechanism of activation.

We must conclude that activation by collisions of the second kind does not now seem possible as a general mechanism of activation for thermal reactions.

331. Activation by the Absorption of Radiation of a Single Frequency.—The very simple postulate has been made by Perrin and W. C. McC. Lewis that the process of activating a molecule consists in the absorption of a quantum of energy from the store of thermal radiation which surrounds the reacting molecules in accordance with the expression

$$h\nu + R \rightarrow R' \tag{623}$$

In the original formulation of the theory only the absorption of a single frequency was considered, corresponding to the existence of one normal and one activated state, so that the energy of activation would be directly connected with the frequency of the radiation by the quantum relation

$$\varepsilon_{\text{act.}} = h\nu \tag{624}$$

Assuming the radiation distributed in accordance with the Planck law, we can write, for the radiation density of frequency ν the expression

$$u_\nu = \frac{8\pi h\nu^3}{c^3} \frac{1}{e^{\frac{h\nu}{kT}} - 1} \tag{625}$$

and hence if B_{na} is the value of Einstein's coefficient of absorption for the passage from the normal state n to the activated state a, we can write for the rate of activation on the basis of this hypothesis

$$\frac{dN_{\text{act.}}}{dt} = B_{na} \frac{8\pi h\nu^3}{c^3} \frac{1}{e^{\frac{h\nu}{kT}} - 1} N \tag{626}$$

Noting that $e^{\frac{h\nu}{kT}}$ is usually large compared with unity, and substituting equation (624) this can be rewritten in the form

$$\frac{dN_{\text{act.}}}{dt} = B_{na} \frac{8\pi h\nu^3}{c^3} e^{-\frac{\varepsilon_{\text{act.}}}{kT}} N \tag{627}$$

In the case of first-order, unimolecular reactions it is to be assumed that every molecule or some definite fraction of the molecules would react as soon as activated and hence the rate of reaction be some definite fraction of the value given by equation (627). Hence the result agrees in form and content with our earlier equation (615) for the rate of first-order reactions. In the case of second-order, bimolecular reactions, it is assumed that the above mechanism of activation is fast enough to maintain the full quota of activated molecules permitted by the Maxwell-Boltzmann distribution law, the rate of reaction then depending on the number of collisions between activated molecules as treated in Section **318.**

Although there are a number of considerations which show, as we shall see, that the radiation theory in this simple form could not give a valid account of the mechanism of all reactions, nevertheless we cannot but admire the boldness and beauty of this conception which would unify our pictures of thermal and photochemical reaction, and account in the simplest possible manner for the orders of reaction by providing a method of activation which is independent of concentration.

332. We must first inquire whether the rate of activation given by equation (627) could be fast enough to account for known rates of reaction. To make this calculation, we must have a value for the coefficient of absorption B_{na}, and for this may make use of the discussion given in Section **221,** Chapter 16, which indicates that values of B_{na} greater than 10^{20} in c.g.s. units are not likely to be encountered. Hence, since everything else in the equation will be known in the case of any actual reaction whose temperature coefficient has been determined, we are at once in a position to test the question.

Again considering the decomposition of nitrogen pentoxide as a typical unimolecular reaction, we find, however, that the rate of activation calculated from equation (627) is smaller by a factor of 10^6 than the actual rate of decomposition experimentally found.[12] Similar relations will hold for any unimolecular reaction whose rate has a high temperature coefficient.

On the other hand, in the case of bimolecular reactions, equation (627) might lead to fast enough rates of activation provided we assume that half the energy of activation is brought into the reaction by each of the two reacting molecules. Thus for the decomposition of nitrous oxide the calculated rate of activation is 800 times the rate of decomposition at the most unfavorable temperature.[13]

333. In connection with such calculations as to the rate of activation by the absorption of radiation, it should of course be noted that in obtaining equation (627) we have assumed the radiation density distributed in accordance with the Planck radiation law. The process of activation, however, would draw on the supply of radiant energy of the particular frequency required and tend to lower the radiation density at

[12] See Tolman, *Jour. Amer. Chem. Soc.*, **47**, 1543 (1925).
[13] *Ibid*, p. 1544.

that point below the equilibrium value prescribed by the Planck law. Furthermore the occurrence of the reaction would presumably lead to the reemission of radiation of some other frequency, higher or lower than the original, depending on whether the total reaction was exothermic or endothermic. How rapidly radiation equilibrium would reestablish itself under these circumstances is difficult to determine. Of course the walls of the container would play their part in maintaining the equilibrium, but in addition all the radiation in the enclosure through interaction with the gas molecules present would strive to readjust itself to equilibrium, and it is not now evident how fast this would take place.

334. In addition to the failure of the simple radiation theory to provide a fast enough rate of activation to account for the rate of unimolecular reactions of high temperature coefficient, the radiation theory is subject to a very special test, since with the help of equation (624) we can calculate from the empirically determined energy of activation the frequency of light which should be photochemically active in bringing about reaction. Thus, in the case of the decomposition of nitrogen pentoxide, the energy of activation is found to be 24,700 calories per mol and this corresponds to a wave-length of 1.61 μ. It has been definitely shown by Daniels and Johnston,[14] however, that nitrogen pentoxide is ***not*** decomposed by light of this wave-length. Furthermore nitrogen pentoxide has no known absorption line or band at this wave-length.

335. Hence as the conclusion of this section it may be definitely stated that the simple radiation theory which assumed activation by the passage from a single normal to a single activated state through the absorption of a quantum of radiation cannot give a satisfactory account of unimolecular reactions, both because the rate of activation would be too small and because in at least one case the substance does not respond to radiation of the predicted wave-length.

336. Elaborated Radiation Theory of Activation.—The simplicity and apparent promise of the original radiation theory of chemical activation has suggested the hope that an elaborated form of the theory might be able to account for the facts. The natural modifications to make would be of course to assume instead of a single normal state and single activated state, a number of chemically active or inactive quantum states thus leading to the possibility of activation by the absorption of other frequencies than those given by equation (624), which makes the quantum of absorbed radiation exactly equal to the energy of activation.

There are two reasons why such a modification would in any case

[14] Daniels and Johnston, *Jour. Amer. Chem. Soc.*, **43**, 72 (1921).

seem inevitable. In the first place the experimental investigation of the quantum properties of molecules has in every case shown the existence of many quantum states, and in the second place the facts of photochemistry show that chemical reactions are in general not induced by a single frequency, but usually by a whole range of frequencies.

337. Such a modification of the radiation theory of reaction would of course immediately remove the difficulty which arises because the frequency of radiation, calculated from the energy of activation by equation (624), is not found experimentally to be absorbed nor to lead to reaction. On the new basis the energy of activation will not be changed, but this energy need not be acquired by a single jump from a single normal to a single activated state.

338. It is more difficult to decide whether a modification of the theory along the above lines would lead to rapid enough rates of activation to account for known reaction rates. To investigate this point it should be noted that the new theory would lead both to the assumption of chemically inactive states of energy intermediate between that of the lower normal states and that necessary for complete chemical activation, and to the assumption of more than a single higher quantum state which would be chemically active. It will be simplest to give separate consideration to the effect produced by the introduction of these two different kinds of states.

339. With regard to the *intermediate* states of less energy than that necessary for full chemical activation it seems evident that the introduction of such states would not lead to any marked increase in the rate of activation, such as would be necessary if we are to account for known rates of unimolecular reaction. This arises from the fact that the greater density of radiation which would be effective in causing transitions from an intermediate state to the activated state, compared with that effective in the full transition from the normal state, would be compensated by a smaller number of molecules in the intermediate state, compared with the number in the normal state. Thus, if ε_a is the total energy of activation and ε_i the energy necessary to raise molecules to the intermediate state, the density of radiation effective in raising molecules from the intermediate to a completely activated state would, in accordance with the Planck radiation law, be greater than that corresponding to the whole jump by a factor of approximately $e^{\frac{\varepsilon_i}{kT}}$; on the other hand, in accordance with the Maxwell-Boltzmann distribution law, the number of molecules in the intermediate state would be less than that in the normal state by the reciprocal factor $e^{-\frac{\varepsilon_i}{kT}}$. The situation may be summed up by the statement that the assumption of intermediate states between the normal and chemically activated states is almost certainly correct, and leads to an explanation of the failure of reactions to respond to calculated frequencies. Nevertheless, it does not seem probable that a mere assumption of such intermediate states would alone lead to the great increase in speed of activation necessary to rescue the radiation theory of activation.

340. Turning now, however, to the assumption of *more than a single state which is chemically active,* it is evident that the condition of affairs is different, since a jump to any one of these states might lead to reaction. Nevertheless, it is also evident from the discussion in Section **332,** that we should need to postulate a great many such activated states in order to account for known rates of reaction, since in the case of unimolecular reactions of normal temperature coefficient the assumption of a single state gives us a rate of activation which is too small by a factor of the order of 10^6. Hence it would seem evident that the chemically active states would need to lie close together in energy content in order to secure sufficient states within the range where the Planck radiation law would permit enough radiation density to permit of frequent transitions.

There are, however, two reasons for believing that such a continuous range of quantum states with gradually increasing energy contents might exist.

341. In the first place the facts of photochemistry show, as already mentioned, that photochemical reaction is by no means always induced by the absorption of a single frequency, but often by the absorption of a continuous range of frequencies. This corresponds no doubt to the existence of a continuous range of chemically active states.

342. In the second place it seems reasonable to assume that the process of chemical activation consists in increasing the energy of vibration of atoms within the molecule. This would correspond with the chemist's rough idea of "loosening a chemical bond" and would presumably make the atoms more free to enter into new chemical combinations. Furthermore, the magnitudes of known energies of activation correspond to frequencies in the near infra-red which we know to be absorbed, when changes in atomic oscillation take place. Hence, we might feel justified in making the hypothesis that the chemically activated condition of the molecule, at least in certain cases, might be one in which the amplitude of atomic oscillation has become so great that the forces holding an oscillating atom in to the molecule are greatly weakened at the exterior limit of its motion.

This hypothesis, however, immediately leads to an important consequence. If the oscillations become so large that the restoring forces at the exterior limit of oscillation are small, it is evident that the frequency of atomic oscillation will also become small in the same way that the frequency of oscillation of any elastic system decreases as it is strained beyond its elastic limit. For any conditionally periodic system of periodicity s, however, the energy E, fundamental frequencies $\omega_1 \ldots . \omega_s$ and phase integrals $I_1 \ldots . I_s$ are connected by the classical equation

$$dE = \omega_1 dI_1 + \ldots . + \omega_s dI_s \qquad (628)$$

Hence, for any oscillation in which a frequency, say ω_r, has become small, the rate of increase of energy E with the corresponding phase integral I_r will be small in accordance with the equation

$$\frac{\partial E}{\partial I_r} = \omega_r \qquad (629)$$

Successive quantized states of the system, however, are those in which the phase integrals differ by successive amounts h. It is immediately evident that the energy levels of successive quantized states will lie near together when the frequency involved becomes small. The close succession of quantized states for a Planck oscillator of small frequency is of course a special case of the above more general result. In other words, as soon as the amplitude of atomic oscillation within a molecule has become large enough so that the bond holding the atom in question has become greatly weakened, the successive quantized states of the molecule, that is, the activated states of the molecule, will have energy levels lying near together. Indeed, if the oscillations become so large that the force restraining the atom at the exterior limit of its motion approaches zero, we obtain a continuous series of activated states. Such a complete weakening of the bond might be necessary for certain reactions and would seem especially reasonable in the case of unimolecular changes.

343. If we proceed on the suggested basis and assume a series of activated states, it is now evident that the rate of activation will be given by a summation of the form

$$\frac{dN_{act.}}{dt} = \frac{\sum_i \sum_a N p_i e^{-\frac{\epsilon_i}{kT}} B_{ia} \frac{8\pi h \nu_{ia}^3}{c^3} \frac{1}{e^{\frac{h\nu_{ia}}{kT}} - 1}}{\sum_i p_i e^{-\frac{\epsilon_i}{kT}}} \qquad (630)$$

where B_{ia} is Einstein's coefficient of absorption for radiation of the frequency necessary for transition from some particular unactive state i

to a particular chemically active state a, and the double summation is to be taken over all such pairs. The summation could of course be replaced by integration in the case of a continuous series of activated states. Whether the increased rate of activation which equation (630) would give as compared with the earlier equation (627) would be sufficient to account for the observed rate of reaction in any special case would of course be dependent on molecular properties of the substance involved which are not now at our disposal.

344. This elaborated radiation theory can also be formulated in a somewhat different manner. On the basis of this theory thermal reaction is to be looked upon as the combined result of all the photochemical reactions going on in the system because of the thermal radiation. Hence if k^ν is the specific photochemical rate of reaction corresponding to the frequency ν, and we assume that the thermal reaction is due to the thermal radiation present in the reaction mixture with the energy density given by the Planck radiation law, we may evidently write for the specific thermal rate k_T of a unimolecular reaction the expression

$$k_T = \int_0^\infty \frac{8\pi h\nu^3}{c^3} \frac{1}{e^{\frac{h\nu}{kT}} - 1} k_\nu \, d\nu \qquad (631)$$

This expression was first published by the author [15] in 1920 and evidently gives the desired extension to the simple radiation theory of reaction.

Although equation (631) gives an undoubted improvement in the formulation of the radiation theory of reaction, nevertheless the main part of our task would be to determine whether this equation actually leads to rapid enough rates to account for known reaction velocities, thus rescuing the radiation theory of chemical reaction from its main difficulty. Unfortunately, the answer to this question cannot now be given, since it would involve a quantitative knowledge of specific photochemical reaction rates, which we do not possess. Certainly the burden of proof must rest on those who would maintain the applicability of equation (631). The difficulties in subjecting this equation to a proper test should not be overlooked. The effect of a wide range of frequencies must be studied, since of course the actual distribution of active wave-lengths cannot be predicted. In particular it is to be remembered, in accordance with the discussion of the preceding sections, that activation may occur in steps and the last steps of the activation process may involve very small energy increases, so that the far infra-red as well as the near infra-red must be searched. Finally, it must not be forgotten that k_ν is the specific photochemical reaction rate *at the tempera-*

[15] Tolman, *Journ. Amer. Chem. Soc.*, **42**, 2506 (1920).

ture in question in the presence of the thermal radiation of all the other frequencies. Hence, since many photochemical reactions have appreciable temperature coefficients, we must determine directly or indirectly the increase in reaction rate when we shine light of the different wave-lengths into the reaction mixture *at* the reacting temperature.

345. To sum up the results of this treatment, we may state that there are experimental and theoretical reasons for believing in the possibility of a series of activated states of nearly continuously increasing energy content, rather than in a single activated state. It cannot be said at the present time, however, whether this leads to a sufficient increase in rates of activation by absorption of radiant energy to account for known reaction velocities.

346. Concluding Remarks.—The discussion of the temperature coefficient of reaction rate given in the foregoing chapter has concerned itself with the high coefficients found in typical homogeneous gas reactions.

Reactions such as the recombination of bromine atoms to form the molecule which appears to have no appreciable temperature coefficient, present of course no difficulty, since we merely have to assume that a certain fraction of the collisions between unactivated bromine atoms lead to recombination with an emission of radiation to remove the excess energy, the process thus being the reverse of the dissociation of bromine by radiation.

As for heterogeneous reactions, as we saw in the last chapter the temperature coefficient may in some cases be dependent on the change with temperature in the fraction of the surface covered with an adsorbed film. On the other hand, changes with temperature in rate of diffusion and actual chance of reaction will have to be considered when the accumulation of experimental data warrant a more complete theory.

347. The theory presented in this chapter has all been based on the Arrhenius picture of activation as the mechanism of chemical reactions that have large temperature coefficients.

A considerable loophole in the theory consists in the fact that independent evidence for the existence of these activated states is mostly lacking. In this connection, however, the considerations of Wulf [16] giving some evidence for their presence in the thermal decomposition of ozone should be noted. Furthermore, we have the certainty that molecules are chemically reactive when they have been excited by the absorption of radiation in photochemical experiments, or excited by collisions of the second kind in experiments on induced photochemical effects,

[16] Wulf, *Proc. Nat. Acad.*, **12**, 129 (1926).

or excited by collision with fast moving particles shot in from radioactive sources.

348. The worst difficulty, however, which has been encountered by the Arrhenius picture is, as we have seen, that of finding mechanisms of activation which would act rapidly enough to keep up with known rates of reaction.

In this connection, neverthless, we must call attention to a certainly demonstrated and definite difficulty of the same kind in connection with the famous Stern and Gerlach[17] experiment on the orientation of silver atoms in a magnetic field. In this experiment a ray of vaporized silver atoms is shot between the poles of a strong magnet, the atoms remaining in the magnetic field only for a time of the order of 10^{-4} seconds. Nevertheless, in this short time interval all the atoms orient themselves so that their own magnetic axis is parallel to the field and, most remarkable of all, approximately half the atoms turn in the direction of lining up with the field and the other half in the opposite direction. Hence half the atoms receive energy at a rate far greater than we can account for on the basis of any known mechanisms of energy transfer.

A similar difficulty is possible in the case of the decomposition of nitrogen tetroxide to form the dioxide. This reaction appears from experiments on the velocity of sound through a mixture of the two gases to take place at a rate so fast as not to be measurable.[18] Hence, here, too, we must inquire into the mechanism by which the tetroxide molecules can acquire energy fast enough to dissociate in a time which is short compared with the period of the sound pulsations. It is to be specially noted in this case, moreover, that here we are dealing, not with a hypothetical energy of activation, but with the known energy necessary for the dissociation of the tetroxide molecule.

349. These remarks as to difficulties in other cases of accounting for the rapidity with which energy is received by atoms or molecules may indicate the necessity of a fundamental modification in our ideas as to the conservation or transfer of energy. Two possibilities would seem to be open.

First we might give up the conservation of energy except as a statistical principle. Then we should never have to inquire where the energy comes from when a molecule is excited, but should merely have to balance our energy account in the long run. In the case of the Stern and Gerlach experiment, since approximately half the silver atoms receive energy and the other half lose it, this balance is immediately made. In the case of the passage of sound waves through a mixture of nitrogen

[17] Stern and Gerlach, *Zeitschr. für Physik.*, **9**, 349 (1922).
[18] Grüneisen and Goens, *Ann. der Phys.*, **72**, 193 (1923).

tetroxide and dioxide, the balance is made at the end of every complete period of the wave. And finally to balance our account in the case of chemical reactions taking place through activation, we should not have to supply the whole calculated energy of activation, but only the much smaller positive or negative energy actually necessary for the reaction itself, since in endothermic reactions most of the energy of activation is immediately again made available by the deactivation of the products, and in exothermic reactions there is even energy to spare.

A second possibility would be always to assume the conservation of energy even for individual molecular processes, but to postulate when necessary a very special coupling by which energy disappearing from one molecule could reappear in another. Thus in the Stern and Gerlach experiment, every silver atom that turned with the field would be "coupled" with one which turned against the field and would supply to it the necessary energy. In the passage of a sound wave through the mixture of tetroxide and dioxide, each molecule of tetroxide dissociating in region of low pressure could receive its energy from a molecule recombining in a region of high pressure. And in the case of chemical reactions taking place through activation, a molecule to be activated would draw its supply directly from that made available by the completion of the reaction undergone by some other molecule. Such immediate transfers, however, would of course take place in ways not now known to us.

Of course it may not be necessary to introduce hypotheses of such a kind in order to preserve the Arrhenius hypothesis of chemical reaction through activation. On the other hand, it must of course be emphasized that the picture itself may be wrong. These are among the most important questions for future chemistry.

Chapter 22.

Photochemical Reactions.

350. Introduction.—In the present chapter we shall consider the yield, rate, and change of rate with temperature in the case of photochemical reactions. Theoretically at least these should be simpler than thermal reactions, since we have definite information as to the nature of the agency responsible for the action. In practise, however, the radiation absorbed may often start a complicated chain of events so that the different factors in the final result are hard to disentangle. In addition, with the experimental methods usually employed, complications are introduced by the fact that a range of frequencies instead of monochromatic radiation is often used and the density of the radiation is usually unknown and different in different parts of the illuminated mixture.

We shall content ourselves with the theoretical consideration of a few cases which are perhaps artificially oversimplified. Nevertheless, the treatment will indicate the nature of the methods to be employed in analyzing more complicated experimental results into the individual processes in terms of which they must be explained.

351. Einstein's Law of Photochemical Equivalence.—It is evident that the primary process in any photochemical reaction must be the absorption of a quantum of radiation by a molecule in some lower quantum state i to go over to a higher state a, as indicated in reaction form by the expression

$$h\nu + A_i \rightarrow A_a \qquad (632)$$

In case complications do not arise, we can then expect each molecule so activated to enter into reaction, so that the total extent of the chemical change will be directly proportional to the number of quanta absorbed. This is *Einstein's law of photochemical equivalence* which can be expressed in the form

$$Q = n\frac{E}{h\nu} \qquad (633)$$

where Q is the number of molecules reacting, E is the radiant energy absorbed, and n is some small integer which gives the number of molecules whose chemical change is brought about by a single process of activation. Tests of this law are of course to be made by measuring the intensities of the incident and transmitted radiation in a photochemical experiment and comparing the energy absorbed with the extent to which the reaction proceeds.

352. Examples of Einstein's Law.—This law of photochemical equivalence is found to be obeyed in the case of a considerable number of reactions, and may be regarded as the ideal state of affairs from which deviations are to be reckoned.

As a simple example of the law we may consider the decomposition of gaseous hydrogen bromide by light, a reaction which has been studied by Warburg.[1] With radiation of wave-length 2090 Å. and 2530 Å. it was found within the experimental error that *two* molecules of hydrogen bromide were decomposed for each quantum absorbed.

This result might be explained by either of the mechanisms

$$h\nu + HBr \rightarrow H + Br$$
$$H + HBr \rightarrow H_2 + Br$$
$$2Br \rightarrow Br_2$$

or

$$h\nu + HBr \rightarrow [HBr]$$
$$[HBr] + HBr \rightarrow H_2 + Br_2$$

where the brackets indicate an activated form of the molecule. The energy contained in one quantum at the frequencies used was as a matter of fact sufficiently great to dissociate the molecule. Nevertheless, molecules are known to have electron jumps which absorb more than the energy of dissociation, without the atoms separating. Hence a decision between the two mechanisms on this basis alone would not be possible. Similar results were also found by Warburg for hydrogen iodide.

As another example of the law, this time involving two reacting substances, we may consider the photochemical reaction between chlorine and trichlor-brom-methane under the influence of light of wave-lengths which are absorbed by the chlorine. In this reaction one molecule of chlorine and two of the trichlor-brom-methane are found to react for each quantum absorbed, the yield being independent of the concentration of the latter substance over a wide range.[2]

The mechanism of the reaction might be pictured as follows:

$$h\nu + Cl_2 \rightarrow [Cl_2]$$
$$[Cl_2] + CCl_3Br \rightarrow CCl_4 + Cl + Br$$
$$Cl + CCl_3Br \rightarrow CCl_4 + Br$$
$$2Br \rightarrow Br_2$$

[1] Warburg, *Sitz.-ber. preuss. Akad.*, p. 314 (1916).
[2] Grüss, *Zeitschr. für Elektrochem.*, **29**, 144 (1923).

Since the yield remains constant even when the concentration of trichlor-brom-methane is very considerably reduced, we must ascribe a long enough life to the activated chlorine molecule $[Cl_2]$ to make sure that it will collide with a trichlor-brom-methane molecule before it has lost its activation. In the experiments under discussion it was merely possible to conclude that the mean life was greater than 2×10^{-9} seconds. This of course is entirely reasonable. (See Chapter 16.)

353. Einstein's Law in Cases of Photochemical Sensitization.—There is of course no reason why the molecule which actually absorbs the radiation should itself be one of those entering into chemical reaction. By collisions of the second kind [3] it can transmit its energy to some other kind of molecule which reacts chemically. This explains the great field of photochemical reactions which only occur in the presence of a suitable added substance which can absorb the radiation employed. They are said to be *photosensitized* by the addition of this *photocatalyst.*

Here, too, in cases which are not complicated by secondary factors, we should expect Einstein's law of photochemcial equivalence to hold, since each molecule of the photocatalyst activated will pass on its energy to a molecule of reactant.

As an example of such a photosensitized process, we may consider the reaction between oxygen and trichlor-brom-methane, which has been found to take place when sensitized by bromine under the action of light of wave-lengths 4360 and 4070 Å. which are absorbed by bromine. Since bromine is itself produced in the process, the reaction is autocatalytic. Two molecules of the trichlor-brom-methane are found to be used up per quantum absorbed, and the yield is found over a wide range to be independent of the concentration both of oxygen and trichlor-brom-methane.[4] The course of the reaction can be indicated as follows:

$$h\nu + Br_2 \rightarrow [Br_2]$$

$$[Br_2] + \{2CCl_3Br + O_2\} \rightarrow Br_2 + 2COCl_2 + Cl_2 + Br_2$$

where it seems wisest not to try to specify the complete mechanism of the second step, which presumably involves the deactivation of the excited bromine by a collision of the second kind. The lack of dependence of yield on the concentration of the reactants sets a limit of the order 10^{-7} seconds for the life of the activated state.

Another example of photosensitization is given by the decomposition of ozone in the presence of chlorine.[5] It is found that two molecules

[3] The *activated* molecule of the photocatalyst might also be able to enter into temporary reaction with the reactant. The formulation would give the same final result as that using the assumption of collisions of the second kind.

[4] Grüss, *loc. cit.*

[5] Bonhoeffer, *Zeitschr. für Phys.*, **13**, 94 (1923).

of ozone decompose for every molecule of chlorine activated by absorption. Since the yield is independent of the concentration of the ozone between 0.1 to 5 per cent, it was possible to calculate a minimum life of the excited chlorine of the order of 1.6×10^{-7} seconds, and to conclude that the excited chlorine could collide with as many as a thousand molecules of oxygen without deactivation. (See Section **330**, Chapter 21.)

354. Deviations from Einstein's Law.—Many photochemical reactions deviate from the law of photochemical equivalence. This, however, is itself to be expected on the basis of the very hypothesis that led to Einstein's law. Although the primary process of absorption will always lead to the activation of one molecule per quantum of energy abstracted from the radiation, a molecule thus activated may not always succeed in entering the chemical change that follows before it loses its energy by reemission or collision of the second kind. In such cases the yield will fall below that predicted by the law. On the other hand, molecules which have been activated by the absorption of radiation may be the starting point for a chain of events involving many further molecules depending on their concentration. In this case we shall find deviations from the law of photochemical equivalence in the direction of more reaction than would be predicted.

As an example of a reaction in which the change is less than would be expected on the basis of the Einstein law, we may take the decomposition of ammonia by radiation in the region 2030 to 2140 Å., where it has been found by Warburg[6] that approximately four quanta are absorbed for every molecule of ammonia decomposed.

As a famous example of a reaction in which more molecules are decomposed than would be predicted by the law of equivalence, we may take the combination of chlorine and hydrogen which has long been known to take place under the action of sunlight. In this reaction the number of molecules reacting may be as large as 10^6 times the number of quanta absorbed. It would be beyond the scope of our present interests to discuss the conflicting mechanisms which have been proposed as giving the specific steps of the chain of processes which must explain such a result.

355. Photochemical Reaction Rate.—So far we have been discussing the *yield* of photochemical reactions, based on a comparison of the number of quanta actually absorbed in a given length of time with the amount of chemical change which results. We must now turn our attention to the rate at which the reactions proceed, with a given intensity of illumination and a given concentration of reactants.

[6] Warburg, *Sitz.-ber. preuss. Akad.*, p. 746 (1911); p. 216 (1912).

In analogy with the use of the term specific reaction rate in ordinary thermal reactions, we shall find it convenient to define the *specific photochemical reaction rate* k_ν for radiation of a given *frequency* ν as the rate at which the reaction proceeds, with unit concentration of the reacting substances present, and in the presence of radiation from an external source of frequency ν and unit radiation density $u_\nu = 1$.

The actual rate of course under any given experimental conditions will be dependent on the radiation density actually present and on the concentrations of the reacting substances employed. To express this fact we may *when convenient* employ an *order equation*, which can be written in the form

$$-\frac{dN_A}{dt} = k_\nu u_\nu (N_A)^a (N_B)^b \dots \qquad (634)$$

where the concentrations of the reacting substances N_A, N_B, etc., are preferably expressed in numbers of molecules per cubic centimeter.

356. In this equation the radiation density u_ν enters to the first power, since so far few if any reactions have been found in which the rate of the change is not proportional to the intensity of illumination. Furthermore, such proportionality is to be theoretically expected unless the intensity of the illumination is raised so high as to lead to the partial exhaustion of molecules in the state which is actually absorbing the radiation, or unless more than one molecule has to be activated in order for the reaction to proceed.

357. As to the dependence of the rate on concentrations, in the case of a reaction which obeys the law of photochemical equivalence, the rate will be dependent only on the concentration of the substance which absorbs the radiation. The rate will furthermore usually be nearly proportional to the first power of this concentration, unless there are large deviations from Beer's law of proportionality between absorption and concentration. In cases where the law of photochemical equivalence fails, the rate may depend on the concentrations of more than one of the reacting substances and in a variety of ways. Thus for the reaction between hydrogen and chlorine in the presence of oxygen which acts as an inhibitor Chapman [7] finds that the rate can be expressed by an equation of the form

$$\frac{d(\mathrm{HCl})}{dt} = u_\nu \frac{k_1(\mathrm{H_2})(\mathrm{Cl_2})^2}{k_2(\mathrm{H_2})^{2-x}(\mathrm{O_2}) + (\mathrm{Cl_2})}$$

where x varies over a small range of the order zero to one half, with the ratio of the concentrations of hydrogen and chlorine.

[7] M. C. C. Chapman, *Journ. Chem. Soc.*, **123**, 3062 (1923).

358. In determining the value of the specific photochemical reaction rate from actual experimental data, it is to be recognized that u_ν is the radiation density of frequency ν in excess of that prescribed by the Planck radiation law. The actual value of u_ν will of course vary from point to point in a mixture experimentally illuminated from an external source. It is related to the intensity of illumination I_ν at any point by the equation

$$u_\nu = \frac{I_\nu}{c} \tag{635}$$

where c is the velocity of light, and I_ν falls off in a homogeneous mixture exponentially with the distance x in the familiar way

$$I_\nu = (I_\nu)_0 e^{-a_\nu x} \tag{636}$$

Furthermore, in determining the value of photochemical reaction rates, it is evident that the total observed rate of change must be corrected when necessary by subtracting the rate of change due to ordinary thermal reaction.

359. Theoretical Expression for a Unimolecular Photochemical Rate.—With the help of our knowledge of the mechanism of radiation absorption, it will now be easy to obtain a theoretical expression for the rate of unimolecular photochemical change.

Let $B_{i\nu}$ be the chance per second that a molecule in the i'th quantum state, surrounded by unit radiation density of frequency ν will absorb a quantum of energy $h\nu$ and hence go over into an excited state.

$B_{i\nu}$ is thus the value of Einstein's coefficient of absorption for the transition in question, and with its help we can of course calculate the number of molecules activated per second and hence the number that would react if every molecule activated should undergo the chemical change. In order, however, to allow for the possibility that some molecules, through deactivation, fail to react we shall take $s_{i\nu}$ as the chance of reaction for a molecule in the excited state of energy content $\varepsilon_i + h\nu$. Hence for the rate of reaction, due to molecules in state i, we obtain

$$-\frac{\partial N}{\partial t} = B_{i\nu} u_\nu s_{i\nu} N_i \tag{637}$$

Since molecules in more than one quantum state may absorb the frequency ν and react[8] we can obtain the total rate of reaction by summing over all states i. Doing this and substituting the Maxwell-Boltzmann expression for the number of molecules in any state i, we obtain for the total rate of photochemical reaction

[8] This will be true in the case of a continuous series of activated states.

$$-\frac{dN}{dt} = u_\nu \frac{\sum_i B_{i\nu} s_{i\nu} p_i e^{-\frac{\epsilon_i}{kT}}}{\sum_i p_i e^{-\frac{\epsilon_i}{kT}}} N \tag{638}$$

or for the specific rate

$$k_\nu = \frac{\sum_i B_{i\nu} s_{i\nu} p_i e^{-\frac{\epsilon_i}{kT}}}{\sum_i p_i e^{-\frac{\epsilon_i}{kT}}} \tag{639}$$

360. Lacking knowledge of the factors $B_{i\nu}$ and $s_{i\nu}$ we cannot of course use this expression for the prediction of actual photochemical rates, but by differentiation with respect to the temperature we can acquire information with respect to the temperature coefficient of photochemical reaction rate.

Taking a logarithmic differentiation with respect to T we obtain [9]

$$\frac{\partial \log k_\nu}{\partial T} = \frac{1}{k_\nu} \frac{\sum_i B_{i\nu} s_{i\nu} p_i e^{-\frac{\epsilon_i}{kT}} \frac{\varepsilon_i}{kT^2}}{\sum_i p_i e^{-\frac{\epsilon_i}{kT}}} - \frac{\sum_i p_i e^{-\frac{\epsilon_i}{kT}} \frac{\varepsilon_i}{kT^2}}{\sum_i p_i e^{-\frac{\epsilon_i}{kT}}} \tag{640}$$

The first term on the right-hand side of this equation is evidently the average value of $\frac{\varepsilon_i}{kT^2}$ for the molecules that actually react, in their states before activation, and the second term the average value for all the molecules, so that we can write

$$\frac{\partial \log k_\nu}{\partial T} = \frac{\overline{\overline{\varepsilon}} - \overline{\varepsilon}}{kT^2} \tag{641}$$

where the double line indicates an average taken for the molecules that enter into reaction and the single line an average for all the molecules. We have thus obtained a very simple equation for temperature coefficient of photochemical reaction rate, which we shall later interpret.

361. Theoretical Expression for a Bimolecular Photochemical Rate.—Let us next consider a reaction between substances of two kinds, having respectively N and N' molecules present per cubic centimeter, and let us assume that reaction is brought about by the photochemical activation of the first of the two substances.

[9] In the differentiation we have taken $s_{i\nu}$ as independent of the temperature. There could be a small tendency for $s_{i\nu}$ to decrease with temperature, owing to more frequent deactivations without reaction. See Section 317, Chapter 21.

If $B_{i\nu}$ is the value of Einstein's coefficient of absorption for molecules of the first kind in any quantum state i, we can then write for the rate at which these molecules are being activated

$$\frac{\partial N_{i\nu}}{\partial t} = u_\nu B_{i\nu} N_i \tag{642}$$

The activated molecules so formed will be removed partly by reaction and, in case the law of photochemical equivalence does not hold, partly by deactivation. For their rate of removal by reaction we can evidently write

$$-\frac{\partial N_{i\nu}}{\partial t} = T^{1/2} \sum_j k_{i\nu,j} N_{i\nu} N_j' \tag{643}$$

where $T^{1/2} k_{i\nu,j}$ is the chance of collision and reaction between a molecule of the first kind in the state $i\nu$ and a molecule of the second kind in the state j. The rate of deactivation can be put proportional to the number of molecules in the state $i\nu$

$$-\frac{\partial N_{i\nu}}{\partial t} = a_{i\nu} N_{i\nu} \tag{644}$$

where $a_{i\nu}$ is the sum of the coefficients of deactivation to all lower states. (See Chapter 21, Section **314**.)

For the steady state of reaction, since the number of molecules in the activated state $i\nu$ is always small, the rate of formation as given by equation (642) must be equal to the total rate of removal as given by equations (643) and (644), so that we can combine these equations and solve for the concentration of activated molecules $N_{i\nu}$. We obtain

$$N_{i\nu} = \frac{u_\nu B_{i\nu} N_i}{T^{1/2} \sum\limits_j k_{i\nu,j} N_j' + a_{i\nu}} \tag{645}$$

The total rate of the reaction can now be obtained by resubstituting this expression into (643) and summing up for all states i. We thus obtain for the rate of reaction

$$-\frac{dN}{dt} = \sum_i \left(\frac{T^{1/2} \sum\limits_j k_{i\nu,j} N_j' u_\nu B_{i\nu} N_i}{T^{1/2} \sum\limits_j k_{i\nu,j} N_j' + a_{i\nu}} \right) \tag{646}$$

362. In case the chance for an activated molecule to react is large compared with the chance of deactivation, the second term in the denominator will be small compared with the first, and the expression will reduce to

$$-\frac{dN}{dt} = \sum_i u_\nu B_{i\nu} N_i = \frac{u_\nu \sum_i B_{i\nu} p_i e^{-\frac{\epsilon_i}{kT}}}{\sum_i p_i e^{-\frac{\epsilon_i}{kT}}} N \tag{647}$$

The rate will then be dependent only on the concentration N of molecules of the first kind, and every molecule that is activated will react in accordance with the law of photochemical equivalence.

On the other hand, in the extreme case that the chance of reaction is very small compared with that of deactivation, equation (646) will assume the form

$$-\frac{dN}{dt} = u_\nu T^{\frac{1}{2}} \sum_i \left(\sum_j \frac{k_{i\nu j}}{a_{i\nu}} B_{i\nu} N_i N_j' \right)$$

$$= u_\nu T^{\frac{1}{2}} N N' \frac{\sum_i \left(\sum_j \frac{k_{i\nu j}}{a_{i\nu}} B_{i\nu} p_i e^{-\frac{\epsilon_i}{kT}} p_j' e^{-\frac{\epsilon_j'}{kT}} \right)}{\left(\sum_i p_i e^{-\frac{\epsilon_i}{kT}} \right) \left(\sum_j p_j' e^{-\frac{\epsilon_j'}{kT}} \right)} \tag{648}$$

so that the rate becomes proportional to both concentrations. Since the relative chance of reaction to deactivation can be increased by increasing the concentration N', we may expect to encounter order equations of form intermediate between (647) and (648) as we vary the concentration, as well as different variations from the yield calculated from the law of photochemical equivalence.

363. The general treatment of the temperature coefficient of a reaction such as we are considering could be obtained by differentiating the general equation (646) with respect to T, but the result is pretty complicated. It will be sufficient for our cases to consider the two limiting cases given by equations (647) and (648).

Referring to equation (647) and taking the specific reaction rate k_ν, which is the factor by which $u_\nu N$ must be multiplied to get the actual rate, we easily obtain by logarithmic differentiation [10] just as in the derivation of equation (641)

$$\frac{\partial \log k_\nu}{\partial T} = \frac{\bar{\bar{\varepsilon}} - \bar{\varepsilon}}{kT^2} \tag{649}$$

where $\bar{\bar{\varepsilon}}$ is the average energy of the molecules of the *photoactive* substance that pick up radiation and react, and $\bar{\varepsilon}$ is the average energy of all these molecules.

[10] The change of $a_{i\nu}$ with temperature is neglected, see Section **317**, Chapter **21**.

Using the value of k_ν coming from equation (648) we obtain by logarithmic differentiation with respect to T[10]

$$\frac{\partial \log k_\nu}{\partial T} = \frac{1}{2T} + \frac{T^{1/2}\sum\limits_i\sum\limits_j \frac{k_{i\nu j}}{a_{i\nu}} B_{i\nu} p_i e^{-\frac{\epsilon_i}{kT}} \frac{\varepsilon_i}{kT^2} p_j' e^{-\frac{\epsilon_j'}{kT}}}{k_\nu\left(\sum\limits_i p_i e^{-\frac{\epsilon_i}{kT}}\right)\left(\sum\limits_j p_j' e^{-\frac{\epsilon_j'}{kT}}\right)}$$

$$+ \frac{T^{1/2}\sum\limits_i\sum\limits_j \frac{k_{i\nu j}}{a_{i\nu}} B_{i\nu} p_i e^{-\frac{\epsilon_i}{kT}} p_j' e^{-\frac{\epsilon_j'}{kT}} \frac{\varepsilon_j'}{kT^2}}{k_\nu\left(\sum\limits_i p_i e^{-\frac{\epsilon_i}{kT}}\right)\left(\sum\limits_j p_j' e^{-\frac{\epsilon_j'}{kT}}\right)}$$

$$- \frac{\sum\limits_i p_i e^{-\frac{\epsilon_i}{kT}} \frac{\varepsilon_i}{kT^2}}{\sum\limits_i p_i e^{-\frac{\epsilon_i}{kT}}} - \frac{\sum\limits_j p_j' e^{-\frac{\epsilon_j'}{kT}} \frac{\varepsilon_j'}{kT^2}}{\sum\limits_j p_j' e^{-\frac{\epsilon_j'}{kT}}} \qquad (650)$$

But by the principles of averaging, so often used before, this evidently becomes

$$\frac{\partial \log k_\nu}{\partial T} = \frac{1}{2T} + \frac{\bar{\bar{\varepsilon}} - \bar{\varepsilon} + \bar{\bar{\varepsilon}}' - \bar{\varepsilon}'}{kT^2} \qquad (651)$$

where $\bar{\bar{\varepsilon}}$ is the average energy of the molecules of the *photoactive* substance that pick up radiation and react and $\bar{\varepsilon}$ the average energy of all the molecules of photoactive substance, while $\bar{\bar{\varepsilon}}'$ is the average energy of the molecules of the other substance entering into the reaction, and ε' the average energy of all of them.

Comparing equations (649) and (651) for the two limiting cases, we see that when the law of photochemical equivalence is followed, the temperature coefficient depends only on the energies of molecules of the photoactive substance, while in the other limiting case we have to include terms depending on the energies of molecules of the other substance as well as the term $1/2T$. For intermediate cases we should expect intermediate results.

364. It is evident that other mechanisms of photochemical reaction besides the unimolecular and bimolecular ones discussed could be theoretically treated by similar methods.

365. The method of treating cases of photochemical steady states in which a thermodynamic equilibrium is displaced by illumination will also be evident. To do this we shall merely have to equate the rates for the different reverse processes which are occurring.

366. Comparison of Theoretical and Experimental Results as to the Temperature Coefficient of Photochemical Reaction Rate.— Examining equations (641), (649), and (651) for the temperature coefficient of photochemical reaction rate, we see that if we neglect the small term $1/2T$, they can all be written in the form

$$\frac{\partial \log k_\nu}{\partial T} = \frac{\bar{\bar{\varepsilon}} - \bar{\varepsilon}}{kT^2} \tag{652}$$

where $(\bar{\bar{\varepsilon}} - \bar{\varepsilon})$ is the difference between the average energy of the molecules that enter into the reaction and that of all the molecules.

The equation has the same form as equation (613) in the last chapter for the temperature coefficient of thermal reactions. The quantity $\bar{\bar{\varepsilon}}$ is, however, quite different in the two cases, since in the equation for thermal reactions $\bar{\bar{\varepsilon}}$ was the average energy of the molecules that react as they exist in the activated state that leads to reaction, while in equation (652) $\bar{\bar{\varepsilon}}$ is the average energy of the molecules that react before they have picked up the radiation which leads to reaction.

With the help of equation (652), we can now interpret existing data on temperature coefficients.

367. The most striking fact concerning the effect of temperature on photochemical reaction rates is the extremely great number of reactions which have nearly zero temperature coefficient. Plotnikow[11] in his collection of existing information on photochemistry lists 21 photochemical reactions for which the maximum value of $(\partial \log k_\nu/\partial T)$ is approximately 0.008.

The existence of such a group of reactions is readily understood on the basis of equation (652), since in accordance with that equation we may write

$$\frac{\partial \log k_\nu}{\partial T} = \frac{\bar{\bar{\varepsilon}} - \bar{\varepsilon}}{kT^2} = 0 \tag{653}$$

and conclude that for this great group of well-known photochemical reactions the average energy of the molecules which enter into the reaction is practically the same as the average energy of all the molecules. This result is entirely understandable; it means that the average molecule has as good a chance of picking up radiant energy and becoming activated as the molecules in some special quantum state far removed in energy content from the average. For reactions of this group, preliminary partial activation is not necessary in order that a molecule of

[11] Plotnikow, "Lehrbuch der Photochemie," de Gruyter and Co., Berlin and Leipzig, 1920, p. 62.

the photoactive substance may absorb its quantum and react, nor is preliminary activation necessary for the molecules of any other substance involved.

368. The above discussion also makes it clear that we might expect to find cases in which preliminary activation of the molecules would be advantageous in promoting reaction. This may arise because molecules in the lower quantum states are not in a condition to absorb radiant energy of the frequency used, or because the energy level which they attain after the absorption is not high enough to lead to chemical reaction. In such cases the average energy $\bar{\bar{\varepsilon}}$ of the molecules which enter into the reaction will be greater than the average energy $\bar{\varepsilon}$ of all the molecules and we should find

$$\frac{\partial \log k_\nu}{\partial T} = \frac{\bar{\bar{\varepsilon}} - \bar{\varepsilon}}{kT^2} > 0 \tag{654}$$

As a matter of fact, Plotnikow lists 17 reactions in which the value of $(\partial \log k_\nu/\partial T)$ varies from about 0.016 to 0.040.

This too is entirely reasonable on the basis of our picture. A value of $(\partial \log k_\nu/\partial T) = 0.040$ leads at room temperature to a value of about 7000 calories per mol. This corresponds to a value of 0.3 volt per molecule which is an entirely reasonable figure for the difference in energy contents between successive quantum levels in a molecule.

A photochemical reaction taking place at room temperature in a homogeneous system with such a temperature coefficient might be very reasonably interpreted by assuming that the molecules in the lowest quantum state were not photoactive, and that molecules had to be in a higher quantum state with a preliminary activation corresponding to 0.3 volt in order to absorb light and react.

The process of reaction itself would then consist in the picking up of a quantum $h\nu$ by molecules in this higher quantum state followed by reaction. If the photoactive light has the wave-length 500 $\mu\mu$ this denotes a further increase in energy level corresponding to an activation of 2.5 volts per molecule, again an entirely reasonable figure.

369. Before this explanation of these small temperature coefficients could be finally accepted, however, it would be necessary to show that we could still account for the absolute magnitude of the rates if it is mainly molecules in higher quantum states that are involved. Moreover, these small temperature coefficients might not be solely due to the above mechanism, since, as the experiments are actually carried out, the velocity of diffusion of molecules into a strongly illuminated zone may be of importance. For aqueous solutions the viscosity decreases in the neighborhood of 20 per cent per 10° rise in temperature. If we assume the same proportional increase in reaction rate this would give us a value of $(\partial \log k_\nu/\partial T)$ of about 0.018, which falls in the region of discussion. Nevertheless small temperature coefficients of the same order occur in gas reactions where the rate of diffusion would be much less affected by temperature, so that the viscosity explanation could not be a general one.

370. If we differentiate equation (652) with respect to the frequency ν, we obtain

$$\frac{\partial}{\partial \nu}\left(\frac{\partial \log k_\nu}{\partial T}\right) = \frac{1}{kT^2}\frac{\partial \bar{\bar{\varepsilon}}}{\partial \nu} \tag{655}$$

Since $\bar{\bar{\varepsilon}}$ is the average energy of the molecules that actually react, this quantity may decrease with frequency, for at higher frequencies the magnitude of the quantum $h\nu$ is increased and molecules from a lower quantum state may be raised to a level where reaction is possible. This means that we may expect to find cases where the second derivative in question is negative, provided $(\partial \log k_\nu/\partial T)$ is not already zero.

Experimental data of the kind in question are not numerous nor certain. Nevertheless, referring once more to Plotnikow's treatise, expressing temperature coefficients as the ratio between the rates at temperatures 10° apart, we find that silver citrate paper has the temperature coefficient 1.19 in the blue and 1.07 in the ultraviolet; the phototropic substance salicylidenenaphthylamine has the values 1.8 in the green, 1.45 in the blue and 1.39 in the violet; and the reaction between chlorine and hydrogen has the values 1.50 for green light (550-530 $\mu\mu$), 1.31 for blue light (490-470 $\mu\mu$) and 1.21 for violet light (400-350 $\mu\mu$). These results if reliable are in complete agreement with the theory presented above.[12]

[12] For the first presentation of the suggested theory of temperature coefficients, see Tolman, *Journ. Amer. Chem. Soc.*, **45**, 2285 (1923).

Chapter 23

The Relation between Statistical Mechanics and Thermodynamics

371. Introduction.—It has been made apparent by the foregoing chapters that statistical mechanics is capable of treating all the problems ordinarily handled by thermodynamics. It is true that the language of thermodynamics is oftentimes simpler than that of statistical mechanics, and its equations better adapted for the substitution of empirical values of the parameters than are the equations of statistical mechanics. For this reason we have not hesitated in Chapters 9 and 13 to show the bearing of some of our results on ordinary thermodynamic calculations. This was due, however, to practical rather than theoretical considerations.

In addition to the possibility of treating what are usually considered as thermodynamic problems by statistical mechanical methods, we have also seen that the latter science permits the theoretical prediction of the values of many quantities which thermodynamics must take as empirical parameters, and furthermore permits the treatment of rates of physical-chemical change, a problem which thermodynamics cannot attack. It seems evident then that we must regard the science of statistical mechanics as the more fundamental, and investigate the possibility of a direct deduction of the axioms of thermodynamics from the principles of statistical mechanics itself. This is the problem which we shall treat in the present chapter.

The most satisfactory method of developing the relations between thermodynamics and statistical mechanics is not as yet a matter of universal agreement. Hence we shall content ourselves in the present chapter with a partially complete account of some of the different modes of attack that have been undertaken. The discussion, however, will be sufficient to show interrelations between different proposals, and to provide a basis for the treatment of certain important physical-chemical problems where it is advantageous to commence the discussion on a statistical mechanical basis, and then translate the results into thermodynamic language so that the discussion can be completed by thermodynamic methods.

372. The Two Laws of Thermodynamics.—The whole of classical thermodynamics may be regarded as obtainable from two axioms known as the first and second laws.

The *first law* may be written in the form

$$dE = dQ - dW \tag{656}$$

and states that the energy increase in a macroscopic system is equal to the heat which flows into it less the work which it does on its surroundings.

The above equation may be regarded as an immediate consequence of atomistics, provided we merely interpret the heat given to the system as energy transferred by random molecular processes, and assume that the law of the conservation of energy applies either directly to individual molecular processes or at the worst applies to the total result of their overall behaviour.

To state the second law, we may define the infinitesimal increase in the entropy of a system dS, accompanying any reversible change in state in terms of temperature T and heat absorbed, by the equation

$$dS = \frac{dQ}{T} \tag{657}$$

The *second law* of thermodynamics now requires that $\frac{dQ}{T}$ shall be a perfect differential for any *reversible* path, so that the entropy S may be regarded as a function of the state for any system which is in equilibrium. The law further requires that T shall be a quantity such that heat will only flow from systems of higher to those of lower temperature, and that for actual *nonreversible* changes in the state of a system, in which the external parameters of the system are varied, dQ being kept zero, the entropy S shall increase, instead of remaining constant in accordance with equation (657), which applies only to reversible processes.

Equations (656) and (657) may now be combined to advantage in the form

$$dE = TdS - A_1da_1 - A_2da_2 - \ldots \tag{658}$$

where the external work done by the system on its surroundings has been expressed in terms of the external forces acting on the system $A_1, A_2, \ldots$ and the corresponding displacements $da_1, da_2 \ldots$.

373. Evidently to obtain a statistical mechanical interpretation of thermodynamics we must secure statistical mechanical analogues for the quantities occurring in equation (658), which will obey an equation of the same form and will otherwise have properties similar to those of the thermodynamic quantities. The fact that more than a single set of pretty satisfactory analogues have been found explains the possibility of a variety of modes of treating the problem.

We shall first describe in outline the three proposals of Gibbs in which analogies were shown between the quantities in equation (658) and quantities which describe the properties of canonical and micro-canonical *ensembles* of systems. We shall then discuss in somewhat more detail the results that can be obtained by the more direct method of Boltzmann in considering those statistical mechanical properties of a *single* system which are analogous to the quantities in equation (658).

374. The First Analogue of Gibbs.—The first analogue of Gibbs was based on the consideration of a *canonical ensemble* of systems of structure identical with that of the system of interest and distributed in the γ-space with the density

$$\varrho = Ne^{\frac{\Psi - E}{\Theta}} \tag{659}$$

where N is the total number of systems in the ensemble, E the energy of a single system and Ψ and Θ are constants, having the dimensions of energy, the latter being called the modulus of distribution.

With this distribution in the γ-space the number of systems having coordinates and momenta in any region $dQ_1 \ldots . dP_m$ will evidently be

$$dN = Ne^{\frac{\Psi - E}{\Theta}} dQ_1 \ldots . dP_m \tag{660}$$

and hence by integration over the whole phase space, we must have the result

$$\int \ldots . \int e^{\frac{\Psi - E}{\Theta}} dQ_1 \ldots dP_m = 1 \tag{661}$$

Furthermore for the average of any property of the systems P we shall evidently have

$$\overline{P} = \int \ldots . \int e^{\frac{\Psi - E}{\Theta}} P dQ_1 \ldots dP_m \tag{662}$$

375. It should also be noted that for such a canonical ensemble of systems, having many degrees of freedom, most of the representative points will be in regions of the phase space where the energy E does not differ greatly from its most probable value. This point has been especially analyzed by Gibbs, but it will be seen qualitatively from the form of equation (660), that very large values of E will not be frequent because of the effect of the exponential factor and very small values of E will not be frequent owing to the small region in the phase space corresponding to low energies. It was hence possible for Gibbs to show that the great majority of the systems in the ensemble have a value of E which does not differ sensibly from the average value $\overline{E}$, and this may be extended to the relation between other properties of the systems and their average values.

376. We are now ready to develop the first analogy of Gibbs. Let us consider the modulus of distribution Θ as a variable parameter, so that by changing its value we can obtain a variety of different canonical ensembles, and let us also consider that the energy E is dependent on certain external parameters $a_1 \ldots . a_n$, of the nature of volume, height, or other macroscopic quantities of an ordinary mechanical nature. Let us now make an infinitesimal (reversible) change in the values of the parameters Θ, $a_1 \ldots . a_n$. This will give us a new canonical distribution, and since equation (661) must then still hold, we may write

$$\int \ldots \int e^{\frac{\Psi - E}{\Theta}} \left[\frac{1}{\Theta} d\Psi - \frac{\Psi}{\Theta^2} d\Theta + \frac{E}{\Theta^2} d\Theta - \frac{1}{\Theta} \left\{ \frac{\partial E}{\partial a_1} da_1 + \ldots + \frac{\partial E}{\partial a_n} da_n \right\} \right] dQ_1 \ldots dP_m = 0 \tag{663}$$

Introducing the principles of averaging given by equation (662) and noting that the external forces corresponding to the parameters $a_1 \ldots . a_n$ will be given by the equations

$$-\frac{\partial E}{\partial a_1} = A_1, \ldots \ldots, -\frac{\partial E}{\partial a_n} = A_n \tag{664}$$

this may be rewritten in the form

$$\frac{d\Psi}{\Theta} - \frac{\Psi}{\Theta^2} d\Theta + \frac{\overline{E}}{\Theta^2} d\Theta + \frac{1}{\Theta} \left\{ \overline{A}_1 da_1 + \ldots . + \overline{A}_n da_n \right\} = 0 \tag{665}$$

Let us now define a quantity $\overline{S}$, called by Gibbs the average value of the negative index of probability, by the equation

$$\overline{S} = \frac{\overline{E} - \Psi}{\Theta} \tag{666}$$

giving us

$$d\overline{S} = \frac{d\overline{E}}{\Theta} - \frac{d\Psi}{\Theta} - \frac{\overline{E}}{\Theta^2} d\Theta + \frac{\Psi}{\Theta^2} d\Theta \tag{667}$$

Combining (665) and (667) and transposing we may then write

$$d\overline{E} = \Theta d\overline{S} - \overline{A}_1 da_1 - \overline{A}_2 da_2 - \ldots . \tag{668}$$

We have thus obtained an equation of exactly the same form as the thermodynamic equation (658), in which the energy E, temperature T, entropy S, and external forces $A_1, A_2 \ldots.$ of a single system, are replaced by the average energy $\overline{E}$, modulus of distribution Θ, average negative index of probability $\overline{S}$, and average values of the external forces $\overline{A}_1, \overline{A}_2 \ldots.$ for all the systems in the ensemble.

To complete the analogy, it was shown by Gibbs that if the systems of two ensembles are brought into a condition analogous to thermal contact, the average result is a passage of energy from the ensemble of the greater modulus to that of the less. And it was further shown that if the external parameters for the systems of a canonical ensemble are subjected to abrupt changes (i.e. nonreversible changes), the ultimate result after sufficient lapse of time will be a new canonical ensemble with a greater value of $\overline{S}$.[1]

We have thus obtained a fairly complete correspondence between the behavior of certain average quantities in an ensemble of systems and the behavior of thermodynamic quantities for a single system. And since we have seen above that these average quantities do not differ sensibly from the individual quantities which we may expect to encounter if we pick systems at random from the ensemble, we may regard the analogy as furnishing a means of predicting the behavior of single systems. It must be remarked, however, that the method seems formal and abstract.

377. The Second Analogue of Gibbs.—The second analogue of Gibbs is based on the consideration of a *microcanonical ensemble* of systems of structure identical with that of interest all having within the narrow limit dE the same energy E as that of the system of interest. Defining V now as the total volume of the phase space corresponding to states of less energy than E, Gibbs was able to obtain the equation

$$dE = \frac{dE}{d\log V}\, d\log V - \overline{A_1}|\, da_1 - \overline{A_2}|\, da_2 - \ldots. \tag{669}$$

where $(dE/d \log V)$ is now the analogue of temperature, $\log V$ of entropy, and $\overline{A_1}|, \overline{A_2}|$, etc., are the average values for the external forces of all the systems in the microcanonical ensemble. It was shown, however, by Gibbs that although the quantity $(dE/d \log V)$ had some of the properties of temperature, it was not true in general that the value of this quantity would remain constant when two systems, having the same value of this quantity, are combined to form a third system.

[1] It must be remarked that it seems doubtful if the rate of approach to the new increased value of $\overline{S}$ could be shown to be fast enough without introducing further hypothesis than those of Gibbs. See P. and T. Ehrenfest, *Encyklop. math. Wiss.*, IV 2, II, Heft 6, page 61.

378. The Third Analogue of Gibbs.—The third analogue of Gibbs was also based on a consideration of a *microcanonical ensemble*. Defining a new quantity ϕ, by the relation

$$\phi = \log \frac{dV}{dE} \tag{670}$$

he obtained the equation

$$dE = \frac{dE}{d\phi}\, d\phi - \overline{\overline{A}}_1 da_1 - \overline{\overline{A}}_2 da_2 - \ldots . \tag{671}$$

where $(dE/d\phi)$ is now the analogue of temperature, ϕ of entropy, and $\overline{\overline{A}}_1$, $\overline{\overline{A}}_2 \ldots .$ the most probable values of the external forces on the systems of the ensembles.

Similar difficulties as to the use of the quantity $(dE/d\phi)$ for temperature were found as in the preceding analogue.

Gibbs apparently considered both these last two analogues as faulty, and regarded the canonical ensemble as superior to the microcanonical for the representation of thermodynamic properties.

379. Boltzmann's Relation between Entropy and Probability.—We may now turn from the attempt to represent thermodynamic quantities by the properties of an *ensemble* of systems, and proceed to the possibility, initiated by Boltzmann, of relating thermodynamic quantities directly to mechanical quantities, which must indeed be studied by statistical mechanical methods, but nevertheless themselves express the properties of a *single* system.

380. Of the three thermodynamic quantities S, Q and T, it is evident that the entropy S is the only one which will need elaborate study. The heat Q absorbed by a single system can be immediately identified on the statistical mechanical basis with the energy which the system receives by molecular impacts or other elementary disordered processes which defy ordinary observational methods. And the temperature T may be regarded as defined by the pressure of a sufficiently dilute gas placed in contact with the system of interest, using the methods of Chapter 4, which permitted us to introduce the temperature immediately into statistical mechanical equations.

Hence we may turn at once to entropy, and consider the relation between the entropy of a system and the probability of its state which was obtained by Boltzmann as the result of the following somewhat formal reasoning.

381. From a thermodynamic point of view, the entropy of an isolated system can only increase if the maximum value has not been reached. On the other hand, from a statistical mechanical point of view, isolated systems may be expected to change from states of lesser probability to those of greater probability. Hence it would seem natural to regard the entropy S of a system as some function of the probability W of its state. We may express this fact by the functional equation

$$S = \phi(W) \tag{672}$$

The properties of entropy and probability, however, permit an immediate solution of the form of this functional equation. If we consider two separate systems of entropies S_1 and S_2 and probabilities W_1 and W_2, we may write

$$\begin{aligned} S_1 &= \phi(W_1) \\ S_2 &= \phi(W_2) \end{aligned} \tag{673}$$

and since entropies are additive and independent probabilities are multiplicative we must have for the combined system

$$S_1 + S_2 = \phi(W_1 W_2) \tag{674}$$

The only solution, however, for equations (673) and (674) is of the form

$$S = k \log W + a \tag{675}$$

where k and a are constants. Of these the constant k merely depends on the interrelation between the scales of measurement for S and W, and it is best to take it the same as the k already introduced. The relation between a and the third law will be discussed later.

The above is Boltzmann's famous equation which serves as a bridge between thermodynamic and statistical mechanical modes of reasoning.

If we consider the change in entropy accompanying a change in the state of a single system we may rewrite equation (675) in the form

$$\Delta S = k \log \frac{W_2}{W_1} \tag{676}$$

382. Entropy Increase and Boltzmann's *H*-Theorem.—The relation of entropy to Boltzmann's quantity H is now immediately evident. In deriving the Maxwell-Boltzmann distribution law, we found that the γ-weight of a given state of gas consisting of N molecules was given by the expression (Chapter 4, equation (38))

$$\log \gamma = N \log N - \sum_i N_i \log N_i + \log d\sigma \tag{677}$$

where N_i is the number of molecules in the i'th region of volume $d\sigma$ into which we have divided the μ-space for the molecules. Hence since probability and γ-weight are proportional we may write

$$S = -k \sum_i N_i \log N_i + \text{const.} \tag{678}$$

In Chapter 18, however, Boltzmann's H was defined by equation (402), as having the value

$$H = \sum_i N_i \log N_i \tag{679}$$

So that we can write

$$S = -kH + \text{const.} \tag{680}$$

The tendency for entropy and probability to increase to a maximum and for Boltzmann's H to decrease to a minimum are thus all seen to be different expressions of the same principle.

383. Different Expressions for Entropy of Systems at Equilibrium.—It will be noticed from the foregoing that Boltzmann's treatment successfully extends the notion of entropy to systems which are not in equilibrium, an extension which could not be made by the Gibbs' method, and one which meets certain difficulties in the field of pure thermodynamics.

In the case of systems which *are* in a state of equilibrium, several different expressions (which are identical for a system with a large number of degrees of freedom), can be used for the Boltzmann result.

Starting with the Boltzmann equation

$$S = k \log W + a$$

since probability is proportional to γ-weight we can write

$$S = k \log \gamma + \text{const.} \tag{681}$$

where γ is the volume in the phase space which corresponds to the state of interest. For a system which has a large number of degrees of freedom, we found, however, in Chapter 4, that practically the whole volume of the γ-space between the limits E and $E + dE$ available to the microcanonical ensemble, corresponded to states of the system which did not differ sensibly from the Maxwell-Boltzmann state of maximum probability. Hence without sensible difference we can replace equation (681) *for a system which is in its equilibrium state by*

$$S = k \log \frac{dV}{dE} + \text{const.} \tag{682}$$

where dV/dE is the change in volume of the phase space with energy at the energy value of interest.

With a system of many degrees of freedom, however, a still further change is possible. The volume V will be that of a many dimensional solid and the logarithm of the surface dV/dE will differ by an inappreciable amount from the logarithm of the whole volume V. This can be seen for example from a consideration of the expressions for many dimensional spheres and ellipsoids given in Sections **158** and **159** of Chapter 12. Hence again without sensible difference we can take *for a system which is in its equilibrium state*

$$S = k \log V + \text{const.} \tag{683}$$

Equations (682) and (683) give the same expressions for entropy as those in the third and second analogues of Gibbs. The present method has the advantage, however, that we do not now need to find properties of the *whole* microcanonical ensemble which are analogous to temperature and external force.

384. Application of the Boltzmann Method to a Specific Case.— It will be desirable to apply the Boltzmann method to a specific simple case.[2] This will have the advantage of showing that the Boltzmann equation for entropy gives an expression which does have the properties of the thermodynamic quantity S, and will also illustrate the procedure that can be undertaken, when statistical mechanical conclusions are for convenience to be translated into thermodynamic language, for the treatment of actual physical chemical problems.

Consider a system of total energy E, composed of N molecules of a single kind, each with n degrees of freedom which may be regarded as fully excited, contained in a vessel of volume V.

For the number of molecules at equilibrium in any region of the μ-space

$$do = dq_1 \ldots . dp_n \tag{684}$$

we may write in accordance with equations (73) and (74) in Chapter 4, the Maxwell-Boltzmann expression in the form[3]

$$dN = e^{\frac{\psi - \epsilon}{kT}} do \tag{685}$$

where the parameter ψ is defined by the equation

$$e^{\frac{\psi}{kT}} = \frac{N}{\int e^{-\frac{\epsilon}{kT}} do} \tag{686}$$

the integration being taken over all regions of the μ-space.

[2] See Tolman, *Journ. Amer. Chem. Soc.*, **44**, 75 (1922).

[3] The quantities ψ and ϵ occurring in this expression for the distribution of *molecules in a system* should be carefully distinguished from the ψ and E in the Gibbs' expression for the distribution of *systems in a canonical ensemble.*

For the average value of any property P of the molecules, which depends on their location in the μ-space, we may evidently write

$$N\overline{P}=\int e^{\frac{\psi-\epsilon}{kT}}\,P\,do \tag{687}$$

where the integration is to be taken over the whole of the μ-space, if we desire the average value for all the molecules of the system.

385. Before proceeding let us call attention to the nature of the quantities occurring in the Maxwell-Boltzmann equation. In accordance with the original way in which ε entered the derivation of the Maxwell-Boltzmann equation in Chapter 4, Section **39,** the quantity ε_i would be the rate of change in the energy of the system per molecule that is added to the i'th region do, *when the system is in its state of maximum probability with energy E, volume V and total number of molecules N,* in accordance with the equation

$$\varepsilon_i=\frac{\partial E}{\partial N_i} \tag{688}$$

Hence, except in the case of perfect gases, ε may be not only a function of the region in the μ-space do, but may also depend on the parameters N, E and V which determine the whole state of the system, since the characteristics of the most probable state will depend on those quantities.

As to the quantities ψ and T in the Maxwell-Boltzmann expression, it is evident that they are functions of the whole state of the system as given by the parameters N, E and V.

386. It will also be advantageous before proceeding to obtain a mathematical identity which will later prove useful. In accordance with the distribution law we may write for the total number of molecules

$$N=\int {}^{\frac{\psi-\epsilon}{kT}}\,do \tag{689}$$

Differentiating with respect to the parameter E, keeping the other two parameters V, and N constant,[4] we obtain

$$\left(\frac{\partial N}{\partial E}\right)_{V,N}=\int\frac{\partial}{\partial E}\left(e^{\frac{\psi-\epsilon}{kT}}\right)do=\int e^{\frac{\psi-\epsilon}{kT}}\cdot\frac{\partial}{\partial E}\left(\frac{\psi-\varepsilon}{kT}\right)do=0 \tag{690}$$

where the expression is equal to zero, since N is constant by hypothesis.

[4] It should be noted that the constancy of V makes it possible to differentiate solely back of the integral sign, since under these circumstances the limits of integration are constant.

387. We are now ready to consider the Boltzmann expression for entropy.

$$S = k \log W + a$$

In accordance with equation (38) in Chapter 4, we may write for the γ-weight of any state of the system

$$\log \gamma = N \log N - \sum_i N_i \log N_i + \log d\sigma \qquad (691)$$

where N_i is the number of molecules falling in the i'th region of the μ-space. Considering now the state of equilibrium or maximum probability, we may substitute the Maxwell-Boltzmann expression for N_i, and remembering that probability is proportional to γ-weight, write in accordance with the Boltzmann expression for entropy

$$S = -k \int e^{\frac{\psi - \epsilon}{kT}} \log \left(e^{\frac{\psi - \epsilon}{kT}} d\sigma \right) d\sigma + \text{const.} \qquad (692)$$

where for convenience we have changed to the language of the integral calculus. In accordance with equation (687) for averages we may now write

$$S = k \log W + \text{const.} = \frac{N\bar{\varepsilon} - N\psi}{T} + \text{const.} \qquad (693)$$

as the Boltzmann expression for entropy.

It will be seen from the development that the constant in this equation can depend on the number of molecules N and on the size chosen for $d\sigma$, as well as on the original undetermined constant a in the Boltzmann equation, which could itself be a function of N. If we drop the constant in (693) we shall later find that the correct dependence of S on N is obtained, and since we are not now interested in absolute values of entropy, it will be most convenient to do this and write

$$S = \frac{N\bar{\varepsilon} - N\psi}{T} \qquad (694)$$

388. We may now investigate to see if this equation for entropy obtained by the Boltzmann method really does have the known properties of the thermodynamic quantity S.

Except for the additive constant, in which we are not now interested, the entropy of a homogeneous system can be completely defined by the thermodynamic equations

$$\left(\frac{\partial S}{\partial E}\right)_{V,N} = \frac{1}{T} \tag{695}$$

$$\left(\frac{\partial S}{\partial V}\right)_{E,N} = \frac{P}{T} \tag{696}$$

$$\left(\frac{\partial S}{\partial N}\right)_{P,T} = \frac{S}{N} \tag{697}$$

The first two of these equations are direct consequences of the fundamental equation

$$dS = \frac{dE + PdV}{T} \tag{698}$$

and the third equation results from the fact that the entropy of a system is always taken as the sum of the entropies of its parts, and hence for a homogeneous system, which has the same pressure and temperature throughout, the entropy will be proportional to the number of molecules.

Let us now show that the quantity given by equation (694) actually does satisfy these thermodynamic equations.

389. We may first investigate the derivative with respect to E. Noting equation (687) for the average value of a quantity, we may write

$$\left(\frac{\partial}{\partial E}\frac{N\bar{\varepsilon} - N\psi}{T}\right)_{V,N} = \frac{\partial}{\partial E}\int e^{\frac{\psi-\varepsilon}{kT}}\frac{\varepsilon-\psi}{T}\,do$$

$$= \int \frac{\partial}{\partial E}\left(e^{\frac{\psi-\varepsilon}{kT}}\right)\frac{\varepsilon-\psi}{T}\,do + \int e^{\frac{\psi-\varepsilon}{kT}}\frac{\partial}{\partial E}\left(\frac{\varepsilon-\psi}{T}\right)do$$

Remembering that ψ and T are independent of o, this may be rewritten in the form

$$\left(\frac{\partial}{\partial E}\frac{N\bar{\varepsilon} - N\psi}{T}\right)_{V,N} = \frac{1}{T}\int\frac{\partial}{\partial E}\left(e^{\frac{\psi-\varepsilon}{kT}}\right)\varepsilon\,do - \frac{\psi}{T}\int\frac{\partial}{\partial E}\left(e^{\frac{\psi-\varepsilon}{kT}}\right)do$$

$$+ \int e^{\frac{\psi-\varepsilon}{kT}}\frac{\partial}{\partial E}\left(\frac{\varepsilon-\psi}{T}\right)do$$

In accordance with equation (690) it is evident, however, that the last two terms in this equation are zero. As to the remaining term on the right-hand side of the above equation, it is evident that the quantity $\frac{\partial}{\partial E}\left(e^{\frac{\psi-\varepsilon}{kT}}\right)do$ is the rate of change in the number of molecules in region do with change in E, while ε is the rate in change of the energy per molecule added to the region do. Hence the value of the definite integral is evidently $\partial E/\partial E = 1$. Our expression thus reduces to

$$\left(\frac{\partial}{\partial E}\frac{N\bar{\varepsilon}-N\psi}{T}\right)_{V,N}=\frac{1}{T} \tag{699}$$

thus agreeing with the first of the three necessary equations (695), (696) and (697).

390. Let us now show that the rate of change of $\frac{N\bar{\varepsilon}-N\psi}{T}$ with respect to volume has the required value. In our previous discussions we have made use of the principle that an isolated system, with a definite energy content E in a container of definite volume V, will assume the state of maximum probability. For our present purposes, we shall make use of the equally valid principle, that the system will assume the state of maximum probability when it is enclosed in a cylinder with walls which do not permit the transfer of energy and provided with a movable piston, thus allowing a simultaneous variation in the parameters E and V, in accordance with the equation

$$\delta E=-P\delta V \tag{700}$$

Now we have already found, in accordance with equation (693), that when the parameters E and V are not allowed to vary, the probability W of the state finally attained, is given by the expression

$$\log W=\frac{N\bar{\varepsilon}-N\psi}{kT}+\text{const.} \tag{701}$$

The variation in this quantity with the parameters E and V will evidently be

$$\delta\log W=\left(\frac{\partial}{\partial E}\frac{N\bar{\varepsilon}-N\psi}{kT}\right)_{V,N}\delta E+\left(\frac{\partial}{\partial V}\frac{N\bar{\varepsilon}-N\psi}{kT}\right)_{E,N}\delta V \tag{702}$$

In the special case that we are considering, W has the maximum possible value not only for changes in internal arrangement with constant values of E and V but also for variations in E and V which do not conflict with the restriction imposed by equation (700). Hence by introducing equation (700) and applying the condition for maximum probability, we obtain

$$-\left(\frac{\partial}{\partial E}\frac{N\bar{\varepsilon}-N\psi}{kT}\right)_{V,N}P\delta V+\left(\frac{\partial}{\partial V}\frac{N\bar{\varepsilon}-N\psi}{kT}\right)_{E,N}\delta V=0$$

Multiplying by k, dividing by δV, and introducing the expression we have already found for the rate of change of $\frac{N\bar{\varepsilon}-N\psi}{T}$ with E, we obtain

$$\left(\frac{\partial}{\partial V}\frac{N\bar{\varepsilon}-N\psi}{T}\right)_{E,N}=\frac{P}{T} \tag{703}$$

and have thus shown the desired agreement with the second of our three necessary equations (695), (696) and (697).

391. In order to show the agreement of the quantity that we have chosen for entropy with the third of the three equations, we may first obtain an auxiliary equation by differentiating the expression for the total number of molecules

$$N=\int e^{\frac{\psi-\epsilon}{kT}}\,d\sigma$$

with respect to the number of molecules N, holding the pressure P and temperature T constant. In carrying out such a differentiation, it is to be noted that we cannot consider the limits of the definite integral as constant, since the volume V of a homogeneous system at a given pressure and temperature is evidently proportional to the number of molecules N, and the volume over which the generalized space $d\sigma$ is to be integrated is proportional to V. Keeping this consideration in mind, we may write

$$\left(\frac{\partial N}{\partial N}\right)_{P,T}=\int\frac{\partial}{\partial N}\left(e^{\frac{\psi-\epsilon}{kT}}\right)d\sigma+\left(e^{\frac{\psi-\epsilon}{kT}}\right)_{\text{lim.}}\frac{\partial\sigma_{\text{lim.}}}{\partial N}$$

Noting further that the last term in this equation is evidently itself equal to $\partial N/\partial N$, and that $\partial\varepsilon/\partial N$ is evidently zero since we have in no way changed the condition inside of any given region $d\sigma$, we obtain

$$\int e^{\frac{\psi-\epsilon}{kT}}\frac{1}{kT}\frac{\partial\psi}{\partial N}\,d\sigma=0$$

or remembering that ψ and its average value for the system are the same, we may write

$$\left(\frac{\partial\psi}{\partial N}\right)_{P,T}=0 \tag{704}$$

We may now differentiate our expression for entropy with respect to N, and obtain

$$\left(\frac{\partial}{\partial N}\frac{N\bar{\varepsilon}-N\psi}{T}\right)_{P,T}=\frac{\bar{\varepsilon}-\psi}{T}+\frac{N}{T}\frac{\partial\bar{\varepsilon}}{\partial N}-\frac{N}{T}\frac{\partial\psi}{\partial N}$$

In accordance with the foregoing, however, the last two terms of this equation are zero. Hence we may rewrite the above equation in the form

$$\left(\frac{\partial}{\partial N}\frac{N\bar{\varepsilon}-N\psi}{T}\right)_{P,T}=\frac{1}{N}\left(\frac{N\bar{\varepsilon}-N\psi}{T}\right) \tag{705}$$

and have thus shown the agreement of the expression which we have chosen for entropy with the third and last of the three necessary equations (695), (696) and (697).

The foregoing, unfortunately but necessarily lengthy investigation illustrates by a special example the fact that the Boltzmann equation does give a correct expression for entropy, and thus permits the translation of statistical mechanical results into thermodynamic language. If, for illustration, we should desire the value of the Helmholtz free energy of the system, we could write

$$F=E-TS=E-N\bar{\varepsilon}+N\psi \tag{706}$$

or in accordance with equation (686)

$$F=E-N\bar{\varepsilon}+NkT\log N-NkT\log\int e^{-\frac{\varepsilon}{kT}}\,d\sigma \tag{707}$$

and for a case such as a perfect gas, in which the difference between the first two terms could be neglected, this would reduce to

$$F=NkT\log N-NkT\log\int e^{-\frac{\varepsilon}{kT}}\,d\sigma \tag{708}$$

The extension to cases of more than one component and to quantized degrees of freedom will be evident. It should be especially noted that a clear idea as to the meaning of ε, as the rate of change of energy per molecule introduced into the region $d\sigma$, will facilitate the application to concentrated systems. Only for dilute systems is ε the energy of an individual molecule.

392. Relation between Statistical Mechanics and the Third Law of Thermodynamics.—In conclusion we must consider the relation between statistical mechanics and the so-called third law of thermodynamics.

The classical thermodynamics was only concerned with the changes in energy ΔE and changes in entropy ΔS which accompany changes in the state of a system, but did not consider the absolute values of these quantities. Since the time of the origin of the classical thermodynamics, the development of the theory of relativity has furnished us with a relation between mass and energy, $E=mc^2$, which gives us a rational starting point for energy values. The third law of thermodynamics endeavors in a similar way to furnish a rational starting point for entropy values.

393. The safest method to attack this problem is to use the methods of Chapter 13, where we showed from a consideration of the vapor pressures of crystals, that a system of crystalline substances at the absolute zero must have the same entropy in any form of chemical combination. This makes it possible to take the entropy as zero for all crystalline substances at the absolute zero of temperature.

394. Another mode of attack which has been largely carried out under the influence of the work of Planck is to return to the Boltzmann equation connecting entropy and probability

$$S = k \log W + a$$

and by dropping the undetermined constant a secure a relation

$$S = k \log W \tag{709}$$

which will give entropy an absolute value provided the probability has such.[5]

In the classical mechanics, of course, no absolute value could have been assigned to the probability W, since the numerical value would obviously depend on the size $d\sigma$ chosen for the regions of equal probability in the phase space. With the advent of quantum theory, however, we might expect an absolute value of W owing to the existence of a finite number of discrete quantum states of the system. This is the essence of the dependence of the third law on quantum theory.

In order to avoid all difficulties as to the designation of equally probable conditions, Planck's latest proposal is to define entropy by the equation

$$S = k \log P \tag{710}$$

where P is the number of all the stationary states which the system can assume without overstepping the energy E which it has.[6] In the investigation in which this definition is considered, it is shown that the equation leads in known cases to the familiar expressions for entropy, and indications as to the importance of the definition for the problem of gas degeneration are presented.

This must conclude our somewhat brief treatment of the relation between statistical mechanics and thermodynamics. It is hoped, however, that sufficient has been presented to make the application to actual problems possible.

[5] For an analysis of this equation, see Schrödinger, *Berlin Acad. of Science*, p. 434 (1925).

[6] Planck, *Zeitschr. für Phys.*, **35**, 155 (1925); compare equations (710) and (683). In obtaining the number of states P, states which differ solely by the interchange of identical atoms or other elements are not to be counted separately. This secures the proper dependence of S on N.

Chapter 24

Concluding Remarks

395. Introduction.—In conclusion we shall make brief mention of several important fields, susceptible to statistical mechanical treatment, which unfortunately cannot be fully treated in the present book. We shall also try to indicate something as to the future outlook for statistical mechanical investigations.

396. Effect of External Fields of Force on Systems of Molecules.—In the applications to physical-chemical systems which we have hitherto considered, we have assumed in general the absence of any external field of force. The methods of statistical mechanics, particularly those furnished by the Maxwell-Boltzmann distribution law, can, however, often be applied to investigate the action of such fields on systems of molecules.

397. The simplest case would be the effect of an external *gravitational field* in producing changes in density with height, but this problem is evidently merely a special case of the general problem of determining the relation between pressure and density, and does not need separate treatment. The direct treatment of the problem for a perfect gas is, however, perhaps so simple as to prove instructive.

Consider a tube of gas parallel to a uniform gravitational field which produces the acceleration g in the x-direction. We can then write the Maxwell-Boltzmann distribution law in the form

$$dN = NCe^{-\frac{\epsilon_0 + mgx}{kT}} dxdydzdq_4 \ldots . dp_n \tag{711}$$

where mgx is the only part of the energy dependent on the height x, and integrating for all values of the variables except x we obtain

$$dN = NACe^{-\frac{mgx}{kT}} dx \tag{712}$$

where A is a constant resulting from the integration. This equation immediately gives, however, the well-known distribution of concentration with height, provided we determine the constants so as to account for all the gas in the tube. In the case of imperfect gases or liquids, the energy ε at a height x would be dependent both on gravity and concentration and the treatment would be more complicated.

398. The effect of an external *electric field*[1] in polarizing a gas can also be treated with the help of the Maxwell-Boltzmann distribution law, and this has been done by Debye[2] on the basis of the classical theory and by Pauli[3] and Pauling[4] on the basis of the quantum theory.[5]

The general nature of the expected effect will be evident. The applied electric field will, in the first place, induce an electric moment in the molecules of the gas by deforming them, and, in the second place, will tend to orient the permanent dipoles, already present, parallel to the field. The total polarization of the gas will be the sum of the polarizations produced in these two ways, the first term being at least approximately independent of the temperature and the second term evidently decreasing with the temperature as the alignment of the dipoles is more and more disturbed.

The Maxwell-Boltzmann distribution law can of course immediately be applied to calculate the relative number of dipoles oriented in different directions, since the energy ε occurring in the Maxwell-Boltzmann expression will contain a term dependent on the angle between the axis of the dipole and the direction of the field. Nevertheless, a difficulty arises until we are certain as to whether the possible angles are to be regarded as continuously distributed in accordance with classical considerations, or confined to the special values given by the theory of spatial quantization. The problem is one of the greatest importance and cannot be solved until we are more informed as to the effect of molecular collisions in disturbing spatial quantization.

Unfortunately we cannot here treat this field in more detail. It may be remarked, however, that the interpretation of electric polarization and dielectric constants by the *classical* method of Debye gives much larger values to the electric moments of the dipoles than does the quantum theory of Pauli using *whole* quantum numbers,[6] and this in turn much larger values than the theory of Pauling using *half* quantum numbers.[7]

399. The effect of an external *magnetic field* in causing the magnetic polarization of a paramagnetic substance can also be similarly undertaken

[1] For the most complete report on the effects of electric and magnetic fields, see Debye in Marx "Handbuch der Radioligie," Volume V, page 597, Leipzig, 1925.
[2] Debye, *Physikal. Zeitung.*, **13**, 97 (1912).
[3] Pauli, *Zeitschr. für Phys.*, **6**, 319 (1921).
[4] Pauling, *Phys. Rev.*, **27**, 568 (1926).
[5] Since the above was written, there has appeared a note on this problem, using the new dynamics of Heisenberg and Born, by Kronig, *Proc. Nat. Acad.*, **12**, 488 (1926).
[6] See Zahn, *Phys. Rev.*, **24**, 400 (1924).
[7] See Pauling, *loc. cit.*

with the help of the Maxwell-Boltzmann distribution law as first shown by Langevin [8] in accordance with the classical point of view, and as treated by Pauli,[9] Epstein [10] and Gerlach [11] using the quantum theory. The results on the whole seem to favor the application of spatial quantization, as it gives magnetic moments corresponding to the Bohr magneton.

400. For a time some light seemed to be thrown on the problem of spatial quantization in the presence of collisions by the work of Glaser,[12] which indicated that the susceptibility of diamagnetic gases increases at low pressures, when the time between collisions becomes of the order of the time of the Larmor precession. The later work of Lehrer,[13] however, as well as work of Hamar in the Norman Bridge Laboratory, indicate that Glaser's findings were not justified. At the present time it seems as if spatial quantization should be applied at ordinary pressures, both in the case of electrical and magnetic fields.

401. The Interionic Attraction Theory.—In recent years the Maxwell-Boltzmann distribution law has been used in a very interesting manner to investigate the distribution of the charged ions in an electrolytic solution. The theory has been specially developed by Milner [14] and Debye,[15] and many experimental tests of the conclusions have been presented by Noyes [16] and others.

The general nature of the expected effect will be evident. Since the electric field of force around an ion falls off only with the square of the distance, it is not possible as in so many applications of statistical mechanics to neutral molecules, to assume that charged molecules have no effect on each other except in the short range of action that we ascribe to ordinary collision. Rather we must believe that each positive ion in a solution tends to concentrate negative ions in its vicinity and vice versa. This change from uniform distribution is undoubtedly one of the factors which prevents ionic substances from acting as perfect solutes, and hence its consideration is of interest in view of the historical importance of the theory of electrolytic solutions.

A satisfactory calculation of the distribution of ions in an electrolytic

[8] Langevin, *Journ. de Phys.*, **4**, 678 (1905); *Am. de Chim. et Phys.*, **5**, 70 (1905).
[9] Pauli, *Phys. Zeitschr.*, **21**, 615 (1920).
[10] Epstein, *Science*, **57**, 532 (1923).
[11] Gerlach, *Phys. Zeitschr.*, **24**, 275 (1923).
[12] Glaser, *Ann. der Phys.*, **75**, 459 (1924); **78**, 641 (1925).
[13] Lehrer, *Zeitschr. für Phys.*, **37**, 155 (1926). Hammar, *Proc. Nat. Acad.*, **12**, 594 (1926).
[14] Milner, *Phil. Mag.*, **23**, 551 (1912); **225**, 742 (1913).
[15] Debye and Hückel, *Physik. Zeitschr.*, **24**, 185 (1923); Debye, *Ibid.*, **24**, 334 (1923).
[16] Noyes, *Journ. Amer. Chem. Soc.*, **46**, 1080 (1924), and a series of following articles the latest being by Sherrill and Noyes, *Ibid.*, **48**, 1861 (1926).

solution is, however, not an easy matter. If we had only a single positive ion and a single negative ion with charges $\pm e$, it is evident, from the Maxwell-Boltzmann distribution law, that the chance of there lying a distance between r and $r + dr$ apart, could be put proportional to $4\pi r^2 e^{-\frac{e^2}{rkT}} dr$. In the case of a solution of moderate concentration, however, the tendency of ions of one sign to cluster around an ion of the other sign greatly changes the effective field of the latter and the calculation of the final distribution has been possible only with the help of approximations. The seriousness of these approximations has recently been emphasized by Bjerrum.[17] Furthermore, it has been pointed out by Zwicky [18] that the most probable distribution of ions will be greatly affected, even at extremely low concentrations, by the potential energy of the solvent, which is subjected to an electrostatic compression which varies with the distribution of the charged ions. It seems possible that a completely satisfactory form of the interionic attraction theory has not yet been given. Further theoretical work on the problem will be of considerable interest.

402. Fluctuations.—The whole application of statistical mechanics to physical-chemical systems is based on the assumption of the atomic structure of matter. Such a discrete structure of matter, however, ever, immediately entails fluctuations in the *condition* of a system even when it is in equilibrium and fluctuations in *rate* when a macroscopic physical chemical change is under way.[19]

When the elementary molecular processes happen to be of such a nature as to permit observation, both these kinds of fluctuations can be directly observed. Thus the particles of a colloidal suspension, which can be seen in the microscope, may be regarded as individual molecules, and their Brownian fluctuations in position and velocity directly observed, with the well-known result that the individual particles are found under equilibrium conditions to have the average kinetic energy $\frac{1}{2}kT$ per degree of freedom, as theoretically predicted by statistical mechanics. Fluctuations in the rate of radioactive decay can also be directly followed by observing the scintillations of the alpha particles which are ejected.

Even, however, when the elementary molecular processes cannot be directly observed, the existence of both kinds of fluctuations may be apparent. Thus, as first appreciated by Smoluchowski,[20] the opalescence of a substance in the neighborhood of its critical point, and of liquid

[17] Bjerrum, *Det Kgl. Danske Videns. Selskab. Math. fys. Med.*, VII, 9 (1926).
[18] Zwicky, *Physik. Zeitschr.*, **27**, 271 (1926).
[19] For a review of the general field of fluctuations, see Fürth, "Schwankungererscheinungen in der Physik," Sammlung Veiweg, Braunschweig, (1920).
[20] Smoluchowski, *Ann. der Phys.*, **25**, 205 (1908).

mixtures in the neighborhood of the conditions for complete mutual solubility, are due to fluctuations from a state of uniform density or concentration. And as first studied by Schottky,[21] fluctuations exist in the rate at which an electric current flows from a hot filament, which on analysis are shown to be due to the fact that the current results from the emission of individual electrons.

403. For the case of the macroscopic fluctuations of a system which is in a state of equilibrium, a very simple and beautiful theory has been developed by Einstein,[22] which we must briefly describe.

Consider the fluctuations of some observable property of a system which is in a state of equilibrium. Let x_0 be the value of the quantity measuring this property when the system is in its state of maximum probability, and let us take

$$dW = Adx \tag{713}$$

as the probability that the quantity will be in the range between x and $x + dx$, in the course of its fluctuations.

The quantity A in this equation will evidently vary with the value of x, becoming rapidly smaller the more x differs from its most probable value x_0. The main factor determining A can be determined, however, from the Boltzmann relation connecting probability and entropy

$$S = k \log W + a \tag{714}$$

In accordance with this equation, we may write for the state of maximum probability

$$S_0 = k \log W_0 + a \tag{715}$$

where S_0 and W_0 are the entropy and probability of this state. Or since the probability W will be of the order of unity for the most probable state, we may write

$$S_0 \simeq a \tag{716}$$

Furthermore for the state of interest we can write

$$S = k \log (dW) + a \tag{717}$$

and combining (716) and (717), obtain

$$dW \simeq e^{-\frac{S_0 - S}{k}} \tag{718}$$

We thus see, since k is very small, compared with appreciable differences between the maximum entropy S_0 and the entropy S of the state of interest, that the main reason, why the quantity A in equation (713)

[21] Schottky, *Ann. der Phys.*, **57**, 514 (1918); **68**, 157 (1922).
[22] Einstein, *Ann. der Phys.*, **33**, 1275 (1910).

decreases as x fluctuates away from the most probable value x_0 is due to the increasing value of the difference $S_0 - S$. Hence if we rewrite equation (713) in the form

$$dW = Be^{-\frac{S_0 - S}{k}}\, dx \tag{719}$$

It is evident that the dependence of B on x will be much less sharp than that of A.

We may now transform equation (719) to advantage with the help of the thermodynamic relations between free energy, entropy and temperature

$$F_0 = E_0 - TS_0$$

and

$$F = E - TS \tag{720}$$

And since the energy of the system is the same in the two states and the temperature is not appreciably affected by the fluctuation, we may substitute

$$S_0 - S = \frac{F - F_0}{T} \tag{721}$$

in equation (719) and write

$$dW = Be^{-\frac{F - F_0}{kT}}\, dx \tag{722}$$

Since now we are interested in states of fluctuation in the neighborhood of the most probable value $x = x_0$, we may develop B and $F - F_0$ as power series of the form

$$B = B_0 + B_1(x - x_0) + \dots. \tag{723}$$

$$(F - F_0) = a(x - x_0) + \frac{b}{2}(x - x_0)^2 + \dots. \tag{724}$$

Owing, however, to the very large factor $1/k$ in the exponent of equation (722), and the fact that we saw above that B does not vary sharply with x, we may neglect, in the neighborhood of $x = x_0$, all but the first term of (723). Furthermore in (724) the first term is zero owing to the fact that we are in the neighborhood of minimum free energy, and higher terms than the second may be neglected owing to the small value of $(x - x_0)$. So that we may rewrite equation (722) in the form

$$dW = B_0 e^{-\frac{F - F_0}{kT}}\, dx = B_0 e^{-\frac{b}{2}\frac{(x - x_0)^2}{kT}}\, dx \tag{725}$$

and by the usual process of averaging obtain for the mean fluctuation of the free energy as dependent on any quantity x

$$(F - F_0)_{\text{av.}} = \left[\frac{b}{2}(x - x_0)^2\right]_{\text{av.}} = \frac{kT}{2} \tag{726}$$

Since $(F - F_0)$ is merely the work that has to be done in order to disturb the system from its most probable state to the state of interest, equations (725) and (726) can easily be applied.

Thus, for example, equation (726) gives at once, that the mean component in any direction of the kinetic energy of a Brownian particle will be $kT/2$, and its mean height above the floor of a vessel will be given by the relation $mg\bar{h} = kT/2$, while equation (725) correctly gives the probability of different kinetic energies and heights.[23]

404. As another example we may calculate the fluctuation in the density of a fluid. If v_0 is the mean *specific* volume of the fluid and v is the specific volume which occurs for a mass m located somewhere within the fluid, as a result of fluctuations, we may write

$$F - F_0 = -m\int_{v_0}^{v}(p - p_0)\,dv \tag{727}$$

Expanding p, however, in the form of a series, we may write

$$p = p_0 + \frac{\partial p}{\partial v}(v - v_0) + \frac{1}{2}\frac{\partial^2 p}{\partial v^2}(v - v_0)^2 + \ldots \tag{728}$$

and neglecting the second order and higher terms obtain by substitution in (727)

$$F - F_0 = -m\left(\frac{\partial p}{\partial v}\right)\frac{(v - v_0)^2}{2} \tag{729}$$

Equation (729) may now be substituted in (725) and (726) and the probability for a given fluctuation in the specific volume and the mean fluctuation in the specific volume determined. Thus for the latter we obtain

$$\overline{\delta^2} = \overline{\left(\frac{v - v_0}{v_0}\right)^2} = -\frac{kT}{m\left(\frac{\partial p}{\partial v}\right)v_0^2} \tag{730}$$

We see that the fluctuations are smaller the greater the mass of fluid m considered.

For the case of a perfect gas, we have

$$pv_0 = \frac{RT}{M}$$

[23] See Fürth, *Ann. der Phys.*, 53, 208 (1916).

where M is the molecular weight, and by substituting in (730), can obtain

$$\overline{\delta^2} = \frac{1}{N} \tag{731}$$

where N is the number of molecules in the region considered. It is thus evident, that the fluctuations in the density of a given region, inside a volume of gas, would be too small to detect, but if actually measurable would furnish a means of determining Avogadro's number.

In case of a fluid in the neighborhood of its critical point, however, the quantities $\frac{\partial p}{\partial v}$ and $\frac{\partial^2 p}{\partial v^2}$ approach zero, so that the fluctuations become measurable and as studied by opalescence measurements are shown to be quantitatively in agreement with the theory,[24] thus furnishing an approximate method of determining Avogadro's number.

405. Conclusion.—It is hoped that the foregoing discussion of statistical mechanics and its applications will be sufficient to assist in the further development of this important science.

It appears certain that chemists, in particular, since they have to deal for the most part with the behavior of systems containing many, many molecules, will need to make increasing use of this science. For the treatment of chemical equilibria, it will be most satisfactory for them to continue the use of the simple language of thermodynamics, substituting into its equations, at will, either empirical values of the parameters or theoretical values obtained by the methods of statistical mechanics. In other fields, however, they will have to depend more and more directly on statistical mechanical considerations.

For the physicists, with their happy facility in devising experiments whose result depends on the immediate action of single atoms, the science of statistical mechanics is of less vital interest. Nevertheless, the interchange between the atomic and macroscopic points of view will continue to be stimulating. With the help of hypotheses as to individual atomic processes we predict the behavior of systems containing many atoms, and by observing this behavior we verify or revise our original starting point. Thus the Schottky effect may be predicted as a result of our knowledge of the atomic subdivision of electricity, or may in turn be used for determining the charge of the individual electrons. Atomic hypothesis can lead to macroscopic predictions, and macroscopic experiments back to atomic knowledge.

In the field of quantum phenomena, we must expect in the near future many changes from the methods of presentation used in this book. To some extent these will consist merely in matters of detail, such as changes

[24] Keesom, *Ann. der Phys.*, **35**, 591 (1911).

in the theoretical methods of predicting quantum levels. When the new quantum dynamics is fully developed, however, it may be desirable to present the foundations of statistical mechanics from a new point of view, basing the science from the very start on the new dynamics, instead of first developing the classical statistical mechanics and then introducing modifications to take account of quantum phenomena.

Among the immediate developments that we can wish for statistical mechanics, will be its increasing application to systems, in which the closeness of molecules or the electrical charges that they carry prevents the use of the common simplifying assumptions, which neglect all but discontinuous interactions. The work of Milner and Debye on interionic attraction is a promising development of this kind. If some general and simple method for the rigorous treatment of such problems could be discovered it would serve as a great step forward.

Further applications to the field of chemical kinetics are also to be desired. The problem of reaction velocities is probably nearer to the heart of most chemists than anything else in their whole range of activity. Rates of reaction are the factors that determine yields, and costs, and possibilities, and their theory must eventually succumb to scientific treatment. Some of the possibilities of attacking the problem of reaction velocity have been presented in Chapters 20, 21, and 22 of this book, but the field still remains one of the most obscure in chemistry. Perhaps new principles as to the behavior of individual atoms will be necessary to the final solution.

Whatever may be, the way is long and the traveller must not tarry.

APPENDIX I

Some Useful Integrals

$$\int_0^\infty e^{-ax^2}\,dx = \frac{1}{2}\sqrt{\frac{\pi}{a}} \qquad \int_0^\infty e^{-ax^2}\,x\,dx = \frac{1}{2a}$$

$$\int_0^\infty e^{-ax^2}\,x^2dx = \frac{1}{4}\sqrt{\frac{\pi}{a^3}} \qquad \int_0^\infty e^{-ax^2}\,x^3dx = \frac{1}{2a^2}$$

$$\int_0^\infty e^{-ax^2}\,x^4dx = \frac{3}{8}\sqrt{\frac{\pi}{a^5}} \qquad \int_0^\infty e^{-ax^2}\,x^5dx = \frac{1}{a^3}$$

. . . .

$$\int_0^\infty e^{-ax^2}\,x^{2k}\,dx = \frac{1.3\ldots.(2k+1)}{2^{k+1}}\sqrt{\frac{\pi}{a^{2k+1}}} \qquad \int_0^\infty e^{-ax^2}\,x^{2k+1}\,dx = \frac{\underline{|k}}{2a^{k+1}}$$

$$\int e^{ax}\,dx = \frac{e^{ax}}{a}$$

$$\int e^{ax}\,x\,dx = \frac{e^{ax}}{a^2}\,(ax-1)$$

$$\int e^{ax}\,x^m\,dx = \frac{x^m\,e^{ax}}{a} - \frac{m}{a}\int e^{ax}\,x^{m-1}\,dx$$

APPENDIX II

Some Useful Constants

$c = 2.9986 \times 10^{10}$ cm. sec.$^{-1}$
$e = 4.774 \times 10^{-10}$ electrostatic units.
$h = 6.554 \times 10^{-27}$ erg sec.
$k = 1.372 \times 10^{-16}$ erg deg.$^{-1}$ per molecule.
$N = 6.061 \times 10^{23}$ molecules per mol.
$R = 8.315 \times 10^{7}$ erg deg.$^{-1}$ per mol.

APPENDIX III

MOLECULAR DIAMETERS CALCULATED FROM VISCOSITY AND VAN DER WAALS, *b*.

Gas	$\sigma \times 10^8$ (viscosity)	$\sigma \times 10^8$ (Van der Waals)
Argon	2.84	2.85
Krypton	3.12	3.14
Xenon	3.47	3.42
Helium	1.89	1.96
Oxygen	2.93	2.89
Hydrogen	2.36	[2.52]
Nitrogen	3.10	{3.54 3.08
Air	3.08	3.30
Carbon Dioxide	{3.27 3.20	{3.40 3.20

(Taken from S. Chapman, *Phil. Trans. Roy. Soc.*, **A 216,** 347 (1916). The calculations are based on a theory of viscosity which takes account of the deviations from the Maxwellian distribution.)

INDEX OF SUBJECTS

(Numbers refer to pages)

INDEX OF AUTHORS

(Numbers refer to pages)

Spotted something that isn't quite right?

Let us know and we'll fix it for future reprints. And as a thank you for your diligence, we'll send you a free Sothis book of your choice.

publisher@sothis.press

This page intentionally left blank

Made in United States
Orlando, FL
05 November 2026

90231831R00196